WIRELESS NETWORKS AND MOBILE COMPUTING, VOLUME 6

ADAPTATION TECHNIQUES IN WIRELESS MULTIMEDIA NETWORKS

WIRELESS NETWORKS AND MOBILE COMPUTING

Yi Pan
Series Editor

Volume 1: Design and Analysis of Wireless Networks
Edited by Yi Pan and Yang Xiao
2005 ISBN 1-59454-186-8

Volume 2: Ad Hoc and Sensor Networks
Edited by Yang Xiao, Jie Li and Yi Pan
2005 ISBN 1-59454-396-8

Volume 3: Security and Routing in Wireless Networks
Edited by Yang Xiao, Jie Li and Yi Pan
2005 ISBN 1-59454-316-X

Volume 4: Wireless LANs and Bluetooth
Edited by Yang Xiao and Yi Pan
2005 ISBN 1-59454-432-8

Volume 5: Resource Allocation in Next Generation Wireless Networks
Edited by Wei Li and Yi Pan
2006 ISBN 1-59454-583-9

Volume 6: Adaptation Techniques in Wireless Multimedia Networks
Edited by Wei Li and Yang Xiao
2006 ISBN 1-59454-883-8

WIRELESS NETWORKS AND MOBILE COMPUTING, VOLUME 6

ADAPTATION TECHNIQUES IN WIRELESS MULTIMEDIA NETWORKS

WEI LI
YANG XIAO
EDITORS

Nova Science Publishers, Inc.
New York

For permission to use material from this book please contact us:
Telephone 631-231-7269; Fax 631-231-8175
Web Site: http://www.novapublishers.com

Library of Congress Cataloging-in-Publication Data

Adaptation techniques in wireless multimedia networks.
p. cm. -- (Wireless networks and mobile computing ; v. 6)
Includes index.
ISBN 13 978-1-59454-883-3
ISBN 10 1-59454-883-8
1. Wireless communication systems--Quality control. 2. Multimedia systems--Quality control. 3. Internetworking (Telecommunication) I. Series.
TK5103.2.A36 2006
621.384--dc22 2006003478

Published by Nova Science Publishers, Inc. ✢New York

CONTENTS

PREFACE

Next generation wireless and mobile communication systems have two important features: heterogeneity and adaptation. Heterogeneous wireless communication platforms require inter-operability among different networks such as wireless cellular networks, wireless local area networks, wireless personal area networks, internet, etc. Adaptation techniques become extremely important in such a heterogeneous environment so that quality management and provision to multimedia traffic over the unreliable wireless channels of the heterogeneous systems can satisfy the demands of various network users. With the development of multimedia compression and coding technologies, more and more real-time applications, such as video and audio can be made adaptive. Adaptation provides an alternative for resource planning, especially for bandwidth allocation/reallocation in wireless multimedia networks, where bandwidth is a scarce resource. The system may need to block incoming users if all of the bandwidth has been used up to provide the highest Quality of Service (QoS) to existing users. However, if these existing users can be degraded to a lower but acceptable QoS level, it is possible to reduce the blocking probability without degrading the QoS of existing users to an "unacceptable" level. Various approaches and algorithms adopting this idea have been proposed. It is therefore critical for wireless network designers to utilize these resources efficiently and effectively. In response to the above demand for next generation wireless and mobile communication systems, this book aims at providing a timely and concise reference of the current activities and findings in the relevant technical fields. The primary goal is to address the key technical issues pertaining to the integrated new systems and present novel technical contributions.

This book contains 15 invited chapters from prominent researchers working in this area around the world. Most chapters were reviewed carefully by two reviewers. The reviewers' reports not only correct some deficiencies of the chapters but also greatly improve the quality of this book. We believe that all of these chapters not only provide novel ideas, new analytical models, simulation and experimental results, and handful experience in this field, but also simulate the future research activities in the area of adaptation techniques in next generation wireless networks. A brief summary of each Chapter is listed as follows.

Chapter 1 addresses rate adaptation and power control, two essential link adaptation techniques to achieve high energy efficiency for wireless local area networks (WLANs). Rate adaptation can improve the good put performance by dynamically selecting the data transmission rate, while power control can reduce the energy consumption. The emerging IEEE 802.11h standard extends the current high-speed rate adaptive 802.11a/g to support power control. This paper gives an overview on rate adaptation and power control schemes for WLANs. Several rate adaptation and power control schemes for WLANs using the carrier-sense multiple access (CSMA) medium access control (MAC) protocol are compared in terms of throughput, energy efficiency, and complexity.

Chapter 2 proposes a new rate adaptation mechanism for multimedia multicasting in IEEE 802.11 WLAN environments, called SARM (SNR-based Auto Rate for Multicast). SARM is based on the RBAR (Receiver Based Auto Rate) protocol. Therefore SARM adapts its transmission rate based on the SNR values of mobile nodes. In order to obtain SNR values, SARM needs an additional signaling method by which the AP can estimate channel quality i.e., SNR, since there are no RTS-CTS messages or MAC layer ACK messages in multicast transmissions. By adapting multicast transmission rate based on SNR, SARM guarantees the best quality of multimedia service depending on the mobile nodes' link conditions. Moreover the proposed protocol minimizes the portion of wireless channel resource for multimedia multicasting, so that it can allocate more bandwidth to other unicast traffic.

Chapter 3 investigates the possibilities for adaptation in wireless terminals. The authors distinguish standard level adaptation, algorithm-selection level adaptation, and algorithm parameter level adaptation. They use a Universal Mobile Telecommunications System (UMTS) receiver to illustrate these adaptation related concepts and show how to exploit adaptation with flexible reconfigurable hardware to save energy in wireless terminals. They also demonstrate the approach with examples from UMTS and HiperLAN/2.

Chapter 4 proposes a novel end-system based adaptive protocol called Wireless Hybrid Adaptation Layered Multicast (WHALM) protocol for layered video multicast over wireless Internet. In WHALM the sender dynamically collects bandwidth distribution from the receivers and uses an optimal layer rate allocation mechanism to reduce the mismatches between the coarse-grained layer subscription levels and the heterogeneous and dynamic rate requirements from the receivers, thus maximizing the Quality of Service (QoS) of all the receivers in a multicast session. Based on the sampling theory and the probability theory, the proposed approaches reduce the required number of bandwidth feedbacks to a reasonable degree and use a scalable feedback mechanism to control the feedback process practically. WHALM is also tuned to perform well in wireless networks by integrating an end-to-end loss differentiation algorithm (LDA) to differentiate error losses from congestion losses at the receiver side. With a series of simulation experiments over Network Simulator Version 2 (NS/2) platform, WHALM has been proved that it greatly improves the QoS of all the receivers while avoiding congestion collapse on the wireless Internet.

Chapter 5 develops a framework for provisioning connection-level QoS in wireless cellular networks that support adaptive multimedia services. The framework operates at the connection-level where the bandwidth of ongoing connections can be dynamically adjusted. This framework supports multiple classes of adaptive multimedia services that have diverse QoS requirements. The authors integrate a threshold-based bandwidth reservation policy, an efficient CAC algorithm, and a Bandwidth Allocation Algorithm (BAA) into a new framework called QoS-Adaptive Multimedia Wireless Access (QoS-AMWA). The objective of QoS-AMWA is to provide the mobile users with seamless handoff communication through the wireless network and to distinguish between handoff calls and new calls for each class of traffic while minimizing New Call Blocking Probability (NCBP) and Hand of Call Dropping Probability (HCDP).

Chapter 6 presents a discussion of selected issues that are related to a robust real-time video application. Expecting no special treatment from the underlying network, the authors focus on the application layer. Specifically, they investigate two problems: 1) how to control the transmission error and eliminate its impact, and 2) how to maintain low end-to-end delay and reasonable intra-stream synchronization.

Chapter 7 proposes a Discrete Cosine Transform (DCT)-based audio decomposition (DAD) scheme for real-time general audio streams. This proposed scheme allows audio quality to be traded off against reductions in bandwidth usage and is appropriate for both unicast and multicast environments. The authors compare DAD performance with two other schemes which also support adaptation to restricted bandwidth in terms of objective and subjective measures.

Chapter 8 discusses a framework for exploring adaptive and self-organizing behaviors in wireless networks that involves a mapping between the problem domain using conceptual analysis models, and simulation domains using models representing both the network as a whole, and individual nodes in the network. It is this ability to model workflow-level, network-level and node-level behaviors in a co-design and co-simulation context that allows us to explore introduction of adaptive protocol changes into the nodes, so that the effects of self-organizing behaviors exhibited across the entire wireless network (as evidenced by appropriate fitness metrics) can be evaluated.

Chapter 9 studies the optimal downlink resource allocation problem for data traffic in a CDMA/TDMA cellular network. The authors use the dynamic pricing approach based on utility functions to formulate the optimal downlink resource allocation as an optimization problem. A utility-function is used as an indicator of the network benefit earned by serving a packet of a user from a given base-station. This utility-function is a function of the corresponding base-station load, the packet experienced delay before its transmission, and wireless channel quality. They then formulate the downlink resource allocation problem in each time-slot as an optimization problem with the objective of maximizing the total network utility that is defined as the aggregation of individual users' utilities in each time-slot. This optimization problem is solved for each time-slot to adaptively exploit wireless link variations, users' quality of service flexibility, and base-station load conditions. The authors then show that this optimization problem is a 0-1 multi-dimensional multiple-choice knapsack problem that simultaneously gives base-station assignment and packet scheduling as the output; therefore, it can be adapted to consider not only wireless link dynamics but also network load variations. They also propose a heuristic algorithm with polynomial-time complexity to solve this problem. Simulation results indicate a significant performance improvement. Specifically, a higher performance gain is achieved in the network coverage area with greater non-uniform user spatial distribution.

Chapter 10 focuses on the common adaptation techniques in multi-carrier systems that improve system performance and capacity. First, adaptation methods specific to multi-carrier systems are given. Then, some parameter measurements that make the adaptation techniques possible are discussed in detail. Finally, adaptive OFDM system design for mobile systems is studied and adaptive features in the new IEEE WMAN standard are presented as a case study.

Chapter 11 discusses stochastic learning control based online optimization of wireless multimedia networks. Specifically, the authors consider the following two problems: (a) joint downlink adaptive rate selection and opportunistic user scheduling for a 3G multi-user network, and (b) wireless link state dependent video source coding rate control.

Chapter 12 introduces a quantified method to minimize the fraction of the number of blocked calls out of the number of total calls under non-stationary handover traffic. The authors develop new mathematical formulations of that fraction and optimization models with efficient exact solution algorithms. Performance analysis shows that the proposed method improves the blocking probabilities of handovers and new calls; this improvement results from a highly adaptive and reactive characteristic to the fluctuating handover traffic condition.

Chapter 13 introduces control strategies from wireline networks and examines them in a mobile environment where the network supports multiple classes of subscribers having different bandwidth requirements. The authors also summarize their previous work, which is a generalization of the Erlang fixed point approximation using multi-rate, multi-class, Erlang loss formulas for each class of traffic. They produce expressions for originating calls lost, handoff calls lost, forced-termination, and mean channel occupancy for each class of traffic in each cell for different control strategies. They then present an iterative scheme for determining the residual time of the call holding time distribution that allows investigating the effects of variance in both the call holding time distribution and the dwell time distribution on network performance. For traffic exhibiting non-stationary behavior, they propose an adaptive control strategy that adjusts the amount of bandwidth assigned to each traffic class. The analytic results are supported by simulation.

Chapter 14 studies the performance of GSM/GPRS networks with the emphasis on the adaptive characteristics exhibited by multi-slot capability. The analytical model generalizes the GPRS data channel requirement assumption (M channels). In addition, the authors propose a channel re-allocation scheme (RAS) with or without buffer mechanism to efficiently utilize the limited bandwidth. Numerical investigation is performed to demonstrate the advantage of RAS as well as the integration of buffer mechanism. The results indicate that RAS not only significantly decreases the voice call blocking probability and the average GPRS packet transmission time, but also slightly increases the channel utilization, with a negligible degradation on the GPRS packet blocking probability. The conventional assumption for small M (e.g., M=2) underestimates the improvements obtained by RAS. The buffer mechanism provides a flexible manner in voice/data integrated wireless networks in helping to tune the system performance tradeoff, and an optimum buffer length is able to provide a good balance among the different performance metrics.

Chapter 15 gives an overview for the state-of-art work on multimedia over multi-hop networks. More specifically, the authors review the recent progress in efficient and resilient video coding and protection solutions, the network routing and transport protocols, link-layer QoS support enhancement, and cross-layer design for media over multi-hop environment.

ACKNOWLEDGMENTS

We are greatly indebted to all authors for making the book enjoyable and inspiring. Their tremendous efforts in providing excellent chapters and in taking as reviewers make the book more comprehensive and very informative. Secondly, we would like to express our appreciation to the staff members at Nova Scientific publishing, Inc. for their efficient work and for getting this book together in a nice shape. Finally, we would like to dedicate this book to our family and are grateful for tolerating our long time for editing of this book.

Wei Li
The University of Toledo, USA
E-mail wli@eecs.utoledo.edu

Yang Xiao
The University of Alabama, USA
E-mail: yangxiao@ieee.org

In: Adaptation Techniques in Wireless Multimedia Networks ISBN 1-59454-883-8
Editors: Wei Li and Yang Xiao, pp. 1-19

Chapter 1

LINK ADAPTATION TECHNIQUES FOR CARRIER SENSED MULTIPLE ACCESS BASED WIRELESS LOCAL AREA NETWORKS*

Li-Chun Wang†
Communication Engineering Department,
National Chiao Tung University, Hsinchu, 300, Taiwan

Abstract

Rate adaptation and power control are two essential link adaptation techniques to achieve high energy efficiency for wireless local area networks (WLANs). Rate adaptation can improve the goodput performance by dynamically selecting the data transmission rate, while power control can reduce the energy consumption. The emerging IEEE 802.11h standard extends the current high-speed rate adaptive 802.11a/g to support power control. This paper gives an overview on rate adaptation and power control schemes for WLANs. Several rate adaptation and power control schemes for WLANs using the carrier-sense multiple access (CSMA) medium access control (MAC) protocol are compared in terms of throughput, energy efficiency, and complexity.

Keywords: Carrier Sense Multiple Access (CSMA); power adaptation; rate adaptation; Wireless local area network (WLAN)

1. Introduction

Wireless local area networks (WLANs) have been used widely in both private and public areas, such as home, office and airport. Current the IEEE 802.11a/g WLANs can achieve date rates up to 54 Mbps and the future IEEE 802.11n can support 320 Mbps (as shown in Table 1). The exponentially growth of the WLANs drives the need of development

---

*This work was supported jointly by Lee and MTI Network Research Center of NCTU, the National Science Council and the Program for Promoting University Excellence of Ministry of Education, Taiwan under the contracts 91-2219-E-009-016 and EX-91-E-FA06-4-4.

†E-mail address: lichun@cc.nctu.edu.tw

Table 1. Different types of IEEE 802.11 WLANs

IEEE standard	Speed	Frequency band
802.11	1, 2 Mbps	2.4 GHz
802.11b	up to 11 Mbps	2.4 GHz
802.11a	up to 54 Mbps	5 GHz
802.11g	up to 54 Mbps	2.4 GHz
802.11n	up to 320 Mbps	2.4 and 5 GHz

Table 2. Eight PHY modes of IEEE 802.11a/g WLAN

Mode	**Data rate (Mbits/s)**	**Modulation**	BpS	**Coding rate,** R
1	6	BPSK	3	$1/2$
2	9	BPSK	4.5	$3/4$
3	12	QPSK	6	$1/2$
4	18	QPSK	9	$3/4$
5	24	16-QAM	12	$1/2$
6	36	16-QAM	18	$3/4$
7	48	64-QAM	24	$2/3$
8	54	64-QAM	27	$3/4$

of advanced link adaptation techniques to improve the goodput and energy efficiency for WLANs.

Because the radio channel condition changes rapidly, rate adaptation to dynamically select the physical layer (PHY) transmission parameters becomes an essential link adaptation technique to improve the goodput performance of the IEEE 802.11 WLANs. For example, the IEEE 802.11a/g WLAN provides eight PHY modes with data rates ranged from 6 Mbps to 54 Mbps, as shown in Table 2. In Section 3., we will discuss several rate adaptation schemes proposed for the IEEE WLANs [1, 2, 3, 4, 5, 6].

Power control is another important link adaptation technique for wireless communications. The effects of power control include two important performance aspects. First, power control can reduce the energy consumption, thereby saving battery energy. Second, power control can also decrease the interference to other co-channel users, thereby improving the signal quality and capacity. Based on 802.11 MAC and 802.11 a/g PHY standards, the IEEE 802.11h specifies a transmit power reporting mechanism to facilitate power control at the MAC layer [7]. In Section 4., we will introduce some joint power and rate adaptation mechanisms suggested for the IEEE 802.11 WLAN [8, 9]. We discuss the way to evaluate the goodput and energy efficiency performances for rate-adaptive and power-controlled CSMA-based WLAN in multipath fading channel [10]. Concluding remarks are given in Section 5..

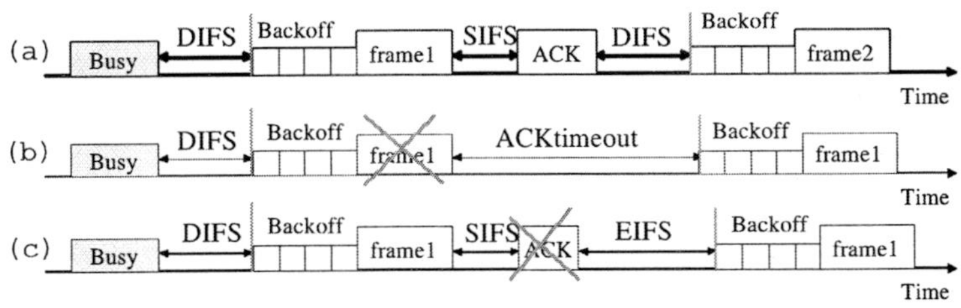

Figure 1. DCF of MAC protocol, i.e. (a)Successful frame transmission. (b)Frame retransmission due to ACK reception error. (c)Frame retransmission due to data frame reception error.

2. Background

We first give some background on the 802.11 MAC protocol and the radio channel modeling.

2.1. IEEE 802.11 MAC

In the IEEE 802.11 distributed coordination function (DCF), all the stations share the medium based on the carrier sense multiple access with collision avoidance (CSMA/CA) MAC protocol. Figure 1 shows three possible scenario when a frame is transmitted.

(1) Successful frame transmission, Fig. 1(a):

Following the CSMA/CA procedure, the transmitter sends a frame after a random backoff duration and a distributed inter-frame space (DIFS) duration. After receiving the frame and waiting a short inter-frame space (SIFS) duration, the corresponding receiver sends an acknowledgment (ACK) control frame back to the sender,

(2) Frame retransmission due to an ACK reception error, Fig. 1(b):

When the transmitter receives a corrupted ACK frame, the transmitter will wait an extended inter-frame space (EIFS) duration and retransmit the data frame. The EIFS is much longer than DIFS and SIFS.

(3) Frame retransmission due to data frame reception error, Fig. 1(c):

When waiting for an ACKtimeout duration, the transmission of the data frame may fail. Thus, the station will transmit the data frame again.

2.2. Generalized Fading Channels

In this paper, we consider the Nakagami fading channel model. The probability density function (pdf) of the signal amplitude in such a channel is given by

$$p_{\alpha}(x) = \frac{2m^m 2^{2m-1}}{\Omega^m \Gamma(m)} exp(-\frac{mx^2}{\Omega}), \tag{1}$$

where α, m, and Ω are the fading amplitude, the Nakagami shape factor, and the average fading power, respectively. When $m = 1$ the Nakagami fading channel becomes the Rayleigh fading channel. As $m \to \infty$, the Nakagami-m fading channel becomes the non-fading AWGN channel.

2.3. Transition Probability Matrix

In [6], a multipath fading channel can be divided into multiple states with E_b/N_o threshold η_i. Denote the transition probability p_{jk} as the probability that the signal changes from the channel state s_j to s_k. Let x_i be the E_b/N_0 of the i-th packet, then p_{jk} can be expressed as:

$$\begin{aligned} p_{jk} &= P(x_{i+1} \in (\eta_{k-1}, \eta_k) | x_i \in (\eta_{j-1}, \eta_j)) \\ &= \frac{P(x_{i+1} \in (\eta_{k-1}, \eta_k), x_i \in (\eta_{j-1}, \eta_j))}{P(x_i \in (\eta_{j-1}, \eta_j))}. \end{aligned} \tag{2}$$

We need to know the joint pdf $p(x_i, x_{i+1})$ to compute $P(x_{i+1} \in (\eta_{k-1}, \eta_k), x_i \in (\eta_{j-1}, \eta_j))$. In the Nakagami fading channel, the joint pdf $p(x_{t-\tau}, x_t)$ can be written as

$$\begin{aligned} p(x_{t-\tau}, x_t) = \frac{4(x_{t-\tau}x_t)^m}{(1-\rho)\Omega(m)\rho^{(m-1)/2}}(\frac{m}{\Omega})^{m+1}. \\ I_{m-1}(\frac{2m\sqrt{\rho}x_{t-\tau}x_t}{(1-\rho)\Omega})\exp(-\frac{m(x_{t-\tau}^2 + x_t^2)}{(1-\rho)\Omega}), \end{aligned} \tag{3}$$

where x_t, $I_{m-1}(\cdot)$, $\Gamma(m)$, ρ, and m are the E_b/N_0 at time instants t, the $(m-1)^{th}$-order modified Bessel function of the first kind, the Gamma function, the channel correlation coefficient, and the Nakagami shape factor, respectively.

Therefore, $P(x_{i+1} \in (\eta_{k-1}, \eta_k), x_i \in (\eta_{j-1}, \eta_j))$ and $P(x_i \in (\eta_{j-1}, \eta_j))$ can be expressed as

$$\begin{aligned} & P(x_{i+1} \in (\eta_{k-1}, \eta_k), x_i \in (\eta_{j-1}, \eta_j)) \\ & = \int_{\eta_{k-1}}^{\eta_k} \int_{\eta_{j-1}}^{\eta_j} p(x_i, x_{i+1}) dx_i dx_{i+1} \end{aligned} \tag{4}$$

and

$$P(x_i \in (\eta_{j-1}, \eta_j)) = \int_0^{\infty} \int_{\eta_{j-1}}^{\eta_j} p(x_i, x_{i+1}) dx_i dx_{i+1}. \tag{5}$$

We apply the transition probability to calculate the energy efficiency and throughput for the CSMA WLAN in the multipath fading channel.

3. Rate Adaptation

Here we discuss different rate adaptation schemes for the IEEE 802.11 WLAN, including automatic rate fallback method (ARF) [1], transmission statistics based rate adaptation [2], received signal strength (RSS) based rate adaptation [4], MAC Protocol Data Unit (MPDU)-based rate adaptation [5], and channel driven rate adaptation (CDRA) [6]. We also discuss the packet shaping issue [3],

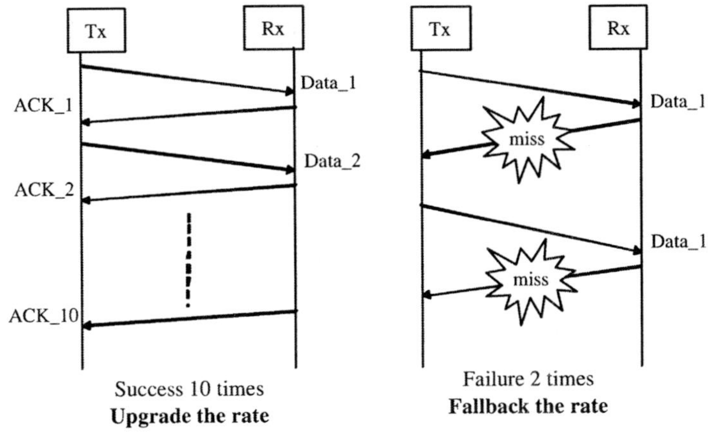

Figure 2. An illustrative example for the automatic rate fallback (ARF) algorithm.

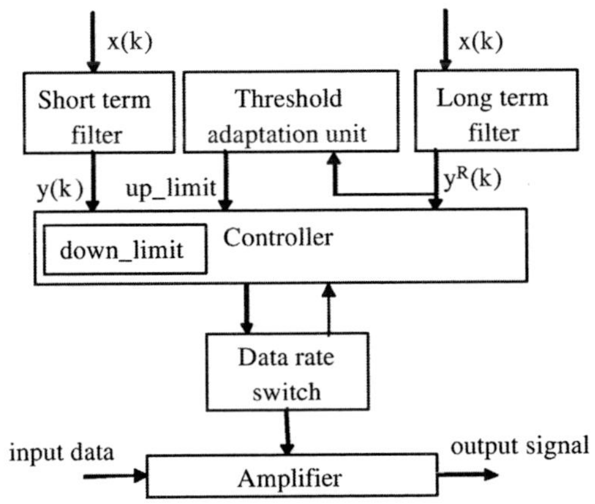

Figure 3. The architecture of the transmission statistics based rate adaptation.

3.1. Automatic Rate Fallback (ARF) Algorithm

One of the well-known rate adaptation schemes is the automatic rate fallback (ARF) algorithm [1] proposed in 1997. The ARF algorithm provides 1 and 2 Mbps rate adaptation for the IEEE 802.11 WLAN. Figure 2 shows the basic idea of rate adaptation. The ARF tracks timing function and the missing ACK frames to alternate between 1 and 2 Mbps PHY modes. If the transmitter does not correctly receive two consecutive ACK frames, the transmitter starts a counter and lowers its data rate in the next retry. When the number of successfully-received ACK frames reaches 10, the transmitter increases the data rate and stops the counter. Because counting the number of ACK frame requires a lot of time, the ARF algorithm may not react fast enough with respect to the fluctuation of the wireless channel.

Table 3. An illustrative example of the transmission statistics based rate adaptation.

Let Short term filter $y(k)$: $c = 0.8$, $d = 0.2$
Long term filter $y^R(k)$: $a = 0.2$, $b = 0.8$
Down_limit $= 0$

k: frame index	Input x(k)	y(k)	$y^R(k)$	up_limit
k = 1 (success)	1	0.8	0.2	0.7
k = 2 (success)	1	0.96	0.36	0.86
k = 3 (success)	1	0.992	0.488	0.98
k = 4 (failure)	-1	-0.6016	0.1904	0.6904
k = 5 (success)	1	0.6796	0.352	0.852

		Rate : {1, 2, 3, 4}
k = 1:	y > up_limit	Rate up to 2
k = 2:	y > up_limit	Rate up to 3
k = 3:	y > up_limit	Rate up to 4
k = 4:	y < up_limit	Rate down to 3
k = 5:	y within limit	Rate hold on 3

3.2. Transmission Statistics Method

In [2], the transmission statistics based rate adaptation scheme was proposed by considering both the long-term and short-term transmission success and failure statistics to determine the next transmission data rate . The functional diagram of the transmission statistics based rate adaptation scheme is illustrated in Table 3. In the figure, input value $x(k)$ to both the short term filter and long term filter is equal to +1 if the transmission attempt of frame k is successful, and -1 if the transmission failed. Thus, the transmission statistics according to the value of $x(k)$ can thus be used for rate adaptation. Let y and y^R represent the output values of the short term filter and the long term filter, respectively. The value of y is reset at each state transition, while y^R is not. By choosing suitable parameters a, b, c, and d, the short term filter and the long term filter can be expressed as

$$y(k) = c \cdot x(k) + d \cdot y(k-1), \tag{6}$$

and

$$y^R(k) = a \cdot x(k) + b \cdot y^R(k-1), \tag{7}$$

respectively. We further denote the parameters up_limit and $down\_limit$ as the upper and lower data rate switching thresholds to which the short term filter value y is compared for adjusting the data rate. Let

$$up\_limit = g + y^R(k), \tag{8}$$

where the parameter g is a constant. To illustrate the basic principles of the statistics based rate adaptation method, we choose $a = d = 0.2$, $b = c = 0.8$, $down\_limit = 0$ and the initial $up\_limit = 0.7$ in an example. At the first transmission, the scheme compares the short term filter $y(k)$ with up_limit. Because y(1)=0.8 is larger than up_limit, the data rate is upgraded to PHY mode 2. Following the same principle, the data rate of the second

Table 4. An illustrative example for the RSS-based rate adaptation.

	Input RSS	RSS_{avg}	Th[i, j]	Data Rate
k = 1 (success)	12 dBm	6 dBm	———	1 Mbps
k = 2 (success)	12 dBm	9 dBm	———	1 Mbps
k = 3 (success)	12 dBm	10.5 dBm	———	2 Mbps
k = 4 (success)	12 dBm	11.25 dBm	———	2 Mbps
k = 5 (failure)	———	———	Th[2,2] = 11.625	1 Mbps

Compare RSS_{avg} with Th[i, j]

k = 1 :	Rate_const	1 Mbps
k = 2 :	Rate_const	1 Mbps
k = 3 :	Rate_up	2 Mbps
k = 4 :	Rate_const	2 Mbps
k = 5 :	Rate_down	1 Mbps

Table 5. The initial thresholds for the RSS-based rate adaptation scheme.

Th[i ,j]	i = 1 (1 Mbps)	i = 2 (2 Mbps)	i = 3 (5.5 Mbps)	i = 4 (11 Mbps)
j = 1 (0-100 bytes)	———	———	———	———
j = 2 (100 -1000 bytes)	5	10	15	20
j = 3 (1000 -2400bytes)	———	———	———	———

transmission is PHY mode 3. Figure 3 lists the results of the short term filter $y(k)$, the long term filter $y^R(k)$, and up_limit in the remaining attempt of frame transmission.

3.3. Received Signal Strength (RSS) Based Method

A transmission rate adaptation method based on the received signal strength (RSS) from the received frames was proposed in [4]. The station selects a transmission rate according to the values of the average RSS (RSS_{avg}), thresholds ($Th[i, j]$) and the frame size. RSS_{avg} is equal to

$$RSS_{avg,new} = a_3 \cdot RSS_{avg,old} + a_4 \cdot RSS \ , \tag{9}$$

where RSS is the current received signal strength; $a_3 + a_4 = 1$, $a_3 \geq 0$, $a_4 \geq 0$. For frame i with size j, $Th[i, j]$ is updated according to the following rule:

$$Th[i, j]_{new} = a_1 \cdot Th[i, j]_{old} + a_2 \cdot RSS \ , \tag{10}$$

where $a_1 + a_2 = 1$, $a_1 \geq 0$, $a_2 \geq 0$.

Table 4 is an example to illustrate the operation of the RSS-based rate adaptation scheme, where $a_1 = a_2 = a_3 = a_4 = 0.5$ for (9) and (10) and the initial thresholds for adjusting the transmission rates in Table 5.

3.4. MAC Protocol Data Unit (MPDU)-based Rate Adaptation

Table 6. Required E_b/N_0 for PER = 10^{-1} on a multipath fading channel with different amount of fading m, where $m = 1$ is equivalent to the Rayleigh fading case.

Mbits/s	6	9	12	18	24	36	48	54
m=1	13.962	56.398	13.962	56.398	16.471	45.757	25.984	63.807
m=2	9.7597	27.016	9.7597	27.016	12.624	28.177	20.927	30.703
m=3	7.4303	17.457	7.4303	17.457	10.602	20.243	17.749	23.747
m=4	6.4016	13.381	6.4016	13.381	9.7048	16.651	16.354	20.529
m=5	5.8631	11.536	5.8631	11.536	9.2052	15.01	15.633	19.04
m=6	5.5178	10.553	5.5178	10.553	8.9484	14.12	15.214	18.217
m=7	5.298	9.9636	5.298	9.9636	8.7146	13.553	14.936	17.693
m=8	5.0805	9.3808	5.0805	9.3808	8.5324	13.187	14.707	17.344
m=9	4.9449	9.2545	4.9449	9.2545	8.3456	12.913	14.566	17.101
AWGN	4.0484	7.253	7.0494	10.262	13.537	17.058	21.164	23.084

In [5], a MAC Protocol Data Unit (MPDU) based link adaptation scheme was proposed for the IEEE 802.11a WLAN. The expected goodput of each PHY mode $\tilde{m}$ for different combination of data payload length ℓ, SNR (denoted by s), and retry count n is computed and compared to select the PHY mode $\tilde{m}^*$ having the maximum goodput. Clearly, the scheme can result in the "optimum" performance of goodput at the cost of complicated dynamic programming calculation. Now we summarize the operation of the scheme in [5]. First, we compute the goodput with the maximum retry n_{max} in each PHY mode:

$$\mathcal{G}(\ell, s, \tilde{m}, n_{max}) = \frac{E[data](\ell, s, \tilde{m}, n_{max})}{E[\mathcal{D}_{data}](\ell, s, \tilde{m}, n_{max})} \quad . \tag{11}$$

In Section 4., we will discuss how to calculate the expected delivered data payload $E[data](\ell, s, \tilde{m}, n_{max})$ and the expected transmission time $E[\mathcal{D}_{data}](\ell, s, \tilde{m}, n_{max})$. Then the "optimum" PHY mode with the maximum transmission attempt will be chosen by

$$\tilde{m}^*(\ell, s, n_{max}) = \arg \max_{1 \leq \tilde{m} \leq 8} \mathcal{G}(\ell, s, \tilde{m}, n_{max}), \tag{12}$$

and its associated goodput is

$$\begin{aligned} \mathcal{G}^*(\ell, s, n_{max}) &= \mathcal{G}(\ell, s, \tilde{m}^*(\ell, s, n_{max}), n_{max}) \\ &= \max_{1 \leq \tilde{m} \leq 8} \mathcal{G}(\ell, s, \tilde{m}, n_{max}). \end{aligned} \tag{13}$$

Similarly, the PHY mode to use for the nth transmission attempt and the corresponding maximum expected effective goodput are

$$\tilde{m}^*(\ell, s, n) = \arg \max_{1 \leq \tilde{m} \leq 8} \mathcal{G}(\ell, s, \tilde{m}, n), \tag{14}$$

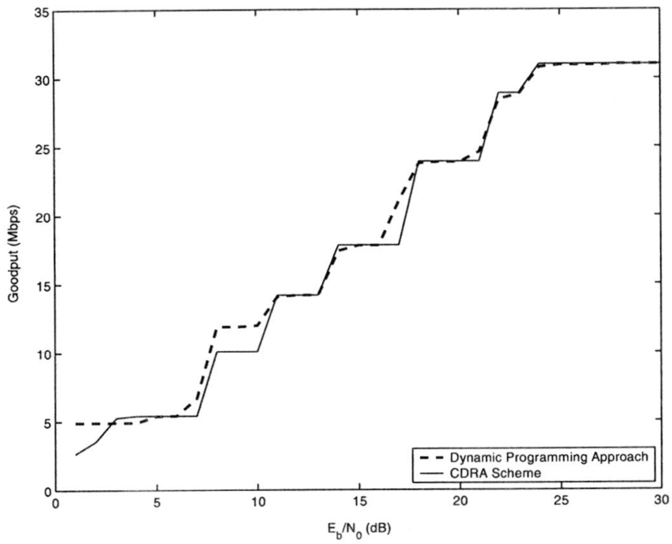

Figure 4. Goodput of our CDRA scheme and the dynamic programming link adaptation scheme over the AWGN channel.

and

$$\begin{aligned} \mathcal{G}^*(\ell, s, n) &= \mathcal{G}(\ell, s, \tilde{m}^*(\ell, s, n), n) \\ &= \max_{1 \le \tilde{m} \le 8} \mathcal{G}(\ell, s, \tilde{m}, n). \end{aligned} \tag{15}$$

By computing the goodput of every combination of ℓ, s, and n for every PHY mode, the system can always find the best PHY mode. However, because there are so many combinations of (ℓ, s, n), the computation can be very complex. Furthermore, the performance of the MPDU-based rate adaptation is related to different types of fading channels. Table 6 shows the required E_b/N_0 for various transmission data rates in the 802.11a/g WLAN under different fading channels, where the fading parameter $m - 1$ is equivalent to the Rayleigh fading channel and $m = \infty$ is for the AWGN channel. In the table, the non-underlined blocks are the "inefficient" PHY modes under the channel and the inefficient modes for different channels are different. Fig. 4 shows the goodput performance of the MDPU-based rate adaptation in the AWGN channel based on the dynamic programming approach. In the figure, it is demonstrated that the MDP-based rate adaptation scheme can choose different seven different PHY modes based on the value of E_b/N_o to provide MAC layer goodput from 5Mbps up to 32 Mbps. In the figure, we also shows the performance of a simpler and effective channel-driven rate adaptation (CDRA) algorithm [6] for comparison, which will be discussed next.

3.5. Channel-Driven Rate Adaptation (CDRA) Algorithm

In [6, 10], a simple and efficient channel-driven rate adaptation (CDRA) algorithm was proposed. One of key ingredients in the CDRA algorithm is the reduced mode rate adaptation. As already shown in Table 6, there are many PHY modes are indeed unnecessary

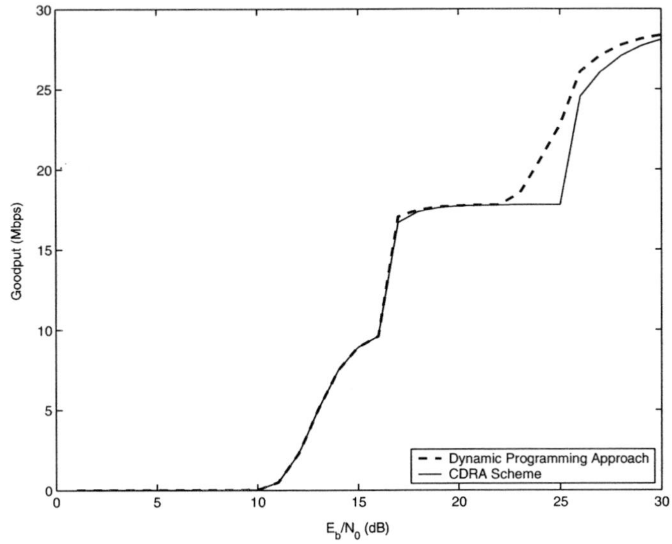

Figure 5. Comparison Goodput Performance of the CDRA scheme and the MPDU-based (dynamic programming) link adaptation scheme in the Rayleigh fading channel (Nakagami parameter m=1).

in different channel condition. Take $m = 1$ Rayleigh fading channel for example. The required E_b/N_0 for data rates of 6 Mbps and 12 Mbps to achieve the same packet error rate performance is indeed the same. Thus, the PHY mode with data rate of 6 Mbps is not needed. Similarly, by comparing the required E_b/N_o for data rates of 9, 18, and 24 Mbps, one can delete the PHY mode with 9 and 18 Mbps in the Rayleigh fading channel. Last, comparing the case for data rates of 36 and 48 Mbps, the PHY mode for 36 Mbps can be neglected. Thus, only four PHY modes for data rates of 12, 24, 48, 54 are necessary from the rate adaptation perspective.

Figure 5 compares the goodput performance of the dynamic programming approach and the CDRA scheme in the Rayleigh fading channel. In the Rayleigh fading channel, only four E_b/N_0 decision regions or three decision boundaries are neccesary. For $E_b/N_0 < 10$ dB, the goodput performances of both schemes are close to zero. For $E_b/N_0 > 22$ dB, the dynamic programming link adaptation scheme performs better than the CDRA scheme by 2 Mbps on average. In general, the goodput performances of the two schemes are very close. The CDRA algorithm can be viewed as the "sub-optimum" rate adaptation scheme, while its computation complexity is much lower due to the less PHY modes in the selection mode and taking advantage of channel information.

3.6. Packet Shaping

After determining the transmission rate based on a rate adaptation scheme, choosing an appropriate MAC service data unit (MSDU) size is another important issue. In [3], the idea of packet shaping was proposed. Because of supporting the multi-rate communications, the packets with the same size may require different transmit duration for different PHY modes. Such a variable transmission durations for different users is undesirable because of the presence of low data nodes reduces overall network capacity. The packet shaping mechanism sets a maximum MSDU size based on the node's data rate so that the maximum transmission time of all the nodes is approximately the same. The main advantage of the packet shaping mechanism is that it can make the network channel resources equally distributed among all the nodes. In such a manner, it is possible to improve network capacity when there are mixed rate nodes in the network. Table 7 is an example of the packet shaping scheme to have about the same maximum transmission time for eight different data rates in the IEEE 802.11a/g WLAN. Note that the MSDU size of packet shaping scheme is adapted according to the transmission data rate. For the transmission data rate of 54 Mbps the MSDU size is between 4096 and 3640 bytes, whereas for the transmission date of 9 Mbps the MSDU size becomes 910 bytes. With the packet shaping scheme, the transmission time of both data rates is about 607 μsec.

Table 7. An example of adjusting the maximum MSDU limits to have the same maximum transmission time by the packet shaping method.

PHY Rate (Mbps)	Maximum MSDU (Bytes)	Max Transmission time (micro second)
54	4096	4096 * 8 / 54 = 606.8
48	3640	3640 * 8 / 48 = 606.6
36	2730	2730 * 8 / 36 = 606.6
24	1820	1820 * 8 / 24 = 606.6
18	1360	1360 * 8 / 18 = 604.4
12	910	910 * 8 / 12 = 606.6
9	680	680 * 8 / 9 = 604.4
6	450	450 * 8 / 6 = 600

4. Power Adaptation

Another important link adaptation is power control, which can extend the lifetime of battery and reduce the interference. Here, we will introduce the IEEE 802.11h standard, path loss estimation based power control, noise margin based power control, the MiSer low-energy transmission strategy and joint channel-driven rate and power adaptation algorithm.

4.1. IEEE 802.11h

The IEEE 802.11h [7] extends the IEEE 802.11a PHY and MAC to provide dynamic frequency selection (DFS) and transmit power control (TPC), in which the newly-defined TPC report element including the transmit power field and link margin field are defined. Figure 6 illustrates the TPC adaptation process defined in the IEEE 802.11h . If the station management entity (SME) of 802.11 station decides to adapt transmit power, the SME will send the request to its MAC sub-layer management entity (MLME). Then, the MLME sends the

TPC request frame and waits for the report of the TPC protocol to adjust transmit power. Table 8 shows the maximum allowable output power.

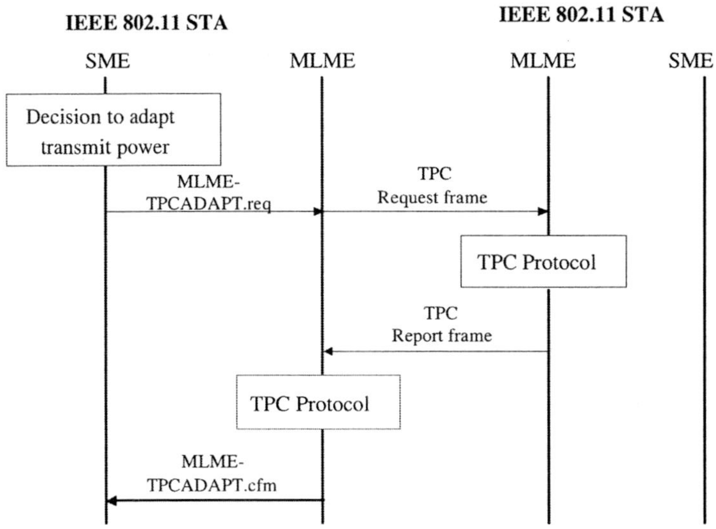

Figure 6. Transmit power control (TPC) adaptation procedures according to the 802.11h.

Table 8. Output transmit power level for IEEE 802.11a WLAN.

Frequency band (GHz)	United States (mW)	CEPT (EIRP)
5.15 -5.25	40 (2.5 mW/MHz)	200 mW
5.25 -5.35	200 (12.5 mW/MHz)	200 mW
5.470 -5.725	———	1 W
5.725 -5.825	800 (50 mW/MHz)	———

4.2. Path Loss Estimation based Power Control

The current IEEE 802.11 does not provide a mechanism to adjust the transmit power dynamically. Typically, each station uses a fixed transmission power level for all the frame transmission throughout its life time. In [8] a simple transmit power control scheme was proposed based on the IEEE 802.11a PHY and/802.h MAC WLAN by using the path-loss estimation. Figure 7 is the flow chart illustrating the power adaptation process. Upon receiving the data frame, the station determines the received power and then estimates the path loss. The path loss can be computed from the difference between the received signal power level and transmit power level. Initially, the transmitting station can measure the path loss by sending an Request-to-Send (RTS) frame and the corresponding Clear-to-Send (CTS) frame shall include the transmission power level in its SERVICE files. Thus, the

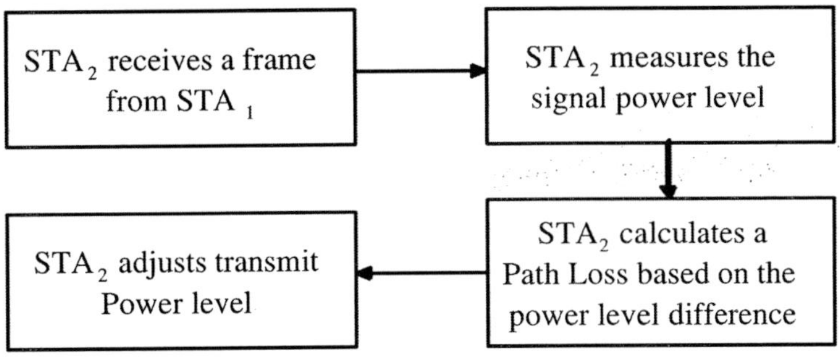

Figure 7. The flow chart of the operation steps of the path loss estimation based power adaptation.

sender of the RTS frame can measure the path loss. After that, the path loss (PL) is updated in every new received data frame by giving a different weight the new and old path loss values:

$$PL = a_1 PL_{new} + a_2 * PL \tag{16}$$

where $a_1 + a_2 = 1$, $a_1 \geq 0$, $a_2 \geq 0$, and PL_{new} represents the estimated path from the new frame reception. Based on the updated path loss information, the transmit power of a receiving station is adjusted. Note that the path loss in (16) is used only when the difference between PL and PL_{new} is smaller than a preset threshold. Otherwise, a new estimation of path loss is performed via an RTS/CTS frame exchange, or the maximum transmission power is used without path loss estimation.

4.3. MiSer

In [9], the MAC Protocol Data Unit (MPDU)-based rate adaptation scheme in Section is extended to incorporate power control aiming to minimize the energy efficiency. The proposed minimum-energy transmission strategy, called MiSer, has two steps. First, the energy efficiency (denoted by ζ) for different rate-power combination (R, P_t) is computed offline, where ζ is defined as the ratio of the expected delivered data payload L to the expected total energy consumption E:

$$\zeta(R, P_t, l, s, SRC, LRC) = \frac{L(R, P_t, l, s, SRC, LRC)}{\xi(R, P_t, l, s, SRC, LRC)} \,. \tag{17}$$

The the quadruplet (l, s, SRC, LRC) represents the data transmission status, where l is data payload length, s is the path loss, SRC is the short retry limit, and LRC is the long retry limit. Second, at runtime the most energy-efficient rate-power pair that maximizes the energy efficiency ζ is selected by table lookup:

$$< R^*, P_t^* >= arg \max_{R, P_t} \zeta(R, P_t, l, s, SRC, LRC). \tag{18}$$

The expected delivered data payload L can be calculated recursively as follows:

$$\begin{aligned} L(R, P_t, l, s, SRC, LRC) &= (1 - P_{c,rts})(1 - P_{e,data}) \cdot l \\ &\quad + (1 - P_{c,rts}) P_{e,data} L(R^*, P_t^*, l, s, SRC, LRC + 1) \\ &\quad + P_{c,rts} L(R^*, P_t^*, l, s, SRC + 1, LRC) \end{aligned} \tag{19}$$

with the boundary conditions:

$$\begin{cases} L(R, P_t, l, s, dot11ShortRetryLimit, LRC) = 0 \ , \\ L(R, P_t, l, s, SRC, dot11LongRetryLimit) = 0 \ , \end{cases} \tag{20}$$

where $P_{c,rts}$ is the RTS collision probability; the data transmission error probability $P_{e,data}$ is a function of R, P_t, l, and s. Similarly, the expected total energy consumption E can be recursively calculated as

$$\begin{aligned} &\xi(R, P_t, l, s, SRC, LRC) = \bar{\xi}(SRC, LRC) + \bar{\xi}_{freeze} + (1 - P_{c,rts})(1 - P_{e,data}) \cdot \\ &[\xi_{rts-sifs-cts-sifs} + \xi_{data} + \xi_{sifs} + \xi_{ack} + \xi_{difs}] + (1 - P_{c,rts})P_{e,data} \cdot \\ &[\xi_{rts-sifs-cts-sifs} + \xi_{data} + \xi_{ack-tout} + \xi(R^*, P_t^*, l, s, SRC, LRC + 1)] \\ &+ P_{c,rts}[\xi_{rts} + \xi_{cts-tout} + \xi(R^*, P_t^*, l, s, SRC + 1, LRC)] \end{aligned} \tag{21}$$

with the boundary conditions:

$$\begin{cases} \xi(R, P_t, l, s, dot11ShortRetryLimit, LRC) = 0 \ , \\ \xi(R, P_t, l, s, SRC, dot11LongRetryLimit) = 0 \ . \end{cases} \tag{22}$$

In (21), the parameter $\bar{\xi}_{freeze}$ (the energy consumption while the backoff counter is frozen) is discussed in more detail in [9], and the average energy consumption while the backoff counter is decrementing can be computed by

$$\bar{\xi}(SRC, LRC) = \frac{\min[2^{SRC+LRC} \cdot (CW_{min} + 1) - 1, CW_{max}]}{2} \cdot T_{slot}P_{r\_mode}, \tag{23}$$

where T_{slot} is the time slot duration and P_{r_mode} is the power consumption in the receive mode. Besides,

$$\begin{aligned} \xi_{rts-sifs-cts-sifs} &= \xi_{rts} + 2\xi_{sifs} + \xi_{cts} \ , \\ &= T_{rts}P_{t\_mode} + 2T_{SIFS}P_{r\_mode} + T_{cts}P_{r\_mode} \ , \end{aligned} \tag{24}$$

where $T_{rts} = 52\mu s$ and $T_{cts} = 44\mu s$ are the slot time for an RTS frame and a CTS frame, respectively; $T_{SIFS} = 16\mu s$ is the SIFT time; P_{t_mode} and P_{r_mode} are the power consumption in the transmit mode and the receive mode, respectively. Also, $E_{difs} = T_{DISF}P_{r\_mood} = 34\mu s P_{r\_mood}$. Moreover, we can compute $E_{data}(l, R, P_t)$ (the energy consumption with the pay load length l and the selected rate R and power P_t as follows:

$$\begin{aligned} \xi_{data} &= P_{t\_mode}(P_t)\Big[T_{PLCP\_Pramble} + T_{PLCP\_SIG} + T_{symbol} \cdot \\ &\left\lceil \frac{MAC\_Header + FCS + (PLCP\_Header + Tailbit)|/8 + l}{BpS(R)} \right\rceil\Big] \ , \end{aligned} \tag{25}$$

where $T_{PLCP\_Pramble} = 16\mu s$, $T_{PLCP\_SIG} = T_{symbol} = 4\mu s$, $MAC\_Header = 24$ bytes, FCS (Frame Checking Sequence) = 4 bytes, $PLCP\_Header = 16$ bits, $Tailbit = 6$ bits, and BpS (bytes per symbol) is dependent on the selected PHY mode or the selected rate. The energy consumed to receive the ACK frame is

$$\begin{aligned} \xi_{ack} &= P_{t\_mode}(P_t)\Big[T_{PLCP\_Pramble} + T_{PLCP\_SIG} + T_{symbol} \cdot \\ &\left\lceil \frac{MAC\_Header\_ACK + (PLCP\_Header + Tailbit)|/8}{BpS(R)} \right\rceil\Big] \ , \end{aligned} \tag{26}$$

where $MAC\_Header\_ACK = 14$ bytes. Furthermore, the energy consumption for a CTS and ACK timeout $E_{cts-tout}$, $\xi_{ack-tout}$ are computed as

$$\xi_{cts-tout} = \xi_{sifs} + \xi_{cts} + T_{slot}P_{r\_mode} \quad , \tag{27}$$

and

$$\xi_{ack-tout} = \xi_{sifs} + \xi_{ack} + T_{slot}P_{r\_mode} \quad . \tag{28}$$

Because all the rate-power pair (R, P) are compared to select the one pair (R^*, P^*) maximizing the energy efficiency $L(R^*, P^*, l, s, SRC, LRC)/\xi(R^*, P^*, l, s, SRC, LRC)$, this approach should give an optimal solution at a cost of high complexity.

4.4. Channel-Driven Rate and Power Adaptation (CDRPA) Algorithm

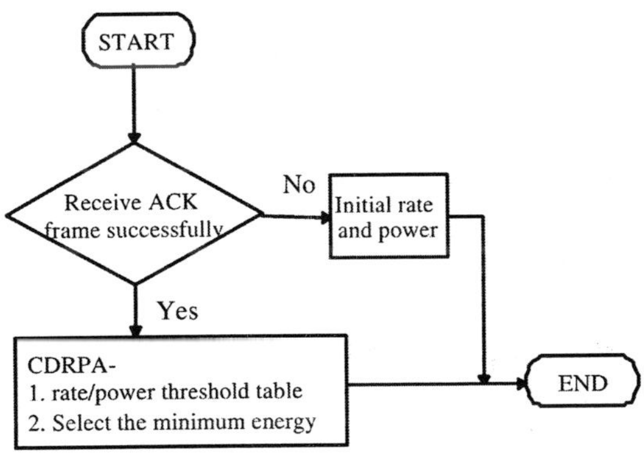

Figure 8. The CDRPA algorithm flow chart.

The CDRA algorithm in Section 3.5. can be modified to a joint rate and power adaptation algorithm for WLANs by measuring the signal power level in the received ACK frame and adapting the power and rate according to the energy efficiency. Figure 8 illustrates the operation steps of the proposed channel driven rate and power adaptation (CDRPA) algorithm. Let $SNR_{(req)}(R(i))$ be the required SNR to achieve the packet error rate (PER) requirement for PHY mode m or rate $R(m)$ with the maximum transmission power P_{t_max}. Upon receiving the ACK frame, based on the measured SNR (denoted by $SNR_{(mea)}$), the lowest possible transmit power $P_t(m)$ to maintain the required PER for each PHY mode m or rate $R(m)$ is determined from the predetermined threshold table of $SNR_{(req)}(R(m))$ as follows:

$$P_t(m) = P_{t\_max} - [SNR_{(mea)} - SNR_{(req)}(R(m))] \quad . \tag{29}$$

Then, the pair $< P_t(m), R(m) >$ with the highest energy efficiency is selected for the next data frame:

$$< R^*, P_t^* >= arg \max_{R(m),P_t} \zeta(P_t(m), R(m), m, l, s, n), \tag{30}$$

where the energy efficiency $\zeta(P_t(m), R(m), m, l, s, n)$ is a function of $P_t(m)$, $R(m)$, the PHY mode (m), the payload length (l), the channel state (s), and the frame retry counts

(n). Since both $P_t(m)$ and $R(m)$ are dependent of the PHY mode m, the energy efficiency $\zeta(P_t(m), R(m), m, l, s, n)$ can be simplified to the function of the quadruplet (m, l, s, n). Define

$$\zeta(l, s, m, n) = \frac{E[L](m, l, s, n)}{E[\xi](m, l, s, n)}, \tag{31}$$

where $E[L](m, l, s, n)$ and $E[\xi](m, l, s, n)$ are the expected data pay load and the energy consumption, respectively. Adopting the concept of the reduced mode rate adaptation in Section 3.5., a multipath fading channel can be classified into multiple states, say N, with the transition thresholds η_i, $i = 1$ to N, and the transition probability p_{jk} for the current frame's SNR $s \in (\eta_{j-1}, \eta_j)$ and the next frame's SNR $r \in (\eta_{k-1}, \eta_k)$. Then, the average delivered data payload can be computed as

$$\begin{aligned} E[L](m, l, s, n) = {} & P_S(m, l, s) \cdot l + [1 - P_S(m, l, s)] \\ & \cdot \sum_{k=1}^{N} p_{jk} \cdot E[L](m(r), l, r, n+1), \end{aligned} \tag{32}$$

where the successful probability of frame transmission $P_S(m, l, s)$ is

$$P_S(m, l, s) = [1 - P_{e,data}(m, l, s)] \cdot [1 - P_{e,ack}(s)] \ , \tag{33}$$

and $P_{e,data}(m, l, s)$ and $P_{e,ack}(s)$ are the error probability of sending a data frame and an ACK frame, respectively. The average energy consumption of a frame transmission, $E[\xi]$ can be expressed as:

$$\begin{aligned} E[\xi](m, l, s, n) = {} & P_{r\_mode} T_{bk}(n) + \\ & P_S(m, l, s) \cdot [P_t(m) T_{data}(l, m) + P_{r\_mode}(T_{sifs} + T_{ack} + T_{difs})] + \\ & P_{e,data}(m, l, s)\{P_t(m) T_{data}(l, m) + P_{r\_mode}[T_{ack-tout} - T_{data}(l, m)]\} + \\ & (1 - P_{e,data}(m, l, s)) P_{e,ack}(s) [P_t(m) T_{data}(l, m) + P_{r\_mode}(T_{sifs} + T_{eifs})] + \\ & [1 - P_S(m, l, s)] \sum_{k=1}^{N} p_{jk} \cdot E[\xi](m(r), l, r, n+1) \\ = {} & P_t(m) T_{data}(l, m) + \\ & P_{r\_mode}\Big\{ T_{bk}(n) + P_S(m, l, s)(T_{sifs} + T_{ack} + T_{difs}) + P_{e,data}(m, l, s) \times \\ & (T_{ack-tout} - T_{data}(l, m)) + (1 - P_{e,data}(m, l, s)) P_{e,ack}(s)(T_{sifs} + T_{eifs}) \Big\} + \\ & [1 - P_S(m, l, s)] \sum_{k=1}^{N} p_{jk} E[\xi](m(r), l, r, n+1) \ , \end{aligned} \tag{34}$$

where $T_{data}(l, m)$, $T_{bk}(n)$, and T_{ack} can be obtained from (25), (23), and (26) , respectively, i.e.,

$$T_{data}(l, m) = 20\mu s + T_{symbol} \cdot \left\lceil \frac{30.75 + l}{BpS(m)} \right\rceil, \tag{35}$$

$$T_{ack} = 20\mu s + T_{symbol} \cdot \left\lceil \frac{16.75}{BpS(1)} \right\rceil \ , \tag{36}$$

and

$$T_{bk}(n) = \frac{min[2^{n-1} \cdot (CW_{min} + 1) - 1, CW_{max}]}{2} \cdot T_{slot} \ . \tag{37}$$

Now we make a simple example to illustrate the principles of the proposed CDRPA algorithm. Referring to Tables 1. and 3.4. under the Nakagami fading channel with the fading parameter $m = 5$, the rate schemes of the CDRA algorithm for the IEEE 802.11a/g WLAN is implemented for the PHY modes 3, 5, 6, 7 and 8 with transmission rates 12, 24, 36, 48, and 54 Mbps and the corresponding SNR thresholds of 5.86, 9.21, 15.01, 15.63, and 19.04 dB, respectively. In this example, we consider the case with the fixed-length payload, i.e. $l = 1500$ bytes and the dynamic range of transmission power is from 16 dBm to 30 dBm (i.e., 15 power levels). Assume that the measured SNR is equal to 16.5 dB. As shown in the table, the minimal energy consumption is the case with $P_t = 20$ dBm and transmission date of 12 Mbps in the PHY mode $m = 3$. The comparison of the CDRPA algorithm to select the best combination of rate and power is only 5 times. However, the MiSer algorithm requires 120 times comparisons to get the best pair of $< P_t, R(m) >$ in the case of eight PHY modes and 15 power levels. Thus the CDRPA algorithm lower the complexity by about 97 % by taking advantage of the channel information and reduced PHY modes in the multipath fading channel. The cost to implement the CDRPA algorithm is to to build up the rate threshold tables, which can be done off-line in the physical layer simulations.

Figure 9 shows the performance for the energy efficiency of CDRPA, MiSer, and the MPDU-based rate adaptation only for various data payloads. As shown in the figure, both the MiSer and the CDRPA can improve the energy efficiency over the MDPU-based rate adaptation. More importantly, we find the CDRPA algorithm can approach the performance of the the MiSer algorithm, while has much lower computation complexity.

Table 9. Required $SNR_{(req)}$ for PER = 10^{-1} in a Nakagami multipath fading channel with the fading parameter $m = 5$. Let SNR_{mea} = 16.5 dB with $P_{t,max} = 30$ dBm and $P_t \in [16 \text{ dBm}, 30 \text{ dBm}]$

PHY mode (m)	3	5	6	7	8
Data rates (Mbits/s)	12	24	36	48	54
$SNR_{(req)}(m)$ dB	5.86	9.21	15.01	15.63	19.04
$\lfloor SNR_{mea} - SNR_{(req)}(m) \rfloor$	10	6	1	0	—
$P_t(m)$, dBm	20	24	29	30	—
$T_{data}(m, l = 1500 \text{ bytes})$, μsec	1044	532	364	276	—
ξ_{min} (Joule $\times 10^{-6}$)	104.4	133.6	289.1	276	—

5. Summary

In this paper, various rate and power adaptation schemes for the IEEE 802.11 WLAN have been introduced. First, rate adaption schemes aim to enhance the throughput. We have discussed rate adaptation schemes, including the ARF, the transmission statistics based, the

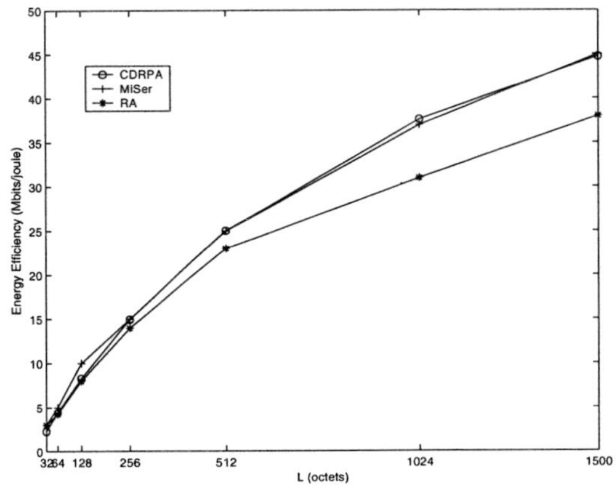

Figure 9. Comparison of energy efficiency the Miser, the CDRPA, and the MPDU-based rate adaptation for versus different data payload length.

RSS-based, the MPDU-based, and the channel driven rate adaptation (CDRA) schemes. The MPDU-based rate adaptation is the optimal scheme in the sense of delivering the highest throughput through the dynamic programming search approach, while the CDRA scheme can approach the performance of the MPDU-based scheme at a lower computation cost by taking advantage of the channel information. Second, power control aims to improve the energy efficiency. We introduced the IEEE 802.11h MAC, the path-loss based power adaptation, the MiSer power and rate adaptation, and the channel driven rate and power adaptation (CDRPA). We show that both Miser and CDRPA can improve the energy efficiency, i.e., the ratio of the data load per energy consumption, compared to the rate adaptation only case. Furthermore, the Miser algorithm has the best performance, whereas the CDRPA algorithm has the sub-optimal performance and the lower computation complexity.

References

[1] A. Kamerman and L. Monteban, "WaveLAN-II: A High-Performance Wireless LAN for the Unlicensed Band," *Bell Labs Technical Journal*, pp. 118–133, Summer 1997.

[2] U. Eckhardt, M. Lenk, and Grell, "Transmitter Adjustment Based Transmission Statistics," Pub. No. US 2004/0086058, Pub. Date: Jun. 19, 2003.

[3] L. J. C. JR., Z. Kostic, K. K. Leung, and H. Yin, "Packet Shaping for Mixed Rate 802.11 Wireless Networks," Pub. No. US 2003/0133427, Pub. Date: Jul. 17, 2003.

[4] J. D. Prado and S. Choi, "Method and System for Generating and Updating Transmission Rate for Link Adaptation in IEEE 802.11 WLAN," Pub. No. US 2004/0017790, Pub. Date: Jan. 29, 2004.

[5] D. Qiao, S. Choi, and K. G. Shin, "Goodput analysis and link adaptation for IEEE 802.11a wirelesslans," *IEEE Trans. on Mobile Computing*, vol. 1, no. 4, pp. 278–292, Oct.-Dec. 2002.

[6] W. C. Liu, L. C. Wang, and Y. W. Lin, "Physical Layer Effects on the MAC Goodput Performance for the Rate Adaptive IEEE 802.11a/g WLAN," *IEEE Wireless Communications and Networking Conference*, vol. 3, pp. 21–25, March 2004.

[7] *Part 11: Wireless LAN Medium Access Control (MAC) and Physical Layer (PHY) specifications: Spectrum and Transmit Power Management Extensions in the 5 GHz band in Europe*, Oct. 2003.

[8] S. Choi and A. A. Soomro, "Updating Path Loss Estimation for Power Control and Link Adaptation in IEEE 802.11h WLAN," Pub. No. US 2002/0168993, Pub. Date: Nov. 14, 2002.

[9] D. Qiao, S. Choi, A. Jain, and K. G. Shin, "MiSer: An Optimal Low-Energy Transmission Strategy for IEEE 802.11 a/h," *The ACM Annual International Conference on Mobile Computing and Networking*, pp. 14–19, September 2003.

[10] Li-Chun Wang, Ya-Wen Lin, and Wei-Cheng Liu, "Cross-layer Goodput Analysis for Rate Adaptive IEEE 802.11a WLAN in the Generalized Nakagami Fading Channel," IEEE ICC pp. pp. 2312-2316, June, 2004.

In: Adaptation Techniques in Wireless Multimedia Networks ISBN 1-59454-883-8
Editors: Wei Li and Yang Xiao, pp. 21-32

Chapter 2

A RATE-ADAPTIVE MULTIMEDIA MULTICASTING MECHANISM IN MULTI-RATE 802.11 WIRELESS LANS*

***Youngsam Park*[1†], *Yongho Seok*[1‡], *Taekyoung Kwon*[1§], *Yanghee Choi*[1¶] *and Jean-Marie Bonnin*[2‖]**
[1]School of Computer Science and Engineering
Seoul National University, San 56-1, Shillimdong, Gwanak-gu, Seoul, Korea
[2]Network and Multimedia Department
ENST Bretagne, 2, rue de la Chataigneraie - CS 17607
35576 Cesson Sevigne Cedex - France

Abstract

IEEE 802.11 WLAN service is broadening its service area rapidly. One of the major driving factors of WLANs is multimedia services such as VOD or HDTV delivered by multicast protocols. However, there have been little research on how to efficiently operate multicast protocols in IEEE 802.11 WLANs. The most widespread variants of IEEE 802.11 standards, 802.11b and 802.11a, are characterized by the rate adaptation mechanism depending on time-varying wireless link conditions. However the current IEEE 802.11 WLAN standard does not utilize the rate adaptation function for multicast transmissions. In this paper we propose a new rate adaptation mechanism for multimedia multicasting in IEEE 802.11 WLAN environments, called SARM (SNR-based Auto Rate for Multicast). SARM is based on the RBAR (Receiver Based Auto Rate) protocol. Therefore SARM adapts its transmission rate based on the SNR values of mobile nodes. In order to obtain SNR values, since there are no RTS-CTS messages or MAC layer ACK messages in multicast transmissions, SARM needs an additional signalling method by which the AP can estimate channel quality (SNR). By

*This work was supported in part by the Brain Korea 21 project of Ministry of Education and in part by the National Research Laboratory project of Ministry of Science and Technology, 2005, Korea.

†E-mail address: yspark@mmlab.snu.ac.kr
‡E-mail address: yhseok@mmlab.snu.ac.kr
§E-mail address: tk@mmlab.snu.ac.kr
¶E-mail address: yhchoi@mmlab.snu.ac.kr
‖E-mail address: jm.bonnin@enst-bretagne.fr

adapting multicast transmission rate based on SNR, SARM guarantees the best quality of multimedia service depending on the mobile nodes' link conditions. Moreover the proposed protocol minimizes the portion of wireless channel resource for multimedia multicasting, so that it can allocate more bandwidth to other unicast traffic.

Keywords: Rate Adaptation, IEEE 802.11, Multicast, Multimedia

1. Introduction

IEEE 802.11 Wireless LAN standards [1][2] are now helping people stay connected to internet while they are away from home and office. With laptops or PDAs, people can easily access internet at HotSpot locations. Currently wireless LAN service providers are actively increasing the number of HotSpot locations to enhance internet accessibility. According to the investigation provided by Gartner, Inc.[3], the number of HotSpot users worldwide will increase to total 30 million in 2004, up from 9.3 million users in 2003. It also states that the number of HotSpot locations in USA has increased rapidly in recent 3 years, to 71,000 in 2003 from 14,753 in 2002 and 1,214 in 2001. In Korea, wireless LAN service called "Nespot" was launched in 2001. Currently it provides more than 16,000 HotSpot locations.

One of the major driving factors of wireless LAN service is multimedia streaming. With the help of several key technologies such as MPEG4 and H.263, wireless LAN service provides various multimedia contents including HDTV or Video On Demand (VOD). The most essential technology in these services is the ability to broadcast or multicast identical multimedia contents to a group of users simultaneously. Multicast is an efficient transmission method since it reduces the demand on bandwidth resources in networks. It will play a key role in delivering multimedia contents shared by many users, such as pay-per-view broadcasts of sporting events, as well as interactive multimedia applications, such as interactive television and video conference.

Compared to traditional application traffic, multimedia traffic requires more bandwidth and QoS (Quality of Service) support. These restrictions are more critical and more difficult to fulfill in wireless environments, since WLANs have relatively low bandwidth and unstable wireless link channel. However, IEEE 802.11 WLAN does not provide any mechanisms to overcome these restrictions. In fact, since multicast transmission shows quite different characteristics from unicast transmission in several ways, it is difficult to provide such mechanisms in IEEE 802.11 WLAN environments.

The most critical component which influences the performance of communication in a wireless environment is the transmission rate adaptation mechanism [4], [5]. Currently, IEEE 802.11b provides channel rate up to 11Mbps and IEEE 802.11a/g provide up to 54Mbps. To utilize this multi-rate capability, AP needs to select adequate channel rate through channel estimation. In unicast transmissions, AP can collect channel state information of MNs (Mobile Nodes) through feedback signals such as MAC layer ACK or RTS-CTS messages and adapt its transmission rate based on those feedback signals. However, in multicast transmissions there are no feedback signals because of the one-to-many communication scenario of multicast transmission. This means that AP cannot easily collect state information of MNs which are participating in a multicast group. Therefore traditional transmission rate adaptation mechanisms such as ARF (Auto Rate Fallback) [6] or

RBAR (Received Based Auto Rate) [7] cannot be used in multicast transmissions since these mechanisms are based on wireless link states of MNs. As the current IEEE 802.11 standards provide up to 11 or 54Mbps channel rate, multicast transmission at the channel rate of basic rate set (e.g., 1 or 2 Mbps in 802.11b) is a significant waste of wireless channel resource.

In this paper, we propose transmission rate adaptation mechanism, called SARM (SNR-based Auto Rate for Multicast). This mechanism aims to maintain adequate quality of multimedia streaming while using minimum wireless channel resource.

The rest of this paper is organized as follows. Section 2. gives a brief overview of traditional transmission rate adaptation mechanisms as well as the motivation behind our work. We describe our transmission rate adaptation mechanism, SARM, in section 3., and present experimental results in section 4.. Finally, we conclude our paper in section 5..

2. Background and Motivation

2.1. Problems of Multicast in IEEE 802.11 Wireless LAN

In this section we briefly investigate the problems of multicast transmission in IEEE 802.11 WLAN environment. Before proceeding, we will describe main features of multicast transmission in IEEE 802.11, which is handled quite differently from unicast transmission.

First, there is no MAC layer ACK and therefore no retransmission mechanism in multicast transmission in IEEE 802.11. This means that all multicast frames have only one chance to be transmitted from AP to MNs. If bit errors occur while transmitting a multicast frame, it is dropped and never recovered. In wireless environments, it is essential to minimize occurrence of bit errors during data frame transmission. Therefore, in the current 802.11 standards, the AP transmits the multicast frame at basic rate set to minimize bit error rate.

Second, multicasting aims at multimedia contents service. Generally, a little distortion caused by a few bit errors in multimedia contents are considered tolerable by service subscribers.. In the case of MPEG video, the quality is measured by PSNR value, where a value over specific level is considered acceptable [8]. This means that absolute error-free transmission is not necessary in multimedia multicast. Therefore in multimedia multicast transmission it is more effective to use error-contained frame instead of just dropping them.

Third, multimedia multicast traffic generally requires relatively higher bandwidth than unicast traffic and it can lead to traffic congestion in AP. It is impossible to apply the over-provisioning policy to IEEE 802.11 WLANs, because it has strictly limited bandwidth resource. Therefore the QoS mechanism in the MAC layer is required for the transmission of multicast frames in IEEE 802.11.

Based on these features, we can draw the following problem of IEEE 802.11 in transmitting multimedia multicast traffic. All commercial APs provide only fixed (relatively very low) transmission rate for the multicast transmission. In fact, the rate adaptation mechanism in IEEE 802.11 is not standardized by any documents, making each manufacturer to choose its own rate adaptation mechanism. In the case of unicast transmission, a few rate adaptation mechanism have been implemented and deployed. However, due to the absence of feedback signal such as MAC layer ACK frame or RTS-CTS message frame in multicast, the rate adaptation mechanism for multicasting has never been proposed. Currently some APs provide manual configuration functionality which enables administrators to select transmission rate for multicasting. On the other hand, some APs only provide transmission rate fixed by manufacturer, in which case the rate is usually the basic service rate, 1Mbps or 2Mbps. Though rate adaptation is one of the critical components to enhance throughput, the natural feature of multicast in IEEE 802.11 limits the method for the channel estimation, which in turn, strictly prohibits the rate adaptation mechanism. According to several researches on multirate capability of IEEE 802.11, it is a well known fact that each device needs to adapt its transmission rate dynamically to achieve high performance in wireless environments. To maximize resource utilization and quality of multimedia contents in wireless environments, it is essential to adapt multicast transmission rate dynamically. Therefore the rate adaptation mechanism for multicast transmissions is the major issue of this paper.

Based on the discussion above, we propose a new rate adaptation mechanism which efficiently supports the multimedia multicast traffic. The proposed mechanism aims to enhance performance of multicast transmission which delivers multimedia contents. It is designed to maintain adequate quality of multimedia contents transmitted through multicast, and also to achieve the throughput gain for unicast transmission by allocating minimum channel resource on multicast transmission.

3. Multicast Transmission Rate Adaptation Mechanism

Transmission rate adaptation mechanism for multicast traffic should be executed differently from traditional mechanism for unicast. In traditional mechanism, an AP collects channel states information of MNs using MAC layer ACK or RTS-CTS messages. Through these feedback signals, AP collects channel states information such as SNR value or the number of successful/failed transmission between AP and specific MN. Based on this information AP dynamically determines the transmission rate for specific MN from 1Mbps up to 11Mbps in IEEE 802.11b or up to 54Mbps in IEEE 802.11a/g. However in multicast, due to absence of traditional feedback signaling an AP cannot collect channel states information of MNs in the same way as in unicast transmission.

In our proposed transmission rate adaptation mechanism, SARM, we add supplementary signaling method through which AP can collect the state information of MNs which are participating in a multicast group. In this scheme, AP uses SNR information between AP and MNs as a criterion of transmission rate selection. By using feedback signal transmitted from MNs, AP collects SNR information for all MNs participating in some multicast group and then determines transmission rate for multicast. This kind of rate adaptation methods based on SNR information has been proved to be feasible in several researches conducted before [7], [12], [14].

3.1. Channel Probing Mechanism in SARM

The key part of rate adaptation mechanism is for each MN to measure the channel state, and to inform the information to the AP. To make this possible, we introduce an additional signaling method. Algorithm 1 describes the scheme of signaling method which is conducted from MN and AP.

Algorithm 1 Channel Probing Mechanism Description

```
Multicast MN
ReceiveBeacon() {
if Beacon Frame Contains No Multicast Information then
    < InitializationPhase >
    MeasureReceivedSNR();
    TransmitFeedback();
else
    < NormalPhase >
    MeasureReceivedSNR();
    if PreviousMinimumSNR > ReceivedSNR then
        TransmitFeedback();
    else if CurrentNode == PreviousMinimumSNRNode then
        TransmitFeedback();
    end if
end if
}

AP
ReceiveFeedback() {
Update Multicast Node Entries;
Pick Out An Entry Which Has The Minimum SNR Value;
Determine Multicast Transmission Rate;
}
Peridocally Transmit Beacon Frame With A Selected Entry;
```

In initialization phase all MNs which are participating in multicast group measure their channel state by measuring received SNR value of beacon frame periodically broadcasted by AP. A periodic time of beacon signal is set as 100 ms in default. If received beacon frame does not contain additional information about multicast transmission rate of its own multicast group, then each MN transmits feedback signal which contains received SNR value. To minimize the possibility of collision between feedback signals from MNs, each MN adjust its backoff time according to the received SNR value. Because the length of backoff time is adjusted according the received SNR value, MN with low received SNR has relatively higher priority than MN with high received SNR. In this mechanism, multicast transmission rate is determined based on the lowest SNR value of MNs.

On receiving feedback signal frame from MNs which are participating multicast group, AP constructs channel-states table for each multicast group. To enable this, AP need to snoop IGMP messages from MNs. Each channel-states table has list of entries of hardware address and received SNR of each MN. After constructing channel-states table, AP picks out an entry with minimum SNR value and adapt multicast transmission rate based on selected minimum SNR value.

After initialization phase, it is necessary for AP to inform the minimum SNR value and hardware address of the MN which has reported the minimum SNR value to all MNs which are participating in multicast group. When AP transmit beacon frame, those information are piggybacked on original beacon frame. Each MN which has received beacon frame measures the channel state repetitively and compares current received SNR value with previous minimum SNR value noted through beacon frame. If currently measured SNR value is greater than previous minimum SNR value then MN does not transmit feedback signal to AP. However if currently measured SNR value is smaller than previous minimum SNR value then this MN need to update minimum SNR value with current one so it transmits feedback signal to AP. Meanwhile the MN which has proved to have the minimum SNR by previous received beacon frame needs to transmit feedback signal whether current received SNR value is smaller than previous one or not. This is for reducing contention overhead on wireless channel caused by transmission of the additional feedback signal frame. If AP does not receive feedback signal frame from the MN with minimum SNR value for specific time, like 3 beacon interval time, then AP considers that the MN has no connection with it any more and restart initialization phase.

3.2. Transmission Rate Selection Criterion in SARM

When AP determines multicast transmission rate based on received SNR values of MNs, it needs specific criterion to adapt its transmission rate adequately. As referred before, proposed mechanism especially aims to maintain adequate quality of multimedia contents transmitted by multicast protocol while using minimum wireless channel resource. Therefore we introduce the concept of PSNR (Peak Signal-to-Noise Ratio) as a criterion for rate selection. In several works which focused on the effect of bit errors on video quality [8][9], PSNR metric has been generally used as a measure of video quality. Moreover, the results state that the PSNR value over 30 should be guaranteed to maintain adequate quality of multimedia contents. Figure 1 shows some examples which show varied quality according to their PSNR values.

(a) 28.2437 PSNR with 10^{-5} BER (b) 19.8604 PSNR with 10^{-4} BER (c) 12.9043 PSNR with 10^{-3} BER

Figure 1. Video Quality at various PSNR values

Therefore, SARM aims to maintain PSNR value of received multimedia contents such as MPEG above 30. To enable this, it is need to construct database which contains the relationship between SNR and PSNR on each transmission rate from 1Mbps to 11Mbps. We conducted preliminary experiment to acquire SNR-PSNR relationship database and this work will be presented in section 4.1.. The result from the experiment is showed in Table 1.

Table 1. SNR-Transmission Rate Mapping Table

	Required SNR		
	SARM with FCS OFF	SARM with FCS ON	RBAR
11Mbps	26	30	30
5.5Mbps	21	24.5	25
2Mbps	17.5	21	21

3.3. Quality Enhancement for Multimedia Contents

With SARM, we also propose one simple method which enhances the quality of multimedia contents and the utilization of wireless channel resource. Due to the absence of the retransmission mechanism in multicast, dropping error-contained frame at MN results in waste of wireless channel resource and also degrades the quality of multimedia contents significantly. In this mechanism, all MNs which receive multimedia multicast traffic do not drop any error-contained multicast frame. This process is possible by turning off FCS (Frame Check Sequence) in MAC layer and setting UDP checksum to zero. With this method our rate adaptation mechanism exploits higher multicast transmission rate at the same channel condition. Details of this method will be presented in section 4.1..

4. Experimental Results

4.1. Experiment for SNR-PSNR Mapping Database

In this section we present and analyze experimental results. We implemented simplified multicast testbed system which delivers MPEG stream from multicast server to MNs. This experiment has two primary goals. First is to acquire the database which shows the relationship between SNR and PSNR values for each transmission rate and second is to analyze the effect on the PSNR of received multimedia contents when we do not drop error-contained frames in a MN.

Main hardware devices used in experiment consist of AP, multicast server and MNs. Multicast server is Pentium 4 desktop PC with RedHat linux 7.3. This server encodes MPEG2 data from original YUV format data at the encoding rate of 1Mbps and transmits MEPG2 data at the rate of 1Mbps. The AP used for our experiment is Orinoco AP-1000 model. This AP can monitor received SNR value for specific MN by transmitting and receiving a kind of probing packet. Also this can set up the multicast transmission rate at fixed rate, by administrator. MN is Pentium 3 Laptop PC with RedHat linux 7.3 and wireless interface card using Intersil chipset[10]. While receiving MPEG2 data from AP it disables any checksum mechanism such as FCS (Frame Check Sequence) in MAC layer or UDP Checksum so that it can use error-contained frames.

Experiment is executed as follows. We locate receiver node so that it can maintain specific level of received SNR. SNR value is monitored and logged by AP. And we set multicast transmission rate of AP at 2, 5.5 or 11Mbps. Then multicast server transmits MPEG2 data to receiver nodes through AP. After receiving MPEG2 data, receiver node decodes MPEG2 to YUV format data and calculates PSNR value by comparing with original YUV format data. Through all the experiment, multicast process which operates beyond MAC layer such as multicast group joining is not considered for convenience. By repeating this procedure with different location of MN each time and different multicast transmission rate of AP, we can get the result of SNR-PSNR relationship at each transmission rate.

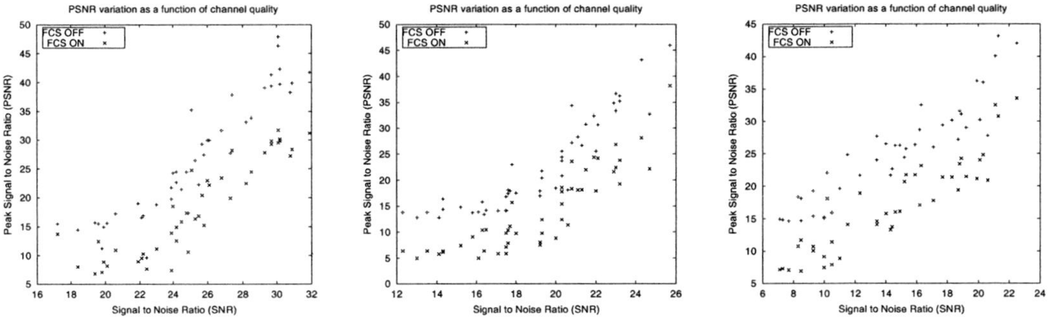

(a) Transmission Rate of 11Mbps (b) Transmission Rate of 5.5Mbps (c) Transmission Rate of 2Mbps

Figure 2. SNR-PSNR Relationship

Figure 2 shows the relationship between SNR and PSNR values. This result shows similar trend with well-known SNR-BER graphs. Based on this result we can acquire required SNR values to maintain PSNR value over 30 at each transmission rate. And ultimately we can construct SNR-Transmission rate mapping database for multicast transmission rate determination in AP. Table 1 shows the SNR-Transmission rate mapping database. Criterion of RBAR[7] based rate adaptation is derived from analytic results from [7], corresponding to the SNR points which meets bit error ratio of 10^{-5}. The result shows that SARM mechanism with FCS OFF exploits almost 1 level higher multicast transmission rate than RBAR and SARM with FCS ON under the same channel condition. This means that wireless channel resource occupied by multicast transmission is significantly reduced.

4.2. NS simulation - Protocol implementation

In this section we evaluate the performance of proposed MAC-layer multicast protocol as compared to fixed-rate multicast transmission and RBAR based multicast transmission. We implemented proposed protocol using NS-2 simulator. All experiments use Ricean propagation model[11][12] as the channel propagation model for IEEE 802.11 PHY and we use NOAH ns-extension[13] to enable infrastructure mode communication. The values for received power thresholds for different data rates were chosen based on the distance ranges specified in the Orinoco 802.11b card data sheet. Also we use the channel error model derived from 4.1. to simulate realistic wireless channel error model of IEEE 802.11b. Figure 3 shows actually measured Packet Error Rate (PER) as a function of SNR for several data rates. Figure 4 (a) and (b) show Probability Density Function (PDF) and Cumulative Density Function (CDR) of bit errors in each transmitted frame from AP, respectively. Through the CDF of bit errors, we can know that the bit errors in each transmitted frame is modeled

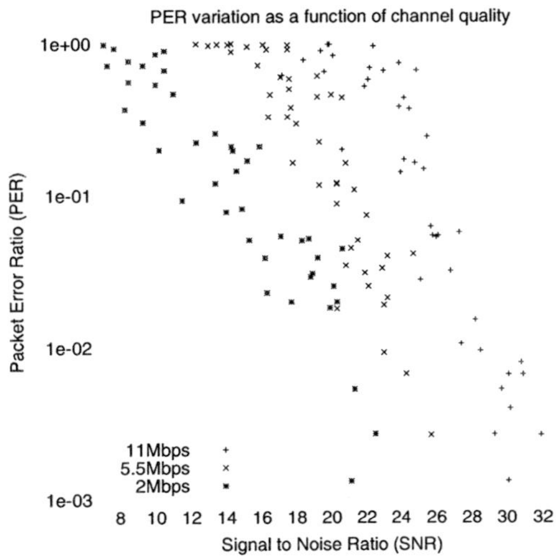

Figure 3. PER Variation as a function of SNR

Simulation topology consists of multicast server and AP which are connected in wired channel, and 3 MN which are connected to AP in wireless channel. Multicast server transmits MPEG2 video data at the rate of 1Mbps and packet sizes are set to 1220 bytes which is exactly same as in 4.1..

Figure 5 shows PSNR variation of each transmission methodologies for different distance from AP, which is corresponding to received SNR. Result shows high fluctuation tendency because of the randomness of channel error model, however proposed protocol with FCS OFF always maintains PSNR value over 30 which is most critical issue in transmitting multimedia multicast traffic. Also proposed protocol with FCS OFF shows outperforms over RBAR based rate adaptation except for the SNR range from 26 to 30. In this SNR range, multicast transmission rate of proposed protocol is 11Mbps and that of RBAR based rate adaptation is 5.5Mbps thus RBAR based rate adaptation show better quality compared to proposed protocol. In the SNR range below 26, proposed protocol with FCS OFF outperforms all other methodologies including 2Mbps fixed rate transmission. This shows that not dropping error-contained packet frame can be better method in the case of multimedia multicast transmission.

Figure 6 shows throughput of background TCP traffic for each multicast rate adaptation methodologies. Proposed protocol outperforms over RBAR based rate adaptation and 2Mbps fixed rate transmission. By adapting multicast transmission rate to satisfy the limitation of PSNR 30, proposed protocol minimized its usage of wireless channel resource. This factor derives higher throughput for TCP based unicast traffic such as FTP file transfer or P2P communication.

5. Conclusion

In this paper we proposed a MAC layer multicast mechanism which utilizes multi-rate transmission capability. By enabling rate adaptation in multicast transmission based on the

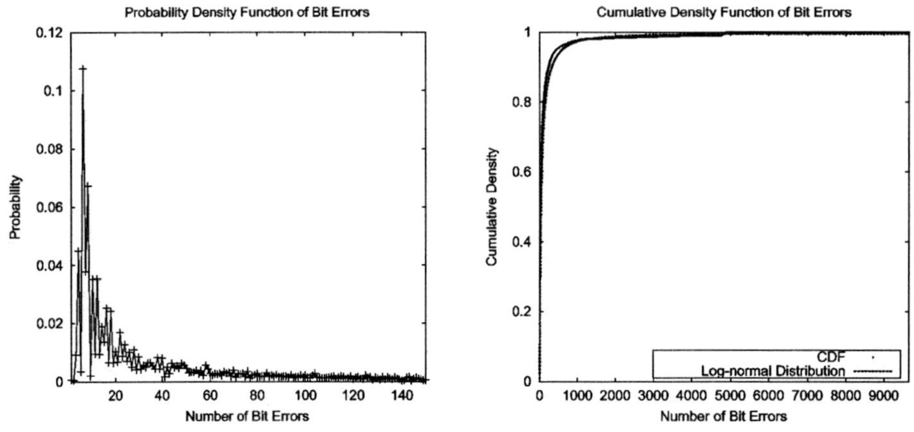

Figure 4. Bit Error in Each Transmitted Frame

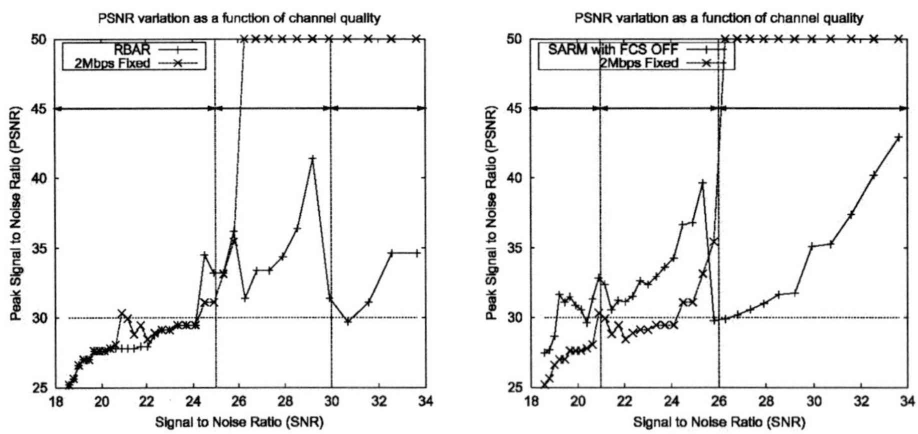

(a) RBAR (b) SARM

Figure 5. PSNR Variation as a function of SNR

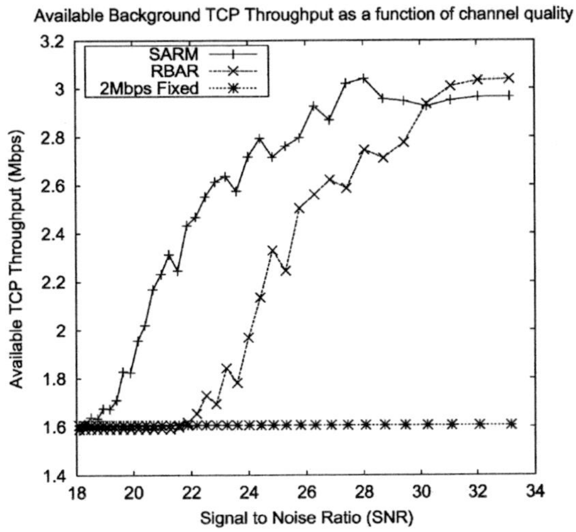

Figure 6. Background TCP Throughput

SNRs of MNs, the proposed mechanism eliminates the waste of wireless channel resource. The proposed mechanism can guarantee the best quality of multimedia contents depending on multicast receivers' wireless link conditions. Moreover it can minimize the usage of wireless channel resource for multicasting compared to the fixed rate transmission or other protocol such as RBAR. Finally we performed experiments using the testbed and simulation using ns-2 simulator. Through experimental and simulation results it is proved that the proposed mechanism outperforms fixed rate transmission and RBAR.

References

[1] IEEE std. 802.11-1999, Part 11: *Wireless LAN Medium Access Control(MAC) and Physical Layer (PHY) specifications*, Reference number ISO/IEC 8802-11:1999(E), IEEE std.802.11,1999 Edition, 1999.

[2] IEEE std. 802.11b, Supplement to Part 11: *Wireless LAN Medium Access Control(MAC) and Physical Layer (PHY) specifications: Higher-speed Physical Layer Extension in the 2.4GHz Band*, IEEE Std. 802.11b-1999,1999.

[3] Gartner Inc., *Press Release*, February 2004

[4] Daji Qiao and Sunghyun Choi, "Goodput Enhancement of IEEE 802.11a Wireless LAN via Link Adaptation," *Proc. IEEE ICC '01*, June 2001, pp. 1995-2000.

[5] Jean-Lien C. Wu, Hunh-Huan Liu and Yi-Jen Lung, "An Adaptive Multirate IEEE 802.11 Wireless LAN," *Proc. 15th International Conference on Information Networking*, 2001, pp. 411-418.

[6] A. Kamerman and L. Monteban, "WaveLAN-II: A High-Performance Wireless LAN for the Unlicensed Band," *Bell Labs Technical Journal*, pp. 118-133, Summer 1997.

[7] G. Holland, N. Vaidya and P. Bahl, "A Rate-Adaptive MAC Protocol for Multi-Hop Wireless Networks," *Proc. of ACM MOBICOM*, Rome, July 2001.

[8] S. Gringeri, R. Egorov, K. Shuaib, A. Lewis, and B. Basch, "Robust Compression and Transmission of MPEG-4 Video," *Proceedings of the seventh ACM international conference on Multimedia* (Part 1), 1999.

[9] Imed Bouazizi, "Size-Distortion Optimized Proxy Caching for Robust Transmission of MPEG-4 Video," LNCS 2899, *Proceedings International Workshop on Multimedia Interactive Protocols and Systems*, November 18-21, 2003, Napoli, Italy.

[10] "HFA3861B; Direct Sequence Spread Spectrum Baseband Processor," January 2000.

[11] R. Punnose, P. Nikitin, and D. Stancil, "Efficient simulation of Ricean fading within a paket simulator," *IEEE Vehicular Technology Conference*, pages 764-767, 2000.

[12] B. Sadeghi, V. Kanodia, A. Sabharwal, and E. Knightly, "OAR : An Opportunistic Auto-rate Media Access Protocol for Ad Hoc Networks," *ACM MOBICOM'02*, Atlanta, GA, Sep 2002.

[13] Jorg Widmer, "Extensions to the ns Network Simulator" (http://www.informatik.uni-mannheim.de/pi4/projects/MobileIP/ns-extension/).

[14] V. Kanodia, A. Sabharwal, and E. Knightly, "MOAR: A Multi-channel Opportunistic Auto-rate Media Access Protocol for Ad Hoc Networks," in *Proceedings of Broadnets 2004*, San Jose, CA, October 2004.

In: Adaptation Techniques in Wireless Multimedia Networks ISBN 1-59454-883-8
Editors: Wei Li and Yang Xiao, pp. 33-56

Chapter 3

ADAPTATION IN THE PHYSICAL LAYER USING HETEROGENEOUS RECONFIGURABLE HARDWARE

Gerard K. Rauwerda, Jordy Potman,
Fokke W. Hoeksema and Gerard J.M. Smit
University of Twente, Department of EEMCS, the Netherlands

Abstract

In this chapter we investigate the possibilities for adaptivity in wireless terminals. We distinguish: *standards level adaptivity*, *algorithm-selection level adaptivity* and *algorithm-parameter level adaptivity*. We use a UMTS receiver to illustrate these adaptivity-related concepts. Next, we show how we can exploit adaptivity with flexible reconfigurable hardware to save energy in wireless terminals. We will demonstrate our approach with examples from UMTS and HiperLAN/2.

1. Introduction

The number of wireless standards has increased significantly over the last number of years. For example, for cellular wireless telephony the GSM and IS-95 standards exist, which in the future will be extended/replaced by the Universal Mobile Telecommunications System (UMTS) and CDMA2000 standards. DECT is a standard used for cordless telephones. Bluetooth is a currently used Personal Area Network (PAN) standard, while 802.15, Zigbee and Ultra Wide Band (UWB) based solutions have already been proposed as future PAN standards. There exist a large number of 802.11 Wireless Local Area Network (WLAN) standards with various data rates. Digital Audio Broadcasting (DAB) and Digital Radio Mondiale (DRM) have been proposed as standards for broadcast radio.

A modern cell phone supports triple-band GSM/GPRS and Bluetooth. If this trend is extended into the future, UMTS cell phones will have to support a number of standards next to UMTS. They will have to support triple-band GSM/GPRS for backwards compatibility. Furthermore, they will also have to support Bluetooth or another PAN standard, to connect to, for example, a headset or a PDA. When the phone is to be used for packet data transfer, it is desirable that it supports one or more of the 802.11 standards to be able to use a WLAN hotspot when it is available. If the phone has to provide broadcast radio reception functionality it will probably have to support DAB and DRM.

The cell phone terminal hardware will have to support all these wireless communication standards. One solution to this problem is to develop terminal hardware that contains specific cores for each communication standard. Another solution is to use a single hardware architecture in the terminal that is flexible enough to support all required standards. In either case, this hardware architecture has to be extremely energy-efficient, as it is used in a battery-operated terminal.

Summarizing, trends that are identified in wireless communication systems are:

- Wireless communication receivers support multiple standards.
- Wireless communication standards evolve quickly.
- Energy-efficiency is of great importance, since mobile terminals are battery-powered.

The Software Defined Radio (SDR) [63] concept perfectly matches with the trends in wireless communication systems. A radio system built using SDR technology extends the utilization of the system to a wide range of applications using different link-layer protocols and modulation/demodulation techniques. SDR provides an efficient and relatively inexpensive solution to the design of multi-mode, multi-band, multi-functional wireless devices that can be enhanced using software upgrades only.

SDR-enabled devices (i.e. mobile terminals) are designed in such a fashion that the characteristics of the device can be reprogrammed. So, the same hardware can be adapted to perform different functions at different points in time. This means that with a single SDR device, different communication standards can be supported, i.e. multi-standard communications.

Another advantage of the SDR concept is the possibility to implement real-time adaptive systems. Traditional algorithms in wireless communications are rather static. The recent emergence of new applications that require sophisticated adaptive, dynamical algorithms based on signal and channel statistics to achieve optimum performance has drawn renewed attention to run-time reconfigurability.

Implementation of SDR requires a flexible hardware architecture. Traditional SDR approaches are implemented on *homogeneous* flexible architectures, like General Purpose Processor (GPP) or Digital Signal Processor (DSP) [63]. Since baseband processing in the wireless receiver is computationally intensive, the terminal's hardware architecture has to be very powerful. Moreover, as wireless terminals are battery-powered, the importance of energy-efficiency of the hardware architecture is emphasized. A common objection to the traditional homogeneous flexible architecture is its relative energy in-efficiency. However, *heterogeneous* reconfigurable hardware, consisting of processing elements with *different* granularities, is designed with these constraints – flexibility, performance and energy-efficiency – in mind.

The idea of heterogeneous processing elements is that one can match the granularity of the algorithms with the granularity of the hardware. Four processor types can be distinguished: *general purpose* processor, *fine-grained* reconfigurable hardware (e.g. FPGA), *coarse-grained* reconfigurable hardware (e.g. MONTIUM [2, 43]) and *dedicated* hardware (e.g. ASIC). Figure 1 depicts a possible heterogeneous reconfigurable hardware platform, consisting of processing elements with different granularities. The processing elements

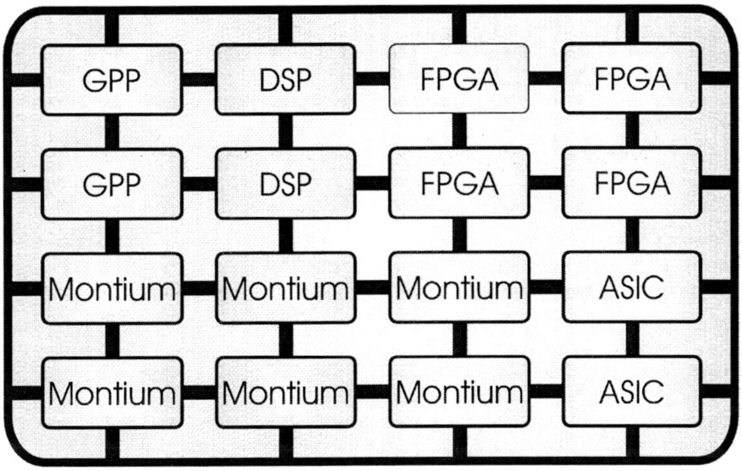

Figure 1. Heterogeneous reconfigurable hardware platform.

are interconnected by a Network-on-Chip (NoC). Matching the granularity of the reconfigurable hardware with the algorithm provides flexibility at the right level:

- *general purpose*
  The general purpose processor is the most flexible hardware architecture. It can be programmed to perform almost any algorithm. General purpose processors are well suited for control-oriented functions. Due to the large overhead in control, these processors are not very energy-efficient.

- *fine-grained*
  Fine-grained reconfigurable devices are bit-level programmable. Because of the configurability at bit-level, the configuration overhead is large. Fine-grained reconfigurable devices are perfectly suited for prototyping and to implement encryption algorithms.

- *coarse-grained*
  Coarse-grained reconfigurable devices are flexible at word-level. Multipliers, adders, etc. are hardwired in these devices. Because only coarse functional blocks have to be configured, the configuration overhead is small. These architectures are more suited for data-oriented functions, like algorithms performed in the DSP domain.

In the remainder of this chapter we will show how heterogeneous reconfigurable hardware can be used to exploit the adaptation of the physical layer of wireless receivers. In Section 2 we will indentify the different levels at which adaptivity is possible in the baseband processing of wireless receivers. Section 3 shows how wireless receivers with these levels of adaptivity can be implemented in heterogeneous reconfigurable hardware. Throughout this chapter a UMTS and HiperLAN/2 receiver are used as an example.

2. Adaptivity Opportunities in the Wireless Receiver

As already stated in the introduction one of the main reasons for introducing reconfigurable hardware in a wireless terminal is to support multiple wireless communication standards. The support of multiple wireless communication standards introduces a first level of adaptivity in the wireless terminal because the terminal can switch between wireless communication standards. For example when packet data transport is performed over UMTS and a WLAN hotspot becomes available the terminal can switch from UMTS to a WLAN standard. This is referred to as *standards level* adaptivity. Standards level adaptivity has an impact on the digital signal processing in the wireless terminal because the wireless communication standard defines the DSP *functions* that have to be performed to implement the standard.

Although a wireless communication standard usually defines the DSP functions which have to be performed to implement the standard, it usually does not define the algorithms that have to be used to implement these functions. So the communication system can therefore select the algorithm that is used to implement a DSP function. This means that the communication system selects an algorithm out of a set of algorithms that implement the same DSP function. Therefore, this second level of adaptivity is referred to as *algorithm-selection level* adaptivity.

For a specific algorithm, there are also opportunities for adaptivity by changing parameters of the algorithm. This third level of adaptivity is called *algorithm-parameter level* adaptivity [36].

An adaptive wireless terminal only makes sense when it better suits the needs of a user at an acceptable complexity and efficiency compared to a traditional non-adaptive terminal. The standards level of adaptivity allows the terminal to adapt the communication standard to the communication environment and the Quality of Service (QoS) requirements. Here the term *communication environment* is used to indicate the available wireless communication standards and wireless channel conditions at a certain location. The algorithm-selection level of adaptivity allows the terminal to select the algorithms that satisfy the QoS requirements in the given communication environment in the most efficient manner. The algorithm-parameter level of adaptivity allows the terminal to do the same with the parameters of a specific algorithm. So a reconfigurable and adaptive terminal is able to track the QoS requirements and communication environment on a finer grained scale than a traditional non-adaptive terminal.

The complexity of an adaptive terminal is determined by the complexity of the standards that it has to support, the complexity of the algorithms that are used to implement the DSP functions of the standard and the complexity of the measurement and control overhead that is required to make the terminal adaptive. The complexity of the algorithms in the set of algorithms that is used for algorithm-selection level adaptivity should therefore be centered around the complexity of the algorithm that is used in a traditional non-adaptive terminal. To minimize the overhead of measuring QoS and channel conditions, measurements that are already required by the wireless communication standard should be (re)used wherever possible. The complexity of any additional measurement algorithm should be kept low. The same is true for the control algorithms.

In Section 2.1, the algorithm-selection and algorithm-parameter levels of adaptivity will be discussed in more detail, based on examples of these levels of adaptivity in the DSP functions of the UMTS terminal receiver. Standards level adaptivity will be discussed in Section 3, where the implementation of a combined UMTS and HiperLAN/2 receiver on a single heterogeneous reconfigurable hardware platform is described.

2.1. Adaptivity in UMTS Terminal Receiver DSP Functions

The Universal Mobile Telecommunications System (UMTS) standard, defined by ETSI, is an example of a Third Generation (3G) mobile communication system. The communication system has an air interface that is based on Code Division Multiple Access (CDMA) [42]. The bit rate of UMTS at the physical level depends on the modulation type and the spreading factor.

In the UMTS base station transmitter roughly the following DSP operations are performed on the user bits that have to be transported over the UMTS physical layer: The user bits are first grouped into code blocks. A CRC checksum is added to each code block and the code block including the CRC checksum is encoded using either a convolutional or Turbo code. The resulting coded bits are then QPSK modulated and spread using the spreading code of the user for which they are destined. At this point in the base station the signals for the different users that are currently in the cell that the base station is servicing can be combined. The combined signal is scrambled with the scrambling code that identifies the base station. The resulting signal is filtered with a root-raised cosine transmit filter so that the transmitted signal conforms to the transmit mask defined in the UMTS standard. The filtered signal is digital-to-analog converted and passed on to the analog front-end where it is carrier-modulated, amplified and finally transmitted.

The wireless channel between base station transmitter and terminal receiver has a number of detrimental effects on the signal. Reflections of the transmitted signal against buildings, hills or other objects cause the arrival of multiple delayed copies of the transmitted signal at the receiver, resulting in a channel with multiple propagation paths. On a smaller scale these reflections cause fading of the received signal strength. The rate at which this fading changes depends on the movements of objects and the velocity with which the terminal is moving.

The main DSP chain of the UMTS terminal receiver roughly performs the reverse of the DSP operations in the transmitter [71]: receive filtering, descrambling and despreading, channel decoding and CRC checking, see Figure 2. The descrambling and despreading function requires a number of supporting DSP functions: the cell search, path search and the channel estimation functions.

The purpose of these functions and how they can be made adaptive will be described in the next sections. Algorithms for implementation of each of these functions found in literature are investigated for opportunities for algorithm-selection and/or algorithm-parameter level adaptivity.

2.1.1. Receive Filter

The most important purpose of the receive filter is to ensure that sufficient adjacent channel attenuation is achieved in order to suppress adjacent channel interference. Adjacent channel

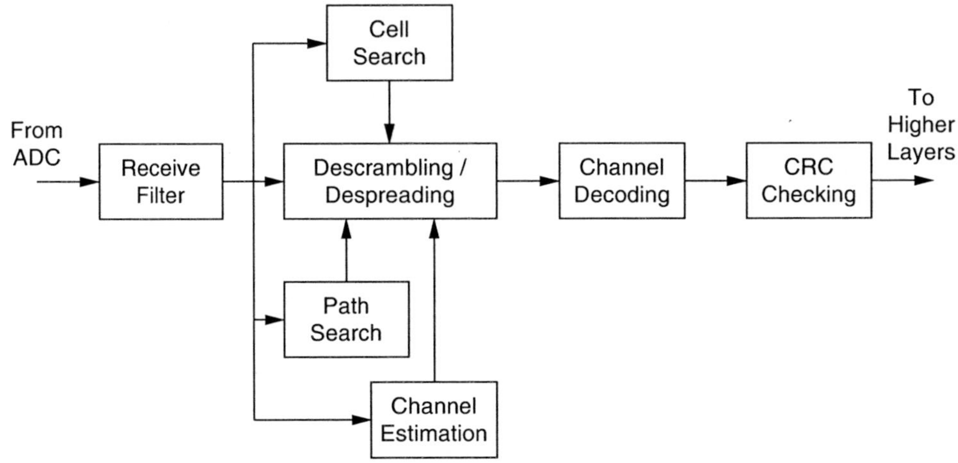

Figure 2. UMTS terminal receiver DSP functions.

interference in the FDD mode of UMTS is caused by the base stations of the networks of other operators transmitting on adjacent channel frequencies. The amount of adjacent channel interference at a specific location varies depending on:

- The distance between the terminal and the base station transmitting on the adjacent channel.
- Whether the adjacent channel is being used or not at that location.

Therefore, the required amount of adjacent channel attenuation varies. In [72] an adaptive digital receive filter for a mobile terminal receiver is proposed that automatically adjusts the adjacent channel attenuation based on the measured in-band and out-of-band powers. So only the required amount of filtering is performed. It does this by adjusting the number of filter coefficients, so a form of algorithm-parameter adaptivity is used.

2.1.2. Descrambling and Despreading

Traditionally UMTS terminals use a RAKE receiver for the descrambling and despreading operation [71]. A RAKE receiver consists of a set of RAKE fingers and a combiner block. A single RAKE finger is an independent correlation receiver that descrambles and despreads the received signal of a single propagation path of a wireless channel. By using multiple RAKE fingers, multiple propagation paths can be handled. The output symbols of the RAKE receiver are constructed in the combiner block by coherently and synchronously combining the finger outputs. This allows the RAKE receiver to recover the energy from multiple propagation paths, which is not possible with a single correlation receiver. The RAKE receiver is, however, suboptimal in UMTS, because its performance is degraded by Multiple Access Interference (MAI). This MAI is caused by the fact that the orthogonality of the spreading sequences of the different users in the system is destroyed by the multipath channel.

In [73] various channel equalizers for implementation of the UMTS terminal receiver descrambling and despreading function are studied. The equalizers provide significant performance improvements compared to a RAKE receiver, due to their superior MAI suppression properties, but only in cells with a large number of active users. The improved performance of the equalizers comes at the cost of an increased computational complexity. Therefore, some energy savings can be achieved if algorithm-selection level adaptivity is used in the terminal to dynamically select between a RAKE receiver and an equalizer implementation of the descrambling and despreading function based on the measured number of active users.

2.1.3. Cell Search

The purpose of the cell search function is to find the scrambling code used by the base station of the cell in which the terminal is currently located. This information is required for the descrambling function. The cell search procedure is clearly defined in the UMTS specification and the algorithms involved do not offer a lot of opportunities for either algorithm-selection adaptivity or algorithm-parameter adaptivity. Thus, this function is not interesting from an adaptive DSP algorithm point of view.

2.1.4. Path Search

A RAKE receiver requires the acquiring and tracking of the delays of the propagation paths associated with each of its fingers [71]. Channel equalizers require acquiring and tracking the delay of the main propagation path [73]. Path delay acquisition is usually performed by a path search algorithm, while path delay tracking is performed by a path tracking algorithm [71].

The ability of the path search function to resolve strong closely spaced paths or paths with a time-variant delay depends on the algorithm that is used to implement the path search function. The improvement in resolving power, however, comes at the cost of an increased computational complexity [74]. So, the path search function is a suitable candidate for algorithm-selection adaptivity.

2.1.5. Channel Estimation

When maximum ratio combining is used to combine the RAKE finger outputs in the RAKE receiver an amplitude and a phase estimate must be known for each symbol to scale and derotate each finger's output prior to combining [71]. These amplitude and phase estimates can be obtained from a channel estimation algorithm. Some of the channel equalization algorithms described in [73] require an estimate of the channel realization at the chip level, which can also be obtained using a channel estimation algorithm.

In the channel estimation function both algorithm-selection adaptivity and algorithm-parameter adaptivity can be used. Algorithm-selection adaptivity is made possible by the fact that there are two sources of pilot symbols that can be used for channel estimation: the pilot symbols on the Common Pilot Channel (CPICH) and the time multiplexed pilot symbols on the Dedicated Physical Channel (DPCH). So it is possible to switch between algorithms that use the CPICH pilot symbols, the DPCH pilot symbols or a combination of both.

Table 1. Overview of adaptivity in UMTS Terminal DSP Functions.

Function	Level of adaptivity	Algorithm reference
Receive filter	algorithm-parameter	[72]
Descrambing and Despreading	algorithm-selection	[73]
Path search	algorithm-selection	[74]
Channel estimation	algorithm-selection algorithm-parameter	[75]
Channel decoding	algorithm-parameter	[76]

Algorithm-parameter adaptivity is possible because there is a trade-off on the length of the filter used in all channel estimation algorithms. Short filters will result in noisy channel estimates, but long filters will reduce the ability of the channel estimator to track fast fading channels [75]. Therefore, it is worthwhile to adjust the filter length based on the channel conditions.

2.1.6. Channel Decoding

In UMTS Turbo or convolutional channel coding is used. Turbo channel coding is decoded using iterative decoding algorithms. The bit-error rate at the Turbo decoder output improves with each decoder iteration. Stopping criteria can be used to stop the decoding iterations when the bit-error rate is below the maximum acceptable bit-error rate specified as a UMTS QoS attribute.

In [76] an approach for adaptively approaching the Viterbi algorithm for decoding convolutional codes is presented. In this approach the pruning threshold and truncation length of a T-algorithm decoder are adapted to the rate of the code, signal-to-noise ratio and the maximum acceptable bit-error rate. This results in considerable energy savings compared to a normal Viterbi decoder.

Both techniques to reduce the complexity of channel decoding described in this section use a form of algorithm-parameter level adaptivity to adapt the channel decoding to the QoS requirements and wireless channel conditions.

Table 1 summarizes the levels of adaptivity that can be applied to the UMTS terminal DSP functions as identified in this section. The table also gives references to algorithms in literature that can be used to implement these levels of adaptivity.

3. Using Flexible Hardware for Exploiting Adaptivity

The concept of a heterogeneous reconfigurable architecture has already been introduced in Section 1. An example of such an architecture is shown in Figure 1 and is used in this section to show the feasibility of implementing adaptive communication algorithms using reconfigurable hardware. The different kinds of adaptivity, as discussed in Section 2, are applied by using the heterogeneous reconfigurable architecture.

Heterogeneous reconfigurable hardware introduces the possibility to map an algorithm to the processing element with the right granularity: e.g. algorithms that operate on bit-level data can be mapped on bit-level reconfigurable hardware and algorithms that operate on word-level data can be mapped on word-level reconfigurable hardware. In this way algorithms can be implemented in a relatively energy-efficient manner while still retaining performance and flexibility.

In Section 3.2 the baseband processing part of a RAKE receiver, used for UMTS communications, is implemented in reconfigurable hardware. The same reconfigurable hardware is also used in Section 3.3 to implement a HiperLAN/2 receiver. Both implementations are based on an implementation in coarse-grained reconfigurable hardware, which is part of the heterogeneous reconfigurable System-on-Chip (SoC). In order to apply the different kinds of adaptivity in the wireless terminal, a configuration manager is needed. This manager controls all the tasks in the SoC and is discussed in Section 3.4. In Section 3.1 an example of a coarse-grained reconfigurable processing element is given.

3.1. The MONTIUM Coarse-Grained Tile Processor

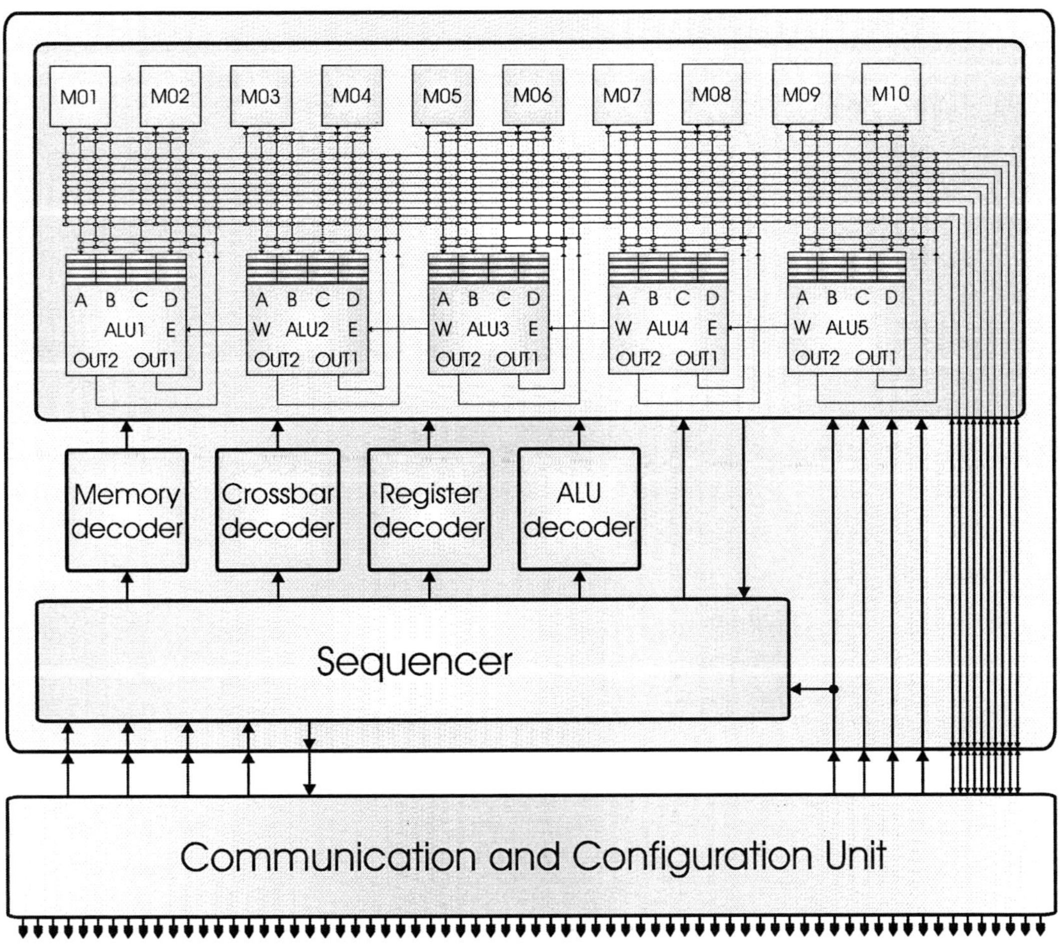

Figure 3. The MONTIUM coarse-grained reconfigurable tile processor.

The MONTIUM is an example of a coarse-grained reconfigurable processing tile. The MONTIUM [2, 43] targets the 16-bit digital signal processing (DSP) algorithm domain. A single MONTIUM processing tile is depicted in Figure 3. At first glance the MONTIUM architecture bears a resemblance to a VLIW processor. However, the control structure of the MONTIUM is very different. For efficiency it is imperative to minimize the control overhead. This can be accomplished by statically scheduling instructions as much as possible at compile time.

The lower part of Figure 3 shows the Communication and Configuration Unit (CCU) and the upper part shows the reconfigurable Tile Processor (TP). The CCU implements the interface for off-tile communication. The definition of the off-tile interface depends on the NoC technology that is used in the SoC.

The TP is the computing part that can be configured to implement a particular algorithm. Figure 3 reveals that the hardware organization of the tile processor is very regular. The five identical ALUs (ALU1 $\cdots$ ALU5) in a tile can exploit spatial concurrency to enhance performance. This parallelism demands a very high memory bandwidth, which is obtained by having 10 local memories (M01 $\cdots$ M10) in parallel. The small local memories are also motivated by the locality of reference principle. The data path has a width of 16-bits and the ALUs support both signed integer and signed fixed-point arithmetic. The ALU input registers provide an even more local level of storage. Locality of reference is one of the guiding principles applied to obtain energy-efficiency in the MONTIUM. A vertical segment that contains one ALU together with its associated input register files, a part of the interconnect and two local memories is called a Processing Part (PP). The five Processing Parts together are called the Processing Part Array (PPA). A relatively simple sequencer controls the entire PPA. The sequencer selects configurable PPA instructions that are stored in the decoders of Figure 3.

Each local SRAM is 16-bit wide and has a depth of 512 positions, which adds up to a storage capacity of 8 Kbit per local memory. A reconfigurable Address Generation Unit (AGU) accompanies each memory. It is also possible to use the memory as a look-up table (LUT) for complicated functions that cannot be calculated using an ALU, such as sine or division (with one constant).

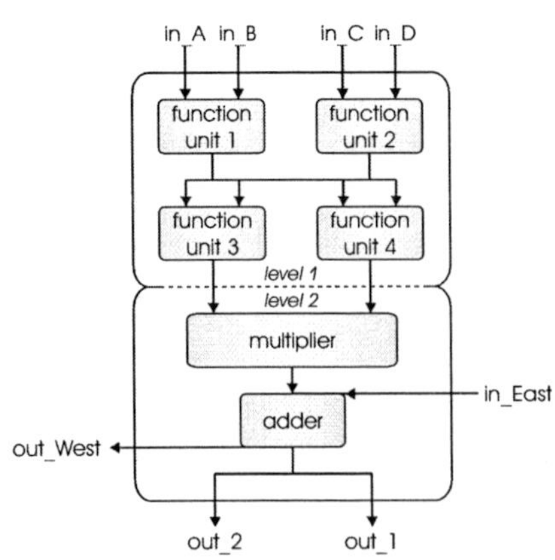

Figure 4. The MONTIUM ALU.

Figure 4 shows the ALU that is used in the MONTIUM. A single ALU has four 16-bit inputs. Each input has a private input register file that can store up to four operands. The input register file cannot be bypassed, i.e. an operand is always read from an input register. Input registers can be written by various sources via a flexible interconnect. An ALU has two 16-bit outputs, which are connected to the interconnect. The ALU is entirely combinational and consequentially there are no pipeline registers within the ALU. Neighbouring ALUs can also communicate directly. The West-output of an ALU connects to the East-input of the ALU neighbouring on the left. The East-West connection does not introduce a delay or pipeline, as it is not registered. Table 2 shows some key performance figures of the current MONTIUM implementation [43].

Table 2. Characteristics of the MONTIUM.

Process	$0.12\ \mu m$
Area	$1.8\ mm^2$
Clock speed	$40\ -\ 150\ MHz$
Energy	$500\ \mu W/MHz$

3.2. UMTS Receiver

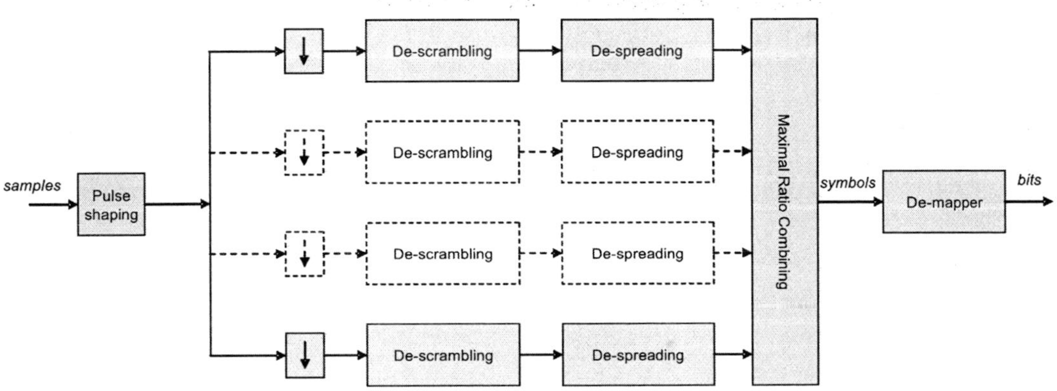

Figure 5. W-CDMA RAKE baseband functions.

Figure 5 shows the baseband processing, performed in the W-CDMA receiver using a RAKE receiver. The W-CDMA receiver has been implemented in heterogeneous reconfigurable hardware. Since most baseband processing consists of multiply-accumulate (MAC) operations, the baseband processing of the receiver has been implemented in coarse-grained reconfigurable hardware, the MONTIUM. The scrambling code in the receiver can be generated with simple combinational logic, consisting of shift-registers and XOR gates. These are typical operations that are performed in fine-grained reconfigurable hardware, like an FPGA. The control-oriented functionality is performed in the GPP and provides the right information to the baseband processing part of the W-CDMA receiver. The control-manager is further discussed in Section 3.4.

The receive filter is commonly implemented with as a pulse-shape filter. The pulse-shape filter, which can be implemented as a FIR filter, is implemented in one MONTIUM tile. The output streams of the pulse-shape filter are the input for the RAKE receiver, which is implemented in a second MONTIUM tile. Figure 6 shows the functional blocks in the W-CDMA receiver that are implemented in the tiles of the heterogeneous reconfigurable SoC.

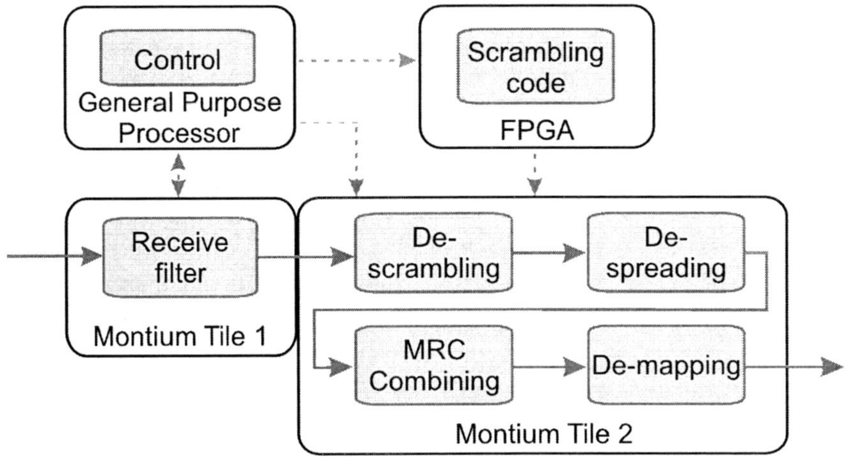

Figure 6. W-CDMA receiver mapped to heterogeneous reconfigurable SoC.

The implemented RAKE receiver can process four individual paths of the received signal. Consequently, the receiver requires four complex-number data streams for the four implemented fingers. All implemented fingers require the same scrambling code. The implemented receiver takes the complex-number scrambling code stream as an input. The spreading code is stored in local memory, because the code is relatively small with a maximum length of 512 samples. Furthermore, the spreading code is assigned to a particular user in the UMTS communication system and, therefore, the spreading code does not change frequently. The received symbols of the individual signal paths – fingers – are combined, while each symbol is scaled with a complex-number coefficient. These coefficients are provided by the channel estimator. So, the implemented receiver takes the complex-number coefficient stream as an input. The receiver outputs a bit stream with the received data.

Figure 3 shows that the CCU is directly connected to the global buses inside the MONTIUM. The CCU implements the interface for off-tile communication and so it guarantees that the correct signals are available for the MONTIUM tile.

3.2.1. Configuration

The configuration size of the flexible RAKE receiver in the MONTIUM is only 858 bytes. One tile can be configured for RAKE receiving in 429 clock cycles. For a configuration clock frequency of 100 MHz this means that a RAKE receiver with 4 fingers can be configured in 4.29 μs.

In case the spreading factor (SF) changes, and thus the spreading code, the new spreading code only has to be loaded in the local memory of the MONTIUM and a constant in the MONTIUM configuration has to be changed. Loading a particular spreading code and reconfiguring the constant costs $SF + 1$ clock cycles.

The signal streams for the different fingers are buffered in local memories inside the MONTIUM. When the delay of one of the paths changes, then the buffering strategy of the local memories has to be changed. The buffering strategy of the memories is configured with 24 bytes. These 24 bytes can be reconfigured in 12 clock cycles. Consequently, the RAKE receiver can update its complete path delay profile in 120 ns, assuming that the configuration clock frequency is 100 MHz.

In total, for reconfiguring the number of fingers from 4 to 2, only 24 bytes have to be reconfigured in the configuration memory of the MONTIUM. Assuming that the clock frequency of the processor tile during reconfiguration is 100 MHz, the RAKE receiver can be reconfigured in 120 ns, which corresponds to 12 clock cycles.

3.2.2. Frequency Scaling

Voltage and frequency scaling is an important measure to control the power consumption of embedded systems. In CMOS design, the power consumption depends quadratically on the supply voltage and linearly on frequency. The main idea of dynamic voltage and frequency scaling is that the supply voltage should be kept as low as possible. Besides, the maximum operating frequency is tightly coupled to the supply voltage level. This means that by downscaling the clock frequency of hardware, the supply voltage can be lowered as well, resulting in a quadratic decrease of the power consumption.

The clock frequency of the MONTIUM during RAKE processing of 4 fingers is about 4 times the chip rate. When the RAKE receiver is reconfigured to 2 finger processing, then the clock frequency of the MONTIUM can be reduced to about 2 times the chip rate.

The dynamic power consumption of a typical multiply-accumulate operation in the MONTIUM is about 0.5 mW/MHz, realized in 0.12 μm CMOS technology. Consequently, the power consumption of the implemented RAKE receiver will be 5 mW in 2-finger mode and 10 mW in 4-finger mode.

3.2.3. Adaptivity Features

The following adaptivity features can be found in the UMTS receiver:

- *standards adaptivity*
  Different communication standards can be implemented using the same reconfigurable hardware. Configuring the reconfigurable processing elements into a W-CDMA receiver requires less than 10 μs configuration time.

- *algorithm-selection adaptivity*
  Within the W-CDMA receiver the functionality of the RAKE receiver can be switched to equalization and the other way around (Section 2.1.2). Reconfiguring the processing element, which was performing the equalizer functionality, to RAKE receiver functionality costs about 4 μs configuration time.

- *algorithm-parameter adaptivity*
  The number of fingers is one of the parameters that can be changed in the RAKE

receiver. The number of fingers in the RAKE receiver of Section 3.2.1 can vary between 1 and 4[1]. Adapting the number of fingers means that the MONTIUM, which implements the RAKE receiver, has to be partially reconfigured with 24 bytes of data. This partial reconfiguration costs 120 ns reconfiguration time. Reducing the number of fingers results in lower clock frequencies of the MONTIUM, which results in reduced energy consumption.

Another parameter that can be changed is the delay profile of the fingers in the RAKE receiver. The reconfiguration time for changing the delay of the fingers is 120 ns, because 24 bytes have to be reconfigured.

The third parameter that can be adapted in the RAKE receiver is the spreading factor. Changing the spreading factor is performed by partial reconfiguration and loading of the new spreading code.

The pulse-shape filter can be implemented as an adaptive filter, which can adapt the number of taps in the FIR filter. Changing the number of taps implies also a partial reconfiguration. Consequently, the clock frequency of the processing element can be adapted to the new filter structure, which results in a change in energy consumption.

3.3. HiperLAN/2 Receiver

In the previous section we saw how a UMTS receiver can be implemented on a heterogenous SoC. In this section we show that the same SoC can be used to implement a HiperLAN/2 receiver (standards adaptivity).

HiperLAN/2 is a wireless local area network (WLAN) access technology and is similar to the IEEE 802.11a WLAN standard. HiperLAN/2 operates in the 5 GHz frequency band and makes use of orthogonal frequency division multiplexing (OFDM) to transmit the analogue signals. The bit rate of HiperLAN/2 at the physical level depends on the modulation type and is either 12, 24, 48 or 72 Mbit/s.

The basic idea of OFDM is to transmit information at a high data rate by dividing the data into several parallel bit streams, and let each one of these bit streams modulate a separate subcarrier. A HiperLAN/2 channel contains 52 subcarriers and has a channel spacing of 20 MHz. 48 subcarriers carry actual data and 4 carry pilots.

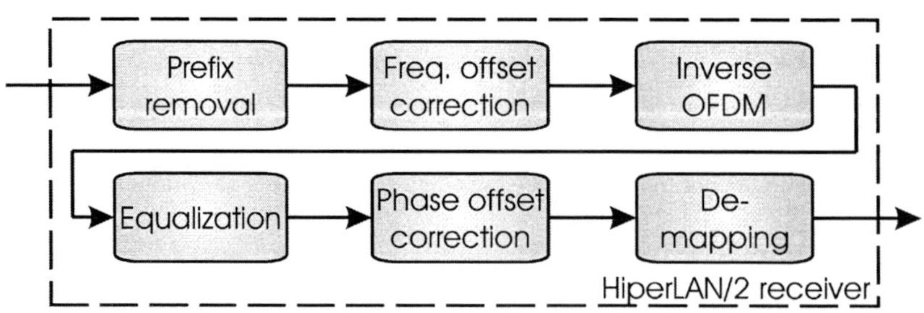

Figure 7. The baseband functions in the HiperLAN/2 receiver.

[1]In the current version of the MONTIUM cannot be dealt with more than 4 fingers because of the available memory bandwidth to store the delayed inputs.

The receiver not only performs the inverse operations of the transmitter, it also has to correct for all the distortions that are introduced in the wireless channel. Figure 7 depicts a model of the HiperLAN/2 receiver. In general, the model can be used for any OFDM-like system. The diffent standards for OFDM-like systems, e.g. IEEE 802.11a/g, HiperLAN/2, DAB, DRM, IEEE 802.16, are generally different in the number of carriers and the transmission bandwidth.

The synchronization of the receiver is performed in two steps. Firstly, coarse-synchronization is performed in order to synchronize the receiver with the frame. During coarse-synchronization the received signal is correlated with known preambles, which indicate the start of a frame. Secondly, the prefix information of an OFDM symbol is used for fine-synchronization. After fine-synchronization, the prefix is removed from the OFDM symbol.

Differences between the oscillator frequencies of the transmitter and the receiver result in frequency offset and cause inter-subcarrier interference. The HiperLAN/2 receiver can compensate for frequency offset by multiplying the data samples of an OFDM symbol with the frequency offset correction coefficient. The frequency offset correction coefficient can be determined by using information from the received preamble sections of the MAC frame.

The inverse OFDM part of the receiver converts the received signal into received subcarrier values. The received subcarrier values may still suffer from distortions that need to be corrected before de-mapping them to a bitstream. The equalizer corrects the distortions caused by frequency-selective fading. The coefficients for the equalizer can be determined by using information from the received preamble sections of the MAC frame. Based on the equalized pilot values, the phase distortion of the received signal is corrected.

Finally, the received complex-number samples will be translated into a useful received bitstream. The de-mapping function assumes that the most likely symbol that was transmitted, was the symbol that maps to the value closest to the received value.

The physical layer of the HiperLAN/2 receiver [30] has been implemented in a combination of general purpose processor and coarse-grained reconfigurable hardware. Figure 8 shows the functional baseband processing blocks in the receiver that are implemented in each tile of the heterogeneous reconfigurable SoC.

Irregular tasks, which are outside the algorithm domain of the MONTIUM, are performed in software (i.e. on a GPP). The irregular processes in the HiperLAN/2 receiver are the estimation of frequency offset and computation of equalization coefficients. These coefficients have to be determined only once per MAC frame, i.e. once per 2 ms.

The frequency offset correction is implemented in one MONTIUM tile. During correction every complex-number sample is multiplied with the frequency offset correction factor. The correction factor is determined with a LUT based on the estimated frequency offset. The frequency offset is estimated in software by the GPP once per MAC frame. One OFDM symbol, containing 64 complex-number samples, can be corrected in 67 clock cycles.

A Fast Fourier Transform (FFT) on a vector of 64 complex-number time samples can perform the inverse OFDM function. The 64-FFT can be performed in 204 clock cycles for one OFDM symbol.

The equalizer, phase offset correction and de-mapping functionality are implemented in one MONTIUM tile in a pipelined fashion. The coefficients for equalization are determined once every MAC frame in software by the GPP. During equalization, the received carriers

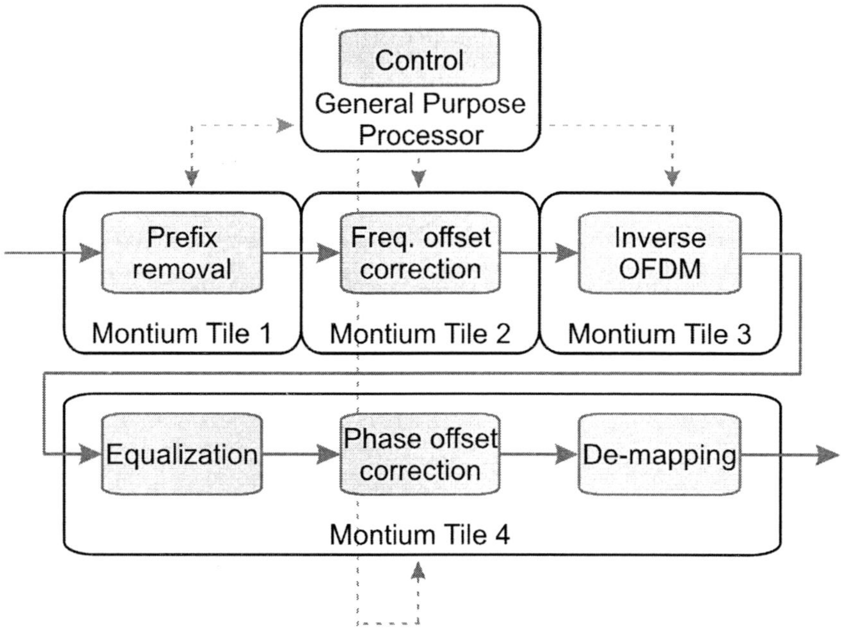

Figure 8. HiperLAN/2 receiver mapped to heterogeneous reconfigurable SoC.

are multiplied with the equalization coefficients. After equalization the pilot values are used to determine the phase offset correction factor. The phase offset correction factor is determined in the MONTIUM, since the phase offset can vary for every OFDM symbol and the correction factor has to be determined on an OFDM symbol basis (once every 4 μs). Hence, determining the phase offset correction factor in software (i.e. GPP) would create large communication overhead between the GPP and the MONTIUM tile. Phase offset correction invokes also a complex multiplication, like equalization. As a consequence the equalizer and phase offset corrector use the same functionality of the MONTIUM. In a pipelined, parallel manner the corrected complex-number samples are translated into a bitstream. Hard-decision de-mapping is implemented with LUT functionality. A parametrizable de-mapper has been implemented, which can be used for QPSK, 16-QAM and 64-QAM modulated signals by only changing the LUT table in the memory of the MONTIUM.

3.3.1. Configuration

The total configuration sizes of the MONTIUM are small for the different functions (Table 3). MONTIUM tile 3, on which the inverse OFDM is performed, requires the largest configuration size. The configuration of tile 3 contains less than 1 Kbyte of data. The configuration data can be written into the configuration memory of the MONTIUM in about 500 clock cycles, since 2 bytes are written in one clock cycle. Suppose that the MONTIUM is running at a clock frequency of 100 MHz, then tile 3 can be configured in 4.73 μs. Notice that the maximum radio turn-around time of the HiperLAN/2 system is 6 μs [33], so the implemented HiperLAN/2 receiver can be considered as a real-time dynamically reconfigurable receiver.

Table 3. Properties of the HiperLAN/2 implementation.

	Frequency offset correction	Inverse OFDM	Equalizer, Phase offset, De-mapper
Execution time [cycles]	67	204	110
Communication time [cycles]	128	116	<100
Minimum system clock with streaming communication [MHz]	17	51	28
Minimum processor clock with block communication (@ 100 MHz) [MHz]	25	72	37
Configuration size [bytes]	274	946	576
Configuration time [cycles]	137	473	288

3.3.2. Frequency Scaling

All operations in the physical layer are performed on OFDM symbols. So, one should assure that each 4 μs a new OFDM symbol can be processed. When a streaming on-chip network between the processors is assumed, the communication time is not a bottleneck and one only has to guarantee that, for example, the data processing for frequency offset correction is performed during 67 clock cycles in 4 μs. Hence, the minimum clock frequency of the MONTIUM tile is 17 MHz, when a streaming on-chip network between the tiles is assumed.

Typically, the clock frequency of the NoC, connecting the reconfigurable processors, will be fixed and only the clock frequency of the reconfigurable processor can be varied. Assuming that the clock frequency of the NoC is fixed at 100 MHz, then the clock frequency of the MONTIUM for frequency offset correction has to be at least 25 MHz (Table 3).

3.3.3. Adaptivity Features

The following adaptivity features can be found in the HiperLAN/2 receiver:

- *standards adaptivity*
  Different communication standards can be implemented using the same reconfigurable hardware. Table 3 shows the size of the configuration files in order to configure the processing elements of the SoC to HiperLAN/2. Configuring the processing elements to the baseband processing functions takes less than 10 μs, assuming that the communication bandwidth of the NoC is high enough.

- *algorithm-parameter adaptivity*
  In the HiperLAN/2 standard different modulation schemes (BPSK, QPSK, 16-QAM and 64-QAM) can be applied. This parameter can be changed by switching the constellation table in the de-mapper. The operation does not require any reconfiguration.

3.4. Controlling the Tasks in the SoC

In Section 3.2 and 3.3 the baseband processing of two different standards have been implemented in the same reconfigurable hardware. Both communication standards serve as a basis for an adaptive UMTS/WLAN receiver. In order to employ an adaptive wireless communication system, which can adapt the baseband processing functions in the physical layer, a centralized control-manager is needed.

The control-manager is implemented in the GPP (e.g. ARM) tile, which is perfectly suited for control-oriented tasks. Tasks of the control-manager are:

- Configuring the processing elements in the SoC.
- Configuring the routing elements of the NoC.
- Administrating the status of the entire SoC and NoC.
- Controlling the different levels of adaptivity:
  - Switching the wireless communication standard based on availability and QoS requirements.
  - Switching the algorithms, used to implement the functions within the receiver for a wireless communication standard, based on channel conditions and QoS requirements.
  - Adapting the parameters of an algorithm based on channel conditions and QoS requirements.

For the adaptive UMTS/WLAN receiver this would mean that the control-manager decides whether the receiver is in UMTS or in WLAN mode. Furthermore, adaptivity within one mode can be applied. Thus, the control-manager can decide whether in the UMTS mode the RAKE functionality or the equalizer functionality will be applied. Moreover, the manager controls the parameters of a certain algorithm; for example the number of RAKE fingers that should be used or the modulation mode in HiperLAN/2.

3.4.1. HiperLAN/2 to UMTS switching

In HiperLAN/2 mode, 4 MONTIUM tiles in the SoC are configured for *Prefix removal*, *Frequency offset correction*, *Inverse OFDM* and *Equalization, Phase offset correction and de-mapping*. Furthermore, the general purpose processor in the SoC is programmed to perform the *Channel estimation*. This task delivers the estimated frequency offset and the equalization coefficients.

Switching to UMTS means that 2 MONTIUM tiles have to be configured as *pulse-shape filter* and *RAKE*. One FPGA tile has to be configured as *scrambling code generator*. Furthermore the *channel estimator* and *path searcher* need to be programmed in the general purpose processor.

After the decision of the manager to switch from HiperLAN/2 to UMTS, the manager has to load the new configurations into the tiles. Furthermore, the communication channels between the different processing elements in the SoC have to be configured. Communication channels are defined between the pulse-shape filter and the RAKE receiver, between the scrambling code generator and the RAKE receiver, and between the general purpose processor and the RAKE receiver. Loading the configuration files into the MONTIUM tiles requires a couple of μs as seen in Section 3.2.1. The manager also adapts the clocks of the processing elements to the required frequencies as discussed in Section 3.2.2.

3.4.2. RAKE versus Equalizer

In Section 2.1.2 we described that the W-CDMA receiver can be implemented in different ways, using a RAKE receiver or an equalizer. Depending on the conditions of the environment, the manager can decide whether the RAKE receiver or equalizer is used in the UMTS receiver. This means that in UMTS mode one tile is configured as *equalizer* and one as *pulse-shape filter*. Now at a certain moment the manager decides to reconfigure the MONTIUM tile that acts as equalizer to RAKE. So, the manager has to load the new configuration in the MONTIUM tile. All communication channels between the tiles remain the same and are not reconfigured.

3.4.3. Adapting the Number of RAKE Fingers

The number of RAKE fingers affects the total signal power that the W-CDMA receiver receives. The environment influences the number of reflected signals, so-called multi-paths, that arrive at the receiving terminal. Based on the number of reflected signals the manager decides how many fingers are applied in the RAKE receiver in order to receive maximum signal power.

After the decision of the manager to change the number of fingers in the RAKE receiver, the manager has to change the configuration of the processing element which implements the RAKE. In Section 3.2.1 the number of RAKE fingers are adapted by dynamic reconfiguration. The reconfiguration time is in the order of ns. By reducing the number of fingers in the RAKE, the clock frequency of the MONTIUM, which implements the RAKE, can be reduced. The manager also handles the settings of the system clock.

4. Summary and Conclusion

Trends in wireless communication systems show that the underlying DSP algorithms become more adaptive. In this chapter three levels of adaptivity in the digital signal processing of wireless receivers have been identified: standards level, algorithm-selection level and algorithm-parameter level. Examples of the last two levels of adaptivity have been investigated for the signal processing functions in the UMTS receiver. Adaptivity requires more flexible hardware architectures. These architectures provide opportunities for adaptation in the digital processing of wireless receivers. Traditional hardware architectures cannot provide the required flexibility efficiently. Therefore, the concept of heterogeneous reconfigurable hardware has been introduced. A heterogeneous reconfigurable architecture has been used to implement an adaptive wireless communication receiver in an energy-efficient way. The implemented receiver can switch between HiperLAN/2 and UMTS. Within the implementation of UMTS or HiperLAN/2 algorithm-selection and algorithm-parameter adaptivity features can be used.

References

[1] Paul M. Heysters and Henri Bouma and Jaap Smit and Gerard J. M. Smit and Paul J. M. Havinga,A reconfigurable function array architecture for 3G and 4G wireless terminals, *Proceedings of 2002 World Wireless Congress*, 2002, P. 399–404, San Francisco, USA.

[2] Paul M. Heysters and Gerard J. M. Smit and Egbert Molenkamp, A Flexible and Energy-Efficient Coarse-Grained Reconfigurable Architecture for Mobile Systems, *Journal of Supercomputing*, 2003, V. 26, No. 3, P. 283–308.

[3] Jan Rabaey, *Silicon Architectures for Wireless Systems*, 2001, Tutorial, Hotchips 2001.

[4] Ralf Niemann, Hardware/Software Co-Design for Data Flow Dominated Embedded Systems, Kluwer Academic Publishers, 1998, Boston.

[5] Bingfeng Mei and Patrick Schaumont and Serge Vernalde, A Hardware-Software Partitioning and Scheduling Algorithm for Dynamically Reconfigurable Embedded Systems, *Proceedings of ProRISC 2000*, 2000, Veldhoven, the Netherlands.

[6] Andras Varga, OMNeT++ – Object-Oriented Discrete Event Simulator, http://www.omnetpp.org, 2003.

[7] Nikolay Kavaldjiev and Gerard J.M. Smit, A Survey of On-Chip Interconnection Networks, University of Twente, faculty EEMCS, Enschede, the Netherlands, 2003.

[8] G. K. Rauwerda and G. J. M. Smit and L. F. W. van Hoesel and P. M. Heysters, Mapping Wireless Communication Algorithms to a Reconfigurable Architecture, *Proceedings of the International Conference on Engineering of Reconfigurable Systems and Algorithms (ERSA'03)*, 2003, Las Vegas, Nevada, USA, P. 242–251.

[9] IST-ADRIATICStudy of existing high-level hardware/software co-design methodologies, Deliverable 2.1, IST-2000-30049/WP/D2.1/IMEC/R, 2002.

[10] OCAPI-xl, http://www.imec.be/ocapi.

[11] SystemC, http://www.systemc.org.

[12] Jordy Potman and Fokke Hoeksema and Kees Slump, DSP Prototyping of Blind Adaptive MMSE Multiuser Detection for Cellular Wireless CDMA Systems, *Proceedings of PRORISC 2002*, 2002, P. 409–418, Veldhoven, the Netherlands.

[13] Lodewijk T. Smit and Gerard J. M. Smit and Paul J. M. Havinga and Johann L. Hurink and Hajo Broersma, Run-Time Control For Software Defined Radio, *Proceedings of PROGRESS Workshop 2002*, 2002, Utrecht, the Netherlands.

[14] Gerard J. M. Smit and Paul J. M. Havinga and Lodewijk T. Smit and Paul M. Heysters and Michel A. J. Rosien, Dynamic Reconfiguration in Mobile Systems, *Proceedings of FPL 2002*, 2002, P. 171–181, Montpellier, France.

[15] Xilinx, Virtex-II Pro Platform FPGAs, http://www.xilinx.com/virtex2pro/.

[16] Altera, *Excalibur Embedded Processor Solutions,* http://www.altera.com/products/devices/excalibur/exc-index.html.

[17] *Chameleon project – Reconfigurable computing in hand-held multimedia computers* , http://chameleon.ctit.utwente.nl.

[18] SDR Forum, http://www.sdrforum.org.

[19] *SDR project - A Bluetooth-HiperLAN/2 SDR receiver*, http://www.sas.el.utwente.nl/home/SDR/.

[20] Pleiades project, http://bwrc.eecs.berkeley.edu/Research/Configurable_Architectures/.

[21] Chameleon Systems, *Reconfigurable Communications Processor*, http://www.chameleonsystems.com.

[22] PACT XPP Technologies, http://www.pactcorp.com.

[23] QuickSilver Technologies, http://www.quicksilvertech.com.

[24] MuMoR project, http://www.mumor.org.

[25] EASY project, http://easy.intranet.gr.

[26] FITNESS project, http://www.ist-fitness.org.

[27] Lodewijk T. Smit and Gerard J. M. Smit and Johann L. Hurink, Energy-efficient Wireless Communication for Mobile Multimedia Terminals, *Proceedings of The International Conference On Advances in Mobile Multimedia*, 2003, Jakarta, Indonesia, P. 115–124.

[28] Anna Berno, *Time and Frequency Synchronization Algorithms for HIPERLAN/2* , University of Padova, Italy, 2001.

[29] J. Khun-Jush and P. Schramm and U. Wachsmann and F. Wenger, Structure and Performance of the HIPERLAN/2 Physical Layer, *Proceedings VTC '99 - Fall*, 1999, Amsterdam, the Netherlands, P. 2667–2671.

[30] ETSI, Broadband Radio Access Networks (BRAN); *HiperLAN Type 2; Physical (PHY) Layer*, 2001, ETSI TS 101 475 v1.2.2 (2001-02).

[31] J. Medbo and P. Schramm, Channel Models for HIPERLAN/2, ETSI/BRAN 3ERI085B, 1998.

[32] R. E. Ziemer and R. L. Peterson, Introduction to digital communication, Second Edition, Prentic Hall, USA, 2001.

[33] ETSI, Broadband Radio Access Networks (BRAN); HiperLAN Type 2; Data Link Control (DLC) Layer Part 1: Basic Data Transport Functions, 2000, ETSI TS 101 761-1 v1.1.1 (2000-04).

[34] Lodewijk T. Smit, Energy-Efficient Wireless Communication, University of Twente, 2004, Enschede, the Netherlands.

[35] Jordy Potman and Fokke Hoeksema and Kees Slump,Tradeoffs between Spreading Factor, Symbol Constellation Size and Rake Fingers in UMTS, *Proceedings of PROR-ISC 2003*, P. 543–548, 2003, Veldhoven, the Netherlands.

[36] Gerard Rauwerda and Jordy Potman and Fokke Hoeksema and Gerard Smit, Adaptive Wireless Networking, *Proceedings of the 4th PROGRESS Symposium on Embedded System*, P. 205–211, 2003, Nieuwegein, the Netherlands.

[37] Paul M. Heysters and Gerard K. Rauwerda and Gerard J. M. Smit, Implementation of a HiperLAN/2 Receiver on the Reconfigurable Montium Architecture, *Proceedings of the 11th Reconfigurable Architectures Workshop (RAW 2004)*, 2004, Santa Fé, New Mexico, USA.

[38] 3rd *Generation Partnership Project, Base Station (BS) radio transmission and reception (FDD)*, 2004, 3GPP TS 25.104 v6.4.0 (2003-12).

[39] 3rd *Generation Partnership Project, Spreading and modulation (FDD),* 2001, 3GPP TS 25.213 v4.1.0 (2001-06).

[40] 3rd *Generation Partnership Project, Physical layer – General description,* 2004, 3GPP TS 25.201 v6.0.0 (2003-12).

[41] Max Nilsson, Efficient ASIC implementation of a WCDMA Rake Receiver, Luleå University of Technology, Sweden, 2002.

[42] H. Holma and A. Toskala, WCDMA for UMTS: Radio Access for Third Generation Mobile Communications, John Wiley & Sons, 2001.

[43] Paul M. HeystersCoarse-Grained Reconfigurable Processors – Flexibility meets Efficiency, University of Twente, 2004, Enschede, the Netherlands.

[44] G. Heidari and K. Lane, Introducing a Paradigm Shift in the Design and Implementation of Wireless Devices, *Proceedings Wireless Personal Multimedia Communications,* 2001, Aalborg, Denmark, P. 225–230".

[45] A. Abnous, Low-Power Domain-Specific Processors for Digital Signal Processing, University of California, 2001, Berkeley, USA.

[46] V. Baumgarte and F. May and A. Nückel and M. Vorbach and M. Weinhardt, PACT XPP – A Self-Reconfigurable Data Processing Architecture, *Proceedings Engineering of Reconfigurable Systems and Algorithms,* 2001, Las Vegas, Nevada, USA, P. 64–70.

[47] Silicon Hive, http://www.siliconhive.com.

[48] K. Masselos and S. Blionas and T. Rautio, Reconfigurability requirements of wireless communication systems, *Proceedings of the IEEE Workshop on Heterogeneous Reconfigurable Systems on Chip*, 2002, Hamburg, Germany.

[49] B. Bougard and S. Pollin and G. Lenoir and W. Eberle and L. Van der Perre and F. Catthoor and W. Dehaene, Energy-Scalability Enhancement of Wireless Local Area Network Transceivers, *Proceedings of the Fifth IEEE Workshop on Signal Processing Advances in Wireless Communication*, 2004, Lisboa, Portugal.

[50] Pascal T. Wolkotte and Gerard J. M. Smit and Gerard K. Rauwerda and Lodewijk T. Smit, An Energy-Efficient Reconfigurable Circuit Switched Network-on-Chip, *Proceedings of the 12th Reconfigurable Architectures Workshop* (RAW 2005), 2005, Denver, Colorado, USA, (to be published).

[51] Nikolay Kavaldjiev and Gerard J. M. Smit and Pierre G. Jansen, A Virtual Channel Router for On-chip Networks, *Proceedings of IEEE International SOC Conference*, 2004, Santa Clara, California, USA, P. 289–293.

[52] Charles M. Rader, Memory Management in a Viterbi Decoder, *IEEE Transactions on Communications*, 1981, V. 29, No. 9, P. 1399–1401.

[53] Andrew J. Viterbi, Error Bounds for Convolutional Codes and an Asymptotically Optimum Decoding Algorithm, *IEEE Transactions on Information Theory,* 1967, V. 13, No. 2, P. 260–269.

[54] Richard D. Wesel, Convolutional CodesWiley Encyclopedia of Telecommunications, John Wiley & Sons, 2003, V. 1, P. 598–606.

[55] Gerhard Fettweis, Algebraic Survivor Memory Management Design for Viterbi Detectors, *IEEE Transactions on Communications*, 1995, V. 43, No. 9, P. 2458–2463.

[56] Shu Lin and Daniel J. Costello, Error Control Coding: Fundamentals and Applications, Prentice-Hall, 1983.

[57] 3rd Generation Partnership Project, Technical Specification Group Radio Access Network; Multiplexing and channel coding (FDD), 2002, 3GPP TS 25.212 v4.3.0 (2001-12).

[58] ETSI, Radio Broadcasting Systems; Digital Audio Broadcasting (DAB) to mobile, portable and fixed receivers, 2001, ETSI EN 300 401 v1.3.3 (2001-05).

[59] ETSI, Digital Radio Mondiale (DRM); System Specification, 2001, ETSI TS 101 980 v1.1.1 (2001-09).

[60] Richard D. Wesel and Xuething Liu and Wei Shi, Trellis Codes for Periodic Erasures, *IEEE Transactions on Communications*, 2000, V. 48, No. 6, P. 938–947.

[61] Smart chipS for Smart Surroundings project, http://www.smart-chips.net.

[62] AMDREL project, http://vlsi.ee.duth.gr/amdrel/.

[63] Roel Schiphorst, Software-Defined Radio for Wireless Local-Area Networks, University of Twente, 2004, Enschede, the Netherlands.

[64] Gerard K. Rauwerda and Gerard J. M. Smit, Implementation of a Flexible RAKE Receiver in Heterogeneous Reconfigurable Hardware, *Proceedings of the 2004 IEEE International Conference on Field-Programmable Technology*, P. 437–440, 2004, Brisbane, Australia.

[65] Sriram Swaminathan and Russell Tessier and Dennis Goeckel and Wayne Burleson, A Dynamically Reconfigurable Adaptive Viterbi Decoder, *Proceedings of the 2002 ACM/SIGDA 10th International Symposium on Field-Programmable Gate Arrays* , P. 227–236, 2002, Monterey, California, USA.

[66] Tobias Gemmeke and Michael Gansen and Tobias G. Noll, Implementation of Scalable Power and Area Efficient High-Throughput Viterbi Decoders, *IEEE Journal of Solid-State Circuits,* 2002, V. 37, No. 7, P. 941–948.

[67] Russell Henning and Chaitali Chakrabarti, An Approach for Adaptively Approximating the Viterbi Algorithm to Reduce Power Consumption While Decoding Convolutional Codes, *IEEE Transactions on Signal Processing,* 2004, V. 52, No. 5, P. 1443–1451.

[68] Michiel Horsman and Thijs Mutter and Marcel van der Veen and Maarten Witteman and Karel Walters, Energy comparison of Viterbi algorithm on ARM and XScale architecture, University of Twente, Enschede, the Netherlands, 2004, Internal report.

[69] Yrjö Neuvo, Cellular Phones as Embedded Systems, Proceedings of the ISSCC 2004, P. 32–37, 2004.

[70] Gerard K. Rauwerda and Paul M. Heysters and Gerard J. M. Smit, Mapping Wireless Communication Algorithms onto a Reconfigurable Architecture, *Journal of Supercomputing,* 2004, V. 30, No. 3, P. 263–282.

[71] Tanner, R. and Woodard, J., *WCDMA Requirements and Practical Design*, John Wiley & Sons, 2004.

[72] Veljanovski, R. and Singh, J. and Faulkner, M., A Proposed Reconfigurable Digital Filter for a Mobile Station Receiver, *Proceedings of GLOBECOM '02*, 2002, P. 524-528.

[73] Hooli, K., *Equalization in WCDMA Terminals* , University of Oulu, 2003.

[74] Damstra, M.C., Path search algorithms for application in W-CDMA systems, University of Twente, 2004.

[75] Choi, J.W. and Lee, Y.H., *Adaptive Channel Estimation in DS-CDMA Systems, ECTI Trans. on Electrical Eng., Electronics*, and Communications, 2004, V. 2, No. 1.

[76] Henning, R. and Chakrabarti, C., An Approach for Adaptively Approximating the Viterbi Algorithm to Reduce Power Consumption While Decoding Convolutional Codes, *IEEE Trans. on Signal Processing*, 2004, V. 52, No. 5, P. 1443-1451.

In: Adaptation Techniques in Wireless Multimedia Networks ISBN: 1-59454-883-8
Editors: Wei Li and Yang Xiao, pp. 57-73

Chapter 4

ADAPTIVE VIDEO MULTICAST OVER WIRELESS INTERNET

***Hao Yin*[†], *Bo Li*,[††] *Chuang Lin*[†], *Geyong Min*[†††] *and Qiang Ni*[††††]**
Institute of Industrial Science and Technical Research,
Kyushu Tokai University, 9-1-1 Toroku, Kumamoto 862-8652, Japan

Abstract

With the rapid growth of wireless networks and great success of Internet video, wireless video services are expected to be widely deployed in the near future. As different types of wireless networks are converging into an all-IP network, i.e., the Internet, it is important to study video delivery over the wireless Internet. This chapter proposes a novel end-system based adaptive protocol called Wireless Hybrid Adaptation Layered Multicast (WHALM) protocol for layered video multicast over wireless Internet. In WHALM the sender dynamically collects bandwidth distribution from the receivers and uses an optimal layer rate allocation mechanism to reduce the mismatches between the coarse-grained layer subscription levels and the heterogeneous and dynamic rate requirements from the receivers, thus maximizing the Quality of Service (QoS) of all the receivers in a multicast session. Based on the sampling theory and the probability theory, we reduce the required number of bandwidth feedbacks to a reasonable degree and use a scalable feedback mechanism to control the feedback process practically. WHALM is also tuned to perform well in wireless networks by integrating an end-to-end loss differentiation algorithm (LDA) to differentiate error losses from congestion losses at the receiver side. With a series of simulation experiments over NS platform, WHALM has been proved able to greatly improve the QoS of all the receivers while avoiding congestion collapse on the wireless Internet.

Keywords: wireless Internet, multicast, scalable video

1. Introduction

The rapid development of the mobile wireless cellular networks and the deployment of new services are changing the ways we live. It is expected that the next generation (3G and beyond 3G) wireless cellular network will be fully capable of delivering multimedia content, in which video multicast might well be the next killer application [1][2][12]. Unfortunately, most of the research on video multicast is performed over Internet. As the different types of wireless networks are converging into all IP networks, i.e., the Internet, it is important to study video multicast over the wireless Internet [4]. Applications of wireless video multicast support group-oriented mobile commerce, military command and control, distance education, and intelligent transportation systems. Many new mobile commerce applications, including mobile auctions, will also gain significant benefit if group communication among mobile users is supported by wireless networks [4][11][14]. Before video multicast can be deployed on wireless Internet, there are many critical issues that need to be examined, i.e. the intrinsic heterogeneity, large scale, error-prone nature, dramatic link status variation, users' mobility and scarce wireless spectrum. In addition, the video streams co-exist and share the resources with the TCP-based data traffic typically in wireless Internet. It is thus important for the video traffic to be adaptive and friendly with the TCP traffic [3]. Lack of bandwidth adaptability will lead to congestion collapse when the aggregate bandwidth of video traffic exceeds network capacity, whereas lack of TCP friendliness will result to compete unfairly with other adaptive traffic, such as TCP [4]. TCP friendliness means, basically, "to achieve a fair share of bandwidth relative to reasonable TCP flows over the same path". The definition still allows for improvements to TCP.

This chapter proposes an end-system based adaptive protocol to support TCP-friendly video multicast in wireless Internet without special support from core network equipments. In the traditional unicast environment, the sender collects the receiver's states via a feedback mechanism and adjusts its transmission rate accordingly. Such an approach faces three major problems in the context of wireless multicast: 1) How to avoid feedback implosion generated by a large number of receivers in a multicasting application? 2) The single sending rate can not satisfy the conflicting requirements of a set of heterogeneous receivers, and layered video transmission has been proposed as an effective approach to support heterogeneous receivers with varying bandwidth requirements [3]. Thus, how should the scarce wireless bandwidth be allocated to different layers? 3) How to design a proper TCP-friendly congestion control protocol in order to handle both the congestion loss and error loss in wireless IP network?

In this chapter we extend the Hybrid Adaptation Layered Multicast (HALM) protocol in [2] [16], which is an end-system based adaptation protocol for layered video multicast over the Internet. HALM adopts a metric called Fairness Index for each receiver, and formulates the joint rate allocation into an optimization problem [2]. It also implements an efficient algorithm to solve the problem. Practical issues about scalable feedback and estimation of available network bandwidth for deploying the optimal algorithm in wireless Internet are solved by using Sender-Adaptive & Receiver-driven Layered Multicast (SARLM) Scheme [5] and TCP throughput formula [6]. One challenge for available bandwidth estimation based on TCP throughput is how to classify the congestion loss and error loss in wireless networks. We refer to the end-to-end loss differentiation algorithm (LDA) [7] to solve the problem. By combining the advantages of HALM, SARLM and LDA, we propose a novel end-system based adaptation protocol for layered video multicast over wireless Internet, which is called Wireless Hybrid Adaptation Layered Multicast (WHALM) protocol.

There are two main wireless network topologies for supporting multimedia applications, namely, networks with last-hop wireless links, and networks with wireless backbones [7]. The former corresponds to cellular networks or satellite modems, and the latter corresponds to high-bandwidth backbones or wireless LAN network such as 802.11 [7]. Taking the characteristics of video multicasting applications into account, WHALM assumes a wireless network topology where the wireless link is the last hop.

The rest of the chapter is organized as follows. Section 2 reviews the related work. Section 3 discusses the framework of WHALM. Section 4 describes the formulation of the optimal rate allocation problem and the integration of its scalable solution into WHALM. Section 5 discusses the available bandwidth estimation for wireless network at the receiver side. Section 6 presents a new scalable feedback mechanism for our protocol. Section 7 evaluates the performance of WHALM through simulation experiments. Finally, Section 8 concludes the chapter and discusses some future directions.

2. Related Work

2.1. Scalable Video Coding

In the coding community, layered coding is often referred to as *scalable coding*. Scalability can be achieved by scaling the frame speed (*temporal scalability*), frame size (*spatial scalability*), and frame quality (*quality or SNR scalability*) [19]. These scalable coding algorithms have been adopted in advanced compression standards, such as H. 264, MPEG-2, and MPEG-4. WHALM does not specify any particular coding algorithm in the application layer. Nevertheless, a coder with a wide dynamic range, fast responsiveness, and fine granularity in terms of rate control is of particular interest. Examples include the *Fine Granularity Scalability* (FGS) [20] or *Progressively FGS* (PFGS) coders [21]. The key technique used here is *bit-plane coding* [20], by which layer rates can be allocated through an assembling/packetization procedure after compression. This is different from the traditional rate control that is performed during compression by adjusting quantizers. Hence, it can speedup response to layer rate adjustment and incur low overhead for layer synchronization. Moreover, the bit-plane coding has been adopted in the MPEG-4 standard.

2.2 TCP-Friendliness

Using TCP for real-time video delivery is not practical, because these applications usually require a smoothed transmission rate and have stringent restrictions on end-to-end delay. However, since a dominant portion of today's Internet traffic is TCP-based, video streaming protocols should have some rate control to ensure that the traffic does not overwhelm the congestion-sensitive TCP flows. This requirement is commonly referred to as *TCP friendliness* [18]. From [18], a flow is said TCP-friendly if its arrival rate does not exceed the arrival rate of a conformant TCP connection in the same network circumstances. A number of TCP-friendly rate control algorithms have been proposed recently, e.g.,TCP-Friendly Rate Control (TFRC) proposed in RFC3448 has a feature of adjusting a transmission rate smoothly while coping with network congestion. Note that, short-term adaptation results in bandwidth oscillations, which is not desirable for video transmission. It is even impossible for a layered video stream to be totally fair to TCP flows, for its adaptation granularity on the receiver's side is at a layer level. Thus our objective is to provide an adaptive protocol that will not starve background TCP traffic and, meanwhile, try to achieve a longterm fair share as close as possible. This loose notion of TCP-friendliness has been widely adopted in existing streaming protocols; see for example [22]. Similar to such protocols, WHALM uses an equation to estimate the longterm throughput of a virtual TCP connection (as if the connection is running over the same path), and adjust the transmission rate accordingly.

2.3. Layered Multicast

McCanne *et al* [1] first introduced a receiver-driven cumulative layered multicast (RLM) congestion control algorithm for heterogeneous users. Vicisano *et al* [23] suggested the coordinated join tests among receivers and sender-initiated probings to reduce the chances of failure of joining tests. Their idea is to try to let the receivers know more about the current network condition without being misled by some false "signals" of network conditions (e.g. congestion and change of package loss ratio). But it can only give users fixed number of layers and each a fixed rate, which leads to low utilization when background traffic varies. Bolot, *et al* [24] presented a sender-driven layered multicast scheme, where the sender adjusts the data rate according to the receiver feedback. However, since they only use one layer, it cannot serve heterogeneous receivers well. Vickers *et al* [25] introduced a source adaptive multi-layered multicast algorithm (SAMM) where the source uses congestion feedback to adjust the number of generated layers and the bit rate of each layer. SAMM is very stable as the high priority packets are always well protected. In addition, since flow-isolation is implemented in routers, TCP-friendliness is not required any more. The prioritized queuing discipline, however, is considerably more complex than the simple FIFO. However they did not address the placement or dynamic instantiation of feedback mergers in the network. They also do not take heterogeneous package loss ratio and fairness problem of receivers into consideration. Liu and Li [2] proposed a Hybrid Adaptive Layered Multicast (HALM) algorithm which performs adaptations on both the sender and the receiver's sides to improve intra-session fairness as well as TCP-friendliness. The main contribution of HALM is a formal study on the sender-based optimal layer rate allocation and its practical use. HALM defined optimization criteria and derived a scalable algorithm to solve the problem. HALM also discussed the implementation issues; specifically, the choice of the layered video coder, the estimation of TCP-friendly bandwidth, and the inference of bandwidth distribution. But HALM does not considers wireless Internet, so its implementation issues are not suitable to work in wireless environment. At the same time, the approach solving feedback implosion in HALM is not very effective. So in this study we extend the HALM into wireless Internet and solve some key problems in the implementation of WHALM.

2.4. Available Bandwidth Estimation

There are three approaches for estimating available bandwidth, namely, Information-share-based approach, Probe-based approach, and Model-based approach [17]. Take the application of the video multicast into account; model-based approach is adopted to estimate end-to-end available bandwidth in WHALM. Model-based approach uses the TCP throughput formula [17], which is obtained from the indicators of the level of network congestion and algorithms of congestion control. The challenge of using model-based approach to estimate available bandwidth of wireless network is how to differentiate congestion loss and error loss in wireless IP network. The typical solution uses a mechanism called end-to-end loss differentiation algorithm (LDA) which differentiates congestion losses from wireless losses by analyzing the variation of packets' transmission delays [7]. The idea is based on the fact that congestion loss is often accompanied by the increased transmission delay while wireless loss does not have this character. LDAs assume that if a packet loss event happens with increased inter-arrival time or increased one-way trip time, it is a congestion loss; otherwise it is a random loss. The advantage of this solution is that it has no dependency on network topology. The transport layers of the sender and receiver view network as a black box. This solution can be applied for networks with either backbone or last-hop wireless links.

3. Overview of WHALM

WHALM works on top of the Real-time Transport Protocol (RTP) [8]. The video stream is delivered by RTP and control messages are exchanged by an application-specific RTP Control Protocol (RTCP) [8]. Its design is motivated by the HALM protocol [2, 16], and yet considers the unique features of wireless links.

3.1. Sender Functionality

In WHALM, the sender encodes the raw video into l cumulative layers using a layered coder. Layer 1 is the base layer and layer l is the least important enhancement layer. The layer rates are given by b_i (i=1,2,…,l). Let c_j denote the cumulative layer rate up to layer j, that is, $c_j = \sum_{i=1}^{j} b_i, j = 1,2,...,l,$ and ρ_l denotes the rate vector, $\rho_l = (c_1, c_2,\ldots, c_l)$. This discrete set offers all the possible video rates of a receiver. In particular, the maximum rate that a receiver with an expected bandwidth r can receive is given by

$$\Gamma(r, \rho_l) = \max\{c : c \le r, c \in \rho_l\} \tag{1}$$

Note that there is a gap between this receiving rate and the expected bandwidth of a receiver. To minimize this gap, the sender also collects the reports of the expected bandwidths from the receivers. Assume the session size is N and the bandwidths expected by receivers are $\{r_1,r_2,\ldots, r_N\}$. The sender will adjust the layer rates based on the distribution of the expected bandwidths with a control period of T_{ctrl} seconds.

The sender multicasts two kinds of report packets to all receivers. One report packet is Sender Control Report (SCR), which is generated and multicasts at the beginning of every control period to distribute the control parameters. SCR has the format of (ID, TS, ρ_l, I, λ, α, T_{ctrl}), where ID is RTP synchronization source identifier (SSRC) [8], TS is a timestamp of the sender's local time, ρ_l is the current rate vector, I is the identification number of the control period, λ and α are parameters of the gamma-distributed timer used by the receivers to avoid feedback implosion, and T_{ctrl} is the interval size of the current control period. Another report packet is Sender Report (SR). SR is multicast every T_{SR} second, where $T_{SR}=T_{ctrl}/k$ for some integer k>1 and is used by the receivers to estimate Round Trip Time (RTT). SR includes SSRC, a timestamp of the sender's local time and the response to receivers' requests. To reduce the control overhead, the sender does not give a response to each request but uses a batch process. An SR multicast at time $t+T_{SR}$ contains the SSRCs of the receivers whose RR packets (described in Section 2.2) arrive in the time slot $[t, t+T_{SR}]$ and their delays t_i^{delay} (the interval between their arrival time and $t+T_{SR}$). Upon receiving the SR, these receivers can use the contained information to calculate their RTTs. Detailed algorithm can be found in [2]. We have depicted these cases in Figure 1.

3.2. Receiver Functionality

A receiver decides whether to join a higher layer, stay at or leave the current layer at the beginning of each control period based on the rate vector in SCR and its expected bandwidth. To be friendly with TCP, a receiver directly uses the following TCP throughput formula [6] to calculate its expected bandwidth that will be sent to the sender:

$$B = \frac{s}{RTT\sqrt{\frac{2p}{3}} + RTO(3\sqrt{\frac{3p}{8}})p(1+32p^2)} \tag{2}$$

This gives the TCP throughput B in bytes/sec, as a function of the packet size s, round-trip time RTT, steady-state loss event rate p, and the TCP retransmit timeout value RTO. The receiver dynamically monitors these parameters to calculate the bandwidth and selectively sends it back to the sender. Specifically, it uses a scalable feedback scheme, as described in Section 5, to decide whether and when to generate report packets in a control period. A feedback packet, named as RR, contains the SSRC of the receiver, the expected bandwidth, and the timer setting z_i of the receiver, which is used by the sender to estimate the number of receivers. It also serves as a request for RTT estimation.

3.3. Operations of WHALM

The sender adjusts its sending rates once during every control period of T_{ctrl} seconds, which varies from 2s to 15s according to the number of receivers. The control process works as follow.

At the beginning of a control period, the sender adjusts its sending rates based on the receivers' bandwidth distribution. It also multicasts a SCR to deliver the new rate vector, the duration of the current control period and the parameters of the gamma-distributed timer (λ and α) used by receivers to avoid feedback implosion.

Upon receiving SCR, each receiver makes decision and then performs its join/leave layer actions. It also generates a gamma distributed random timer to decide whether and when to send feedback in the current control period. It also collects network status from the SR packets and data packets sent by the sender to estimate its available bandwidth.

Based on the feedback packets, the sender estimates the number of receivers, N. The size of N determines the number of feedbacks (denoted by n) that will have to be collected in next control period. Given the limited control bandwidth and n, the sender can derive the duration of next control period. The sender also calculates the new λ and α for the next feedback loop based on *n*.

Figure 1 shows the control process. Note that $receiver_i$' RR packet not only feedbacks its expected bandwidth, but also serves as a request for closed-loop RTT estimation. Each SR serves as the response for all the RR packets that have arrived at the sender since the last SR was multicast.

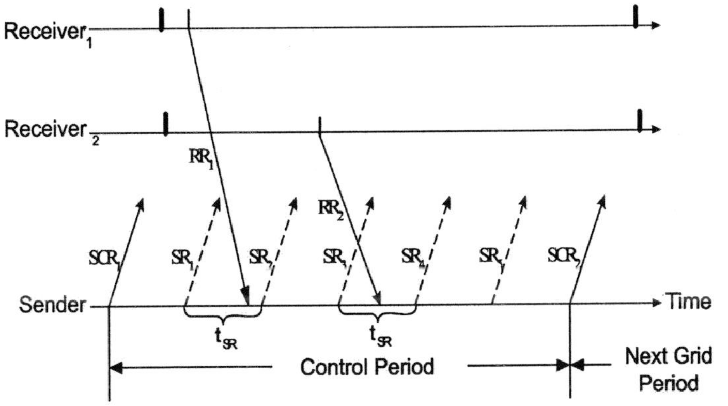

Figure 1. Timing diagram for the control loop

4. Sender-Based Dynamic Rate Allocation

We define a *Fairness Index F*(.) [13]to help establish the optimization objective for rate allocation. The fairness index *F*(.) for a receiver with expected bandwidth *r* is as follows:

$$F(r,\rho_l)=\frac{\Gamma(r,\rho_l)}{r} \tag{3}$$

This definition can be used to assess the satisfaction degree of a receiver when there is performance loss incurred by a mismatch between the discrete set of possible rates and the expected bandwidth. Since the expected bandwidth is estimated as the throughput of a TCP connection over the same path, this index also reflects the degree of fairness between video traffic and TCP traffic. Specifically, the fairness index of 1 is optimal in that it allows the receiver to fully exploit the expected bandwidth, that is, to fairly share the bandwidth with TCP connections.

For a multicast session, a natural optimization objective is to maximize the expected fairness index, $\overline{F}(r,\rho_l)$, for all the receivers by choosing an optimal rate vector. We state the optimization problem as follows:

$$\text{Maximize } \overline{F}(r,\rho_l)=\frac{1}{N}\sum_{i=1}^{N}F(r_i,\rho_l), \tag{4}$$

Subject to $l \le \mathrm{L}$,

$$0 < c_{i\text{-}1} < c_i,\ i=2,3,\ldots,l.$$

where L is the maximum number of layers that the sender can manage.

The complexity of this problem can be further reduced by considering some characteristics of a practical layered codec. First, there are only a finite number of possible rates for any given source. These rates, called *operational rates* [9], depend only on the compression algorithm and source features. Secondly, to avoid the undesired situation where a receiver cannot join any layer, the base layer should adapt to the minimum expected bandwidth. However, the dynamic range of a layered coder is limited which usually places a lower bound to the rate of the base layer. Taking these two characteristics into account, we assume there are M operational points. The set of operational rates is given by $\pi=\{R_1, R_2, \ldots, R_M : R_i < R_{i+1}\}$, and R_1 is the lower bound for the base layer rate. We can then re-formulate the optimization problem as follows:

$$\text{Maximize } \overline{F}(r,\rho_l) = \frac{1}{N}\sum_{i=1}^{N} F(r_i,\rho_l), \tag{5}$$

Subject to $\quad l \le \mathrm{L}$,

$$c_1 = \max_j\{R_j : R_j \le \min_i\{r_i : r_i \ge R_1\}\},$$

$$\mathrm{c_i} \in \pi,\ \mathrm{c_{i\text{-}1}} < \mathrm{c_i},\ \mathrm{i}=2,3,\ldots,\mathrm{l}.$$

A scalable algorithm for this problem with time complexity $O(LM^2)$ and auxiliary storage space O(LM) is derived in [2]. As the complexity does not depend on the number of receivers, the algorithm is highly scalable. Moreover, it relies only on the bandwidth distribution of all the receivers, therefore sampling can be used to reduce collection time for bandwidth reports. From the statistical theory, let n be the number of samples needed to calculate the expected fairness index within confidence interval ε and confidence level 1-α. The smallest n that satisfies:

$$P(|\overline{F}_A^n - \overline{F}_A| < \varepsilon) \ge 1-\alpha, \tag{6}$$

can be calculated by:

$$n_0 = \left(\frac{Z_{\frac{\alpha}{2}} S}{\varepsilon}\right)^2 \quad \textit{and} \quad n = \frac{n_0}{1+\frac{n_0}{N}} \tag{7}$$

where $\overline{F}_A$ is the average fairness based on the distribution of all the receivers, $\overline{F}_A^n$ is the one based on n reports, $Z_{\alpha/2}$ is the upper $\alpha/2$ percentage point of the standard normal distribution, and S is the standard deviation of the fairness indices, which can be estimated once during every control period. Given a fixed average control bandwidth (e.g., 20Kbps), the interval time needed to collect the feedbacks (i.e., the control period T_{ctrl}) can be determined by n. . Note that the more are the receivers, the more samples need to be collected, and the longer is the control period, and vice-versa. Since a very short or long control period may result in inaccurate bandwidth reports, and a short control period also may cause highly oscillative adaptation behavior, we confine T_{ctrl} in the range of [2s, 15s].

5. Available Bandwidth Estimation in Wireless IP Networks

As depicted in Section 2.2, each receiver estimates its available bandwidth using a TCP throughput formula (see Equation 2) which needs the estimation of *RTT*, *RTO* and *p*. But how to estimate *RTT, RTO* and *p* in wireless IP networks brings great challenges [10]. First of all, the varying wireless environment results in dramatic fluctuation of the end-to-end RTT over wireless Internet. Thus the bandwidth estimation counted on RTT may be inaccurate and fluctuate greatly. To solve this problem, the receivers use the following RTT estimation to measure the "average" round trip time during a period of time:

$$\overline{RTT_n} = \overline{RTT_{n-1}} * \alpha + RTT_n * (1 - \alpha) \tag{8}$$

where α is a weighted parameter that is set to 0.9 in our protocol, $\overline{RTT_{n-1}}^{*}$ is the average round trip time at the (n-1)-th measurement interval, and RTT_n * is the estimated round trip time at the n-th measurement interval. As a result, the bandwidth estimation performs more smoothly.

Another parameter RTO can be estimated from RTT. Practically, the simple heuristic of RTO=max {1,4RTT} works reasonably well to provide fairness with TCP [2].

Moreover, in wireless IP networks, the end-to-end packet loss can be caused by either congestion loss in the wired network or the erroneous loss in the wireless part. Traditional TCP and TCP-friendly protocols treat every loss event as a signal of congestion and correspondingly reduce the transmission rate. However, this rate reduction is unnecessary if the loss is due to the error in wireless network. In WHALM we incorporate a loss differentiation algorithm (LDA) called spike [7] on the receiver side to classify the loss type and estimate the packet loss rate using only congestion loss. Since only congestion losses are used as congestion signals, and wireless losses do not restrict the sending rate, WHALM can achieve considerable performance optimization.

Spike scheme [*Reference required] uses the Relative One-way Trip Time (ROTT) to identify the state of the current connection. ROTT is a measure of the time a packet takes to travel from the sender to the receiver. Since the sending and receiving times are measured at the sender and receiver separately and there exists skewness between the two clocks, thus the name "relative" is used.

Spike scheme works as follows. On receipt of a packet with sequence number i, if the connection is currently not in the spike state, and the ROTT for packet i exceeds the threshold Bspikestart, then the state enters the spike state. Otherwise, if the connection is currently in the spike state and the ROTT for packet i is less than a second threshold Bspikeend, the state changes out of the spike state. When the receiver detects a loss because of a gap in the sequence number of received packets, if the connection is in the Spike state, losses are assumed to be due to congestion. Otherwise, losses are assumed to be wireless transmission errors, which are illustrated in Figure2.

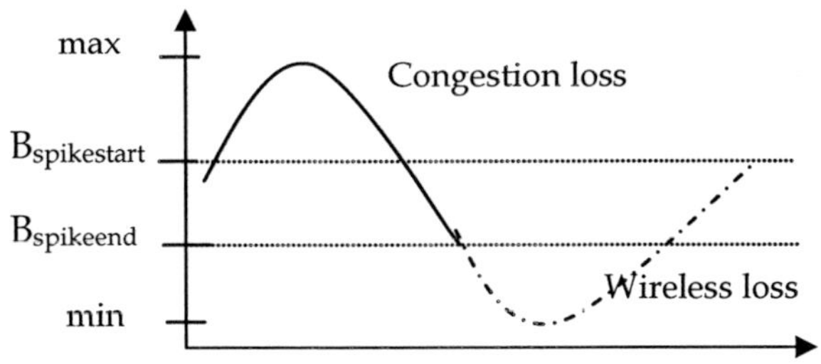

Figure 2. Spike Scheme

6. Scalable Feedback Mechanism

In WHALM the receivers need to send bandwidth feedbacks and RTT estimation request to the sender, thus a feedback control mechanism is needed to suppress feedbacks. Several solutions exist for implosion avoidance based on hierarchy, parameterized and randomly delayed timers. Nonnenmacher *et al* [*Reference required] suggested a truncated exponentially distributed timer, but it requires a multicast feedback channel for every receiver, which leads to additional overheads. We only use a unicast feedback channel from receiver back to the sender. We introduce a devised Gamma-distributed timer from *0* to T_{ctrl} for each receiver, which outperforms other distributed timers when the number of users is large. The density of the Gamma distributed timer is:

$$f_{Z_i}(z_i)=\begin{cases}\frac{1}{(e^{\lambda}-1)}\cdot\alpha\frac{\lambda}{T_{ctrl}^{\alpha}}z_i^{\alpha-1}e^{\frac{\lambda}{T_{ctrl}^{\alpha}}z_i^{\alpha}} & ,0\le z_i\le T_{ctrl}\\ 0 & ,otherwise\end{cases} \tag{9}$$

where T_{ctrl} is the control period, λ and α are factors related to the number of receivers.

At the beginning of a control period, the sender multicasts a SCR packet. SCR contains the parameters of the gamma-distributed timer, i.e., T_{ctrl}, λ and α. Upon receiving the SCR, receiver i schedules a gamma distributed random timer $z_i \sim [0,T_{ctrl}]$. Only the receivers that get the timer between (0,0+c) can send feedback, other receivers are suppressed by this way, where c is the receiver-sender delay. Figure3 shows the Gamma distributed timer setting of z_i .When the timer z_i expires, receiver i sends a RR packet back to the sender.

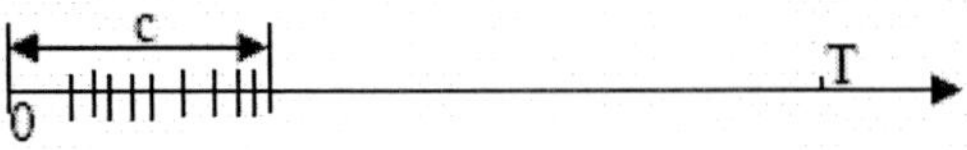

Figure 3. Gamma Distributed Timer Setting z_i

At the end of a control period, the sender computes N^*, the estimated number of the receivers,

$$N^* = \frac{X(1-F_Z(m))}{F_Z(m+c)-F_Z(m)} \quad (10)$$

using the knowledge about the timer settings z_i of all the receivers that have returned RR packets during current control period. The sender then computes the new λ and α for the next control period based on N^* and n, the desired number of RRs:

$$\lambda = 1.1 \cdot \ln N^* + 0.8 \quad (11)$$

$$\alpha = \frac{\ln\left[\frac{1}{\lambda}\ln[\frac{n(e^{\lambda}-1)+N^*}{N^*}]\right]}{\ln(\frac{c}{T_{ctrl}})} \quad (12)$$

In this way the sender can control the number of feedback packets during each control period and thus avoid feedback explosion.

7. Simulation Results

We simulate WHALM and HALM using network simulator ns-2 [15]. We refer to the network topology that is used in [2] and modify it to evaluate the effectiveness of our protocol by replacing some wired receivers with wireless ones. Figure 4 illustrates the network topology. There is a WHALM sender and *6m* receivers belonging to six LANs, each LAN having *m* receivers. The first three LANs are wireless LANs, where the receivers are connected to the switches through wireless links. The rest LANs are wired LANs whose bottleneck links are (SW0, SWi), i=4, 5, 6, respectively. A TCP connection modeled as a FTP flow shares (SW0, SWi), i=1, 2, …, 6, with the corresponding video streams. The cumulative layer rates of the video source are initialized to {128, 384, 896} Kbps, and the lower bound of base layer rate is 100 Kbps.

It is important to accurately estimate available bandwidths of receivers. We also calculate the optimal allocation based on the exact and instant bandwidth distribution of all the receivers. Assume the expected fairness index under this allocation is F^*, and the one under a practical algorithm (in WHALM and HALM) is F', the accuracy of the practical algorithm is defined as [2]:

$$Accuracy = \frac{F'}{F^*} \quad (12)$$

Figure 5(a) illustrates the performance comparison of WHALM and HALM.

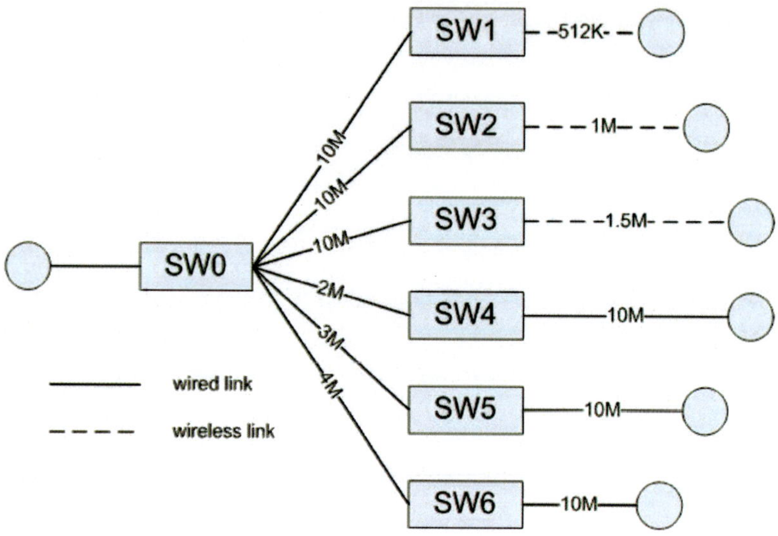

Figure 4. Topology of simulation experiments

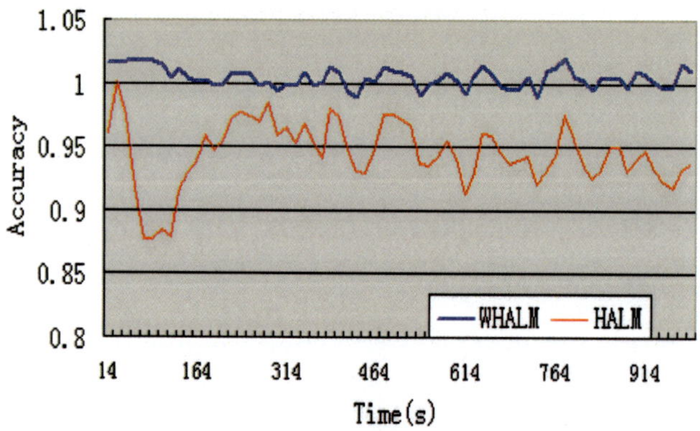

(a) Accuracy comparison

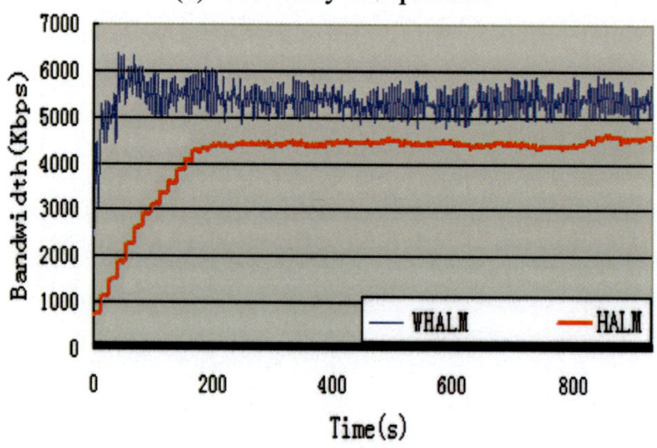

(b) Throughput comparison

Figure 5. Performance Comparison of WHALM and HALM

It can be seen that WHALM achieves better accuracy than HALM. This improvement results from: (1) HALM collects feedbacks in a fixed control period of 15s, which is long enough for sampling up to 5000 receivers. Such a long collection time is unnecessary when the number of receivers is smaller. Moreover, it causes serious skewness between a receiver's current expected bandwidth and its recent report in a highly dynamic environment such as wireless network. While in WHALM, the control period varies with the number of receivers to reduce the skewness. (2) WHALM incorporates spike scheme which differentiates between congestion and wireless losses, thus it uses relatively accurate packet loss rates when estimating the available bandwidths.

Figure 5(b) shows throughput comparison between WHALM and HALM. Here the throughput is the cumulative one of all the receivers. Evidently WHALM achieves higher throughput than HALM. This is because wireless losses do not restrict the sending rate in WHALM.

Figure6 shows the bandwidth distribution between the competing video streams using WHALM and TCP flows at switches 1 - 4. Compared to TCP flows, WHALM not only exhibits smoother behavior but also has a higher bandwidth usage in wireless environments. At the same time, experimental results also show that the TCP friendliness is preserved. Figure 6(a)-(c) illustrate the obvious TCP friendliness, while Figure 6(d) shows that the WHALM consume more bandwidth than TCP traffic. The reason of this phenomenon is that the traditional TCP protocol does not take the loss differentiation into account, which will decrease the bandwidth utility. On the contrary, WHALM can differentiate congestion loss and wireless loss effectively, so it has better bandwidth utility. According to the definition of TCP friendliness that achieves a fair share of bandwidth relative to reasonable TCP flows over the same path. The definition still allows for improvements to TCP. It is even acceptable to perform better than TCP. So we can also conclude that WHALM still preserves the TCP friendliness.

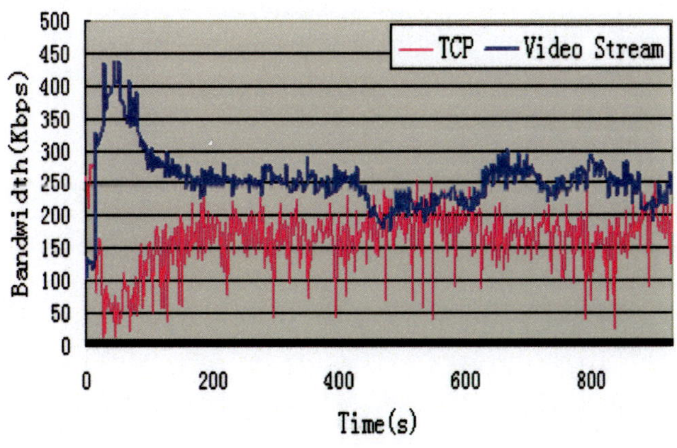

(a) Switch 1

Figure 6. (Continued)

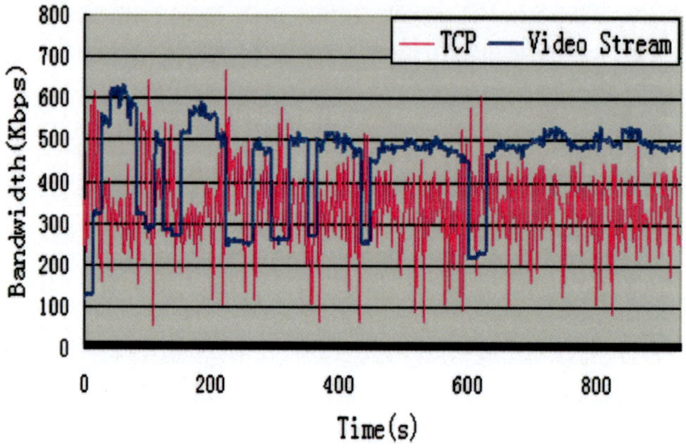

(b) Switch 2

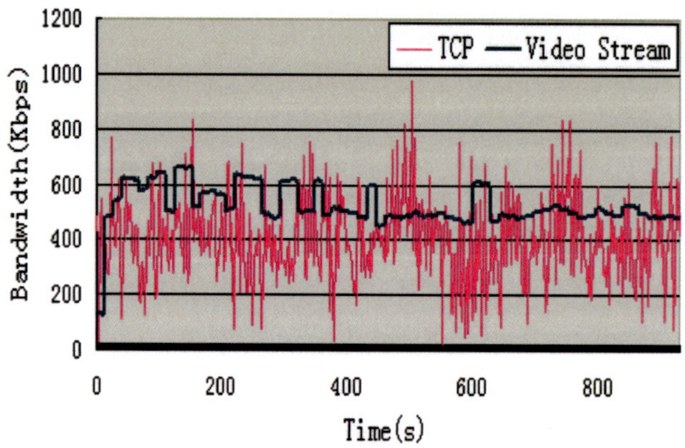

(c) Switch3

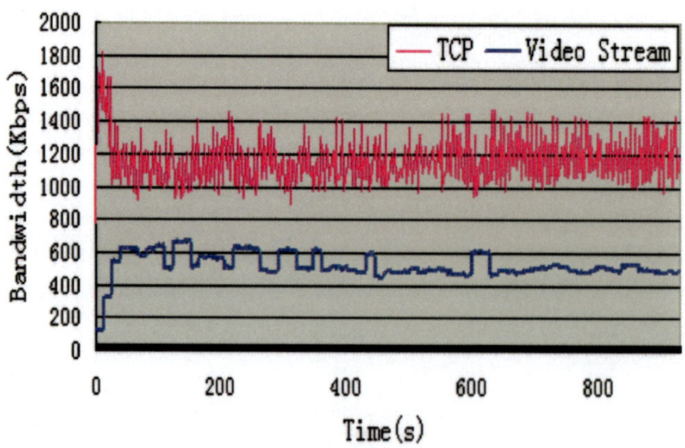

(d) Switch 4

Figure 6. Bandwidth distribution between WHALM and TCP at switches 1–4

8. Conclusions

This chapter has proposed an end-system based TCP friendly adaptation protocol, named WHALM, for layered video multicast over wireless Internet. WHALM integrates an optimal rate allocation algorithm and a scalable feedback mechanism with the aim to dynamically adapt to the heterogeneous requirements from the receivers. In addition it efficiently avoids the well-known performance degradation in wireless Internet by incorporating a loss differentiation scheme. Using a series of simulation experiments over NS platform, we can conclude the following unique advantages for WHALM: 1) By developing a metric called Fairness Index for each receiver, and formulating the joint rate allocation into an optimization problem, WHALM greatly improves the QoS of receivers with heterogeneous bandwidth requirements; 2) WHALM not only maintains high bandwidth utility but also preserves smoothness and TCP friendliness of output rate, which are very important for the QoS improvement of multimedia transmission and network stability. 3) WHALM is feasible to be deployed in wireless Internet for adopting our proposed scalable feedback control mechanism and end-to-end loss differentiation algorithm.

Our future work is to conduct more simulations and real experiments with advanced layered coding algorithms. This also enables more extensive and realistic comparisons with other layered multicast protocols over wireless Internet. Other potential work includes how to improve the error resilience of video multicast over wireless channels by using error control algorithms, and how to improve the accuracy of the estimation of available network bandwidth.

Acknowledgment

This work was supported in part by a grant from the Projects of Development Plan of the State Key Fundamental Research (No.2003CB314804), the National Natural Science Foundation of China (No. 60372019, No.60473086, No. 60429202).

References

[1] S.McCanne, V.Jacobson and M.Vetterli, "Receiver-driven Layered Multicast," in *Proceedings of ACM SIGCOMM 96*, August 1996.

[2] J. Liu, B.Li, and Y.-Q.Zhang, "An End-to-End Adaptation Protocol for Layered Video Multicast Using Optimal Rate Allocation," *IEEE Transaction on Multimedia*, Vol.6, No.1, February 2004.

[3] B. Li and J. Liu, "Multi-Rate Video Multicast over the Internet: An Overview," *IEEE Networks Special Issue on Multicast*, Vol. 17, No. 1, January/February 2003.

[4] Upkar Varshney, "Multicast over Wireless Networks," *Coomunication of the ACM*, Vol.45,No.12,December 2002.

[5] Q.Ni, Q.Zhang, W.W.Zhu. "SARLM: Sender-adaptive and Receiver-driven Layered Multicast for Scalable Video". *IEEE ICME 2001 (IEEE International Conference on Multimedia and Expo)*, Japan,Aug. 22-25,2001.

[6] M.Handley, S.Floyd, J.Padhye and J.Widmer, "TCP Friendly Rate Control (TFRC): Protocol Specification", *RFC* 3448, January 2003.

[7] S.Cen, P.C.Cosman, and G.M.Voelker, "End-to-End Differentiation of Congestion and Wireless Losses," *IEEE Transaction on Networking.*

[8] H.Schulzrinne, S.Casner, R.Frederick and V.Jacobson, "RTP: A Transport Protocol for Real-Time Applications," *RFC* 1889, January 1996.

[9] G. Schuster and A. Katsaggelos, *Rate-Distortion Based Video Compression*, Kluwer Academic Publishers, 1997.

[10] F.Yang, Q.Zhang, W.Zhu, and Y.-Q.Zhang, "End-to-End TCP-Friendly Streaming Protocol and Bit Allocation for Scalable Video over Wireless Internet," to appear in *IEEE Journal on Selected Areas in Communications Special Issue on All-IP Networks*, 2004.

[11] J. Liu, B. Li, H.-R. Shao, W. Zhu, and Y.-Q. Zhang, "A Proxy-Assisted Adaptation Framework for Object Video Multicasting," *IEEE Transactions on Circuits and Systems for Video Technology*, March 2005.

[12] J. Liu, B. Li, and Y.-Q. Zhang, "Adaptive Video Multicast over the Internet," *IEEE Multimedia*, Vol. 10, No. 1, pp. 22-31, January/February 2003.

[13] J. Liu, B. Li, Y.-T. Hou, and I. Chlamtac, "Dynamic Layering for Multi-Session Video Broadcasting with General Utility Functions," *IEEE INFOCOM'03*, San Francisco, USA, March 2003.

[14] J. Liu, B. Li, Y.-T. Hou, and I. Chlamtac, "On Optimal Layering and Bandwidth Allocation for Multi-Session Video Broadcasting," *IEEE Transactions on Wireless Communications*, February 2004.

[15] "The Network Simulator (NS-2)", http://www.isi.edu/nsnam/ns/.

[16] J. Liu, B. Li, and Y.-Q. Zhang, "A Hybrid Adaptation Protocol for TCP-friendly Layered Multicast and Its Optimal Rate Allocation," *IEEE INFOCOM'02*, pp. 1520-1529, New York City, USA, June 2002.

[17] H. Yin, G. Zhu, Y.Zhu, X.Qin, "Estimation of Network Bandwidth: Challenges and Approaches," *The Proceedings of the International Conference on Fundamentals of Electronics, Communications and Computer Sciences*, Mar. 2002, Tokyo, Japan

[18] J. Widmer, R. Denda, M. Mauve, "A Survey on TCP-Friendly Congestion Control," *IEEE Network Magazine*, **15** (3), pp. 28-37, May/June 2001.

[19] D.-P. Wu, Y.-T. Hou, W. Zhu, Y.-Q. Zhang, and J. Peha, "Streaming video over the internet: approaches and directions," *IEEE Trans. Circuits System Video Technology*, vol. 11, no. 3, pp. 282–300, Mar. 2001.

[20] W. Li, "Overview of the fine granularity scalability in MPEG-4 video standard," *IEEE Trans. Circuits Syst. Video Technol.*, vol. 11, pp. 301–317, Mar. 2001.

[21] S. Li, F. Wu, and Y.-Q. Zhang, "Experimental results with Progressive Fine Granularity Scalable (PFGS) Coding," ISO/IEC JTC1/SC29/WG11, MPEG99/m5742, Mar. 2000.

[22] D. Sisalem and A. Wolisz, "MLDA: a TCP-friendly congestion control framework for heterogeneous multicast environments," in *Proc. 8th Int. Workshop on Quality of Service* (IWQoS), June 2000.

[23] L. Vicsano, etc. "TCP-like Congestion Control for Layered Multicast Data Transfer," in *Proceedings of IEEE INFOCOM'98*.

[24] J. Bolot, T. Turletti, and I. Wakeman, "Scalable feedback control for multicast video distribution in the internet," *Comput. Commun. Rev.*, vol.24, no. 4, pp. 58–67, Oct. 1994.

[25] B. Vickers, C. Albuquerque, and T. Suda, "Source adaptive multi-layered multicast algorithms for real-time video distribution," *IEEE/ACM Trans. Networking*, vol. 8, no. 6, pp. 720–733, Dec. 2000.

In: Adaptation Techniques in Wireless Multimedia Networks ISBN 1-59454-883-8
Editors: Wei Li and Yang Xiao, pp. 75-90

Chapter 5

ADAPTIVE QOS-BASED BANDWIDTH ALLOCATION FOR MULTIMEDIA WIRELESS NETWORKS

***Nidal Nasser*[1]* *and Hossam Hassanein*[2]†**
[1]Department of Computing and Information Science,
University of Guelph, Guelph, Ontario, N1G 2W1 Canada
[2]School of Computing, Queen's University,
Kingston, Ontario, K7L 3N6, Canada

Abstract

New upcoming wireless cellular infrastructures such as 3G and 4G are deemed to support a broad spectrum of multimedia services with Quality of Service (QoS) guarantees. One side of the spectrum is real-time services requiring strong bandwidth guarantees. The other side of the spectrum is services which are adaptive in nature and can operate over a wide range of bandwidth. An example could be a video streaming service with three bandwidth levels: low, medium and high. In a wireless cellular network, a mobile user can freely roam within a networks coverage area, and may undergo a large number of handoff events during a typical session. When a user handoffs to a new cell, there may not be sufficient bandwidth to support his call. Under such a situation, there are two possibilities based on the nature of the service. In the case of a non-adaptive service, i.e., the bandwidth of a call is fixed throughout its life time, needing strict bandwidth guarantees, the call will be dropped. Whereas in the case of an adaptive service, the call will not be dropped but will suffer bandwidth degradation. In earlier studies [1, 2], we developed an adaptive admission control framework to provide connection-level QoS guarantees to mobile users that are running non-adaptive multimedia service. In this chapter, we develop a novel framework for provisioning connection-level QoS in wireless cellular networks that support adaptive multimedia services.

Keywords: QoS provisioning; multimedia traffic; call admission control; bandwidth adaptation; wireless cellular networks

AMS: 68M10, 68M20

*E-mail address: nasser@cis.uoguelph.ca
†E-mail address: hossam@cs.queensu.ca

1. Introduction

Adaptive multimedia networking becomes very attractive due to the scarcity of wireless resources, the availability of a wide range of bandwidth granulites, and their ability to mitigate the highly fluctuating link bandwidth in wireless networks. In multimedia wireless networks using an adaptive framework, it is possible to overcome the link overload condition by dynamically adjusting the bandwidth of individual ongoing calls, which we call bandwidth adaptation. With the help of such adaptive framework, the dropping probability of handoff calls can be reduced to a negligible level under normal traffic load. However, bandwidth adaptation changes handoff dropping into a problem of bandwidth degradation caused by adaptation. Thus, in addition to connection-level QoS parameters: New Call Blocking Probability (NCBP) and Handoff Call Dropping Probability (HCDP), we introduce a new QoS parameter called the call degradation ratio. This ratio represents the average ratio of the number of degraded calls to the number of calls at any state of the system. We refer to a call as degraded if its assigned a bandwidth below its requested bandwidth.

Two major components designed in adaptive multimedia wireless networks contribute to reducing handoff drops due insufficient resources in the new cell while efficiently maintaining bandwidth utilization. The first, related to network resource management, is the Call Admission Control (CAC) algorithm. The second component is a Bandwidth Adaptation Algorithm (BAA), which is related to bandwidth management of ongoing calls in the system. BAA is needed to adjust the calls bandwidth in a cell.

Several CAC algorithms and BAAs have been proposed in wireless cellular networks using the adaptive framework [3, 4, 5, 6, 7, 8, 9]. In [3, 5, 8, 6], it is assumed that all calls belong to a single class of adaptive multimedia traffic and, similar to [11] and [4, 12], receive varying bandwidth values from a discrete set of integer bandwidths. Chou et al [3] derived an analytical model for wireless networks with adaptive bandwidth reservation and traffic restriction CAC. In [5, 8], new QoS parameters for adaptive multimedia in wireless networks are introduced. Kwon et al [5] proposed the cell overload probability parameter, while Xiao et al [8] proposed two QoS parameters: the degradation ratio and the degradation degree. Both schemes suggested a new CAC and BAA that satisfy the application QoS requirements in terms of the proposed parameters. Kwon et al [6] identified the possible objectives of BAAs, based on which they proposed two BAAs. The first algorithm considers the service provider objectives of low cost and high revenue while the second algorithm aims to provide good quality and to maintain fairness from the standpoint of the service user.

Multiple classes of adaptive multimedia services in cellular wireless networks have been rarely introduced in the literature without considering the prioritization between a new call arrival and handoff arrival for each class of traffic [7, 9, 4]. This motivates us to provide a bandwidth reservation mechanism that takes into account the separation between incoming traffic for each class and prioritizes the handoff calls over the new calls. Xiao et al in [7] and [9] proposed fair adaptive bandwidth algorithms for multiple classes of connections. Fairness among classes is achieved by partitioning the bandwidth according to arrival rates. The disadvantage of these algorithms is that they cause all ongoing connections to receive reduced bandwidth. To alleviate this problem, we design a bandwidth adaptation algorithm that aims to minimize the number of ongoing degraded connections. A prioritization in the process of bandwidth adaptation among multiple classes of multimedia services is presented in [4] where the bandwidth of calls with lower priority is preferably adapted. However, in this approach the authors assume no handoff dropping which makes their work impractical. This is not the case in our work; in order to make our contribution more realistic, a handoff call can be dropped if it does not satisfy the adaptation condition.

2. Scheme Outline and Objectives

The main contribution of this chapter is the introduction of a novel adaptive multimedia framework for multimedia wireless networks. The framework operates at the connection-level where the bandwidth of ongoing connections can be dynamically adjusted. This framework supports multiple classes of adaptive multimedia services that have diverse QoS requirements. Here, we integrate a threshold-based bandwidth reservation policy, an efficient CAC algorithm and a BAA into a new framework called QoS-Adaptive Multimedia Wireless Access (QoS-AMWA). The objective of QoS-AMWA is to provide the mobile users with seamless handoff communication through the wireless network and to distinguish between handoff calls and new calls for each class of traffic while minimizing the NCBP and the HCDP. This objective is essential from the point of view of the user. The service provider, on the other hand, is more interested in increasing revenues. For this purpose, we design a BAA which allows the system to offer services whenever there is insufficient amount of bandwidths by intelligently adjusting bandwidth of ongoing calls which results in a near zero HCDP. Therefore, more calls are able to complete their services and as a result better utilization is obtained which translate into more revenue to the service provider.

A block diagram of our proposed QoS-AMWA system architecture is depicted in Figure 1. The QoS bandwidth allocation controller implements the threshold-based bandwidth reservation policy that prioritizes calls according to their QoS constraints by assigning a threshold value to each traffic class. The policy is based on reserving bandwidth for aggregate handoff calls, thus giving them priority over new calls for each class of traffic and providing them with low HCDP. We require that during call setup, a mobile terminal (running a user multimedia service) defines its requirements in a traffic profile before it is transmitted to the CAC at the base station. The traffic profile consists of the traffic type (new or handoff) and the bandwidth requirements. The traffic profile is then sent to the CAC, which determines the acceptance or rejection of a call based on traffic type and the amount of available bandwidth in the system. If the available bandwidth is not sufficient to satisfy the new call requests, the call will be blocked. However, for a handoff call if the available bandwidth is not enough to accommodate this call, then the QoS bandwidth adaptation controller is invoked to execute a Bandwidth Adaptation Algorithm (BAA). BAA sensibly reduces the bandwidth of ongoing calls in the system and assigns the saved bandwidth to the incoming handoff call. Hereafter, a call of any class is referred to as degraded if and only if the assigned bandwidth of the call is lower than its requested bandwidth. Note that the QoS bandwidth adaptation controller will also invoke and execute BAA when a call (new or handoff) completion occurs in the system or handoff to adjacent cell. In this case, the BAA will use the released bandwidth to upgrade degraded calls if applicable.

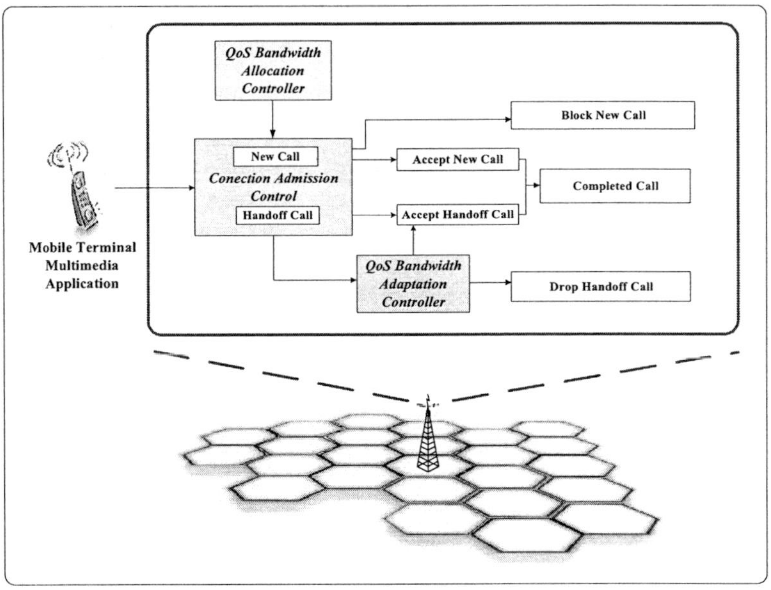

Figure 1. QoS-AMWA system architecture

In Section 3, we describe the adaptive multimedia model and the QoS parameters. Section 4 presents a detailed description of the adaptive multimedia framework with its main components. Section 5 provides the simulation model and the performance evaluation. A chapter summary is provided in Section 6.

3. System Model

We consider adaptive multimedia wireless services within a cellular network infrastructure. The network comprising a wired backbone and a number of base stations (BSs) which reside at the centers of the cells. A mobile, while staying in a cell, communicates with another party, which may be a node connected to the wired network or another mobile, through the BS in the same cell. Mobile users can freely roam within a networks area of coverage, and experience a number of handoffs during a typical session. When a mobile moves into an adjacent cell in the middle of a communication session, a handoff will enable the mobile to maintain connectivity to its communication partner, i.e., the mobile will start to communicate through the new BS, hopefully without noticing any difference.

Here, we are concerned with the connection-level QoS guarantees for new and handoff calls in each cell. Therefore, we decompose the wireless cellular network into individual sub-systems, each comprising a single cell. The correlation between these sub-systems (models), results from handoff calls among corresponding cells. Under this formulation, each cell can be modeled and analyzed individually. Similar model is used for all cells in the network, but the model parameters may be different, reflecting the mobility and traffic conditions in individual cells. Thus, each model can support homogeneous traffic and heterogonous traffic.

We assume the system uses Fixed Channel Allocation (FCA), which means each cell has a fixed amount of capacity. No matter which multiple access technology (FDMA, TDMA, or CDMA) is used, we could interpret system capacity in terms of effective or equivalent bandwidth [10]. Hereafter, whenever we refer to the bandwidth of a call, we mean the number of basic bandwidth units (bbu) that is adequate for guaranteeing the desired QoS for this call with certain traffic characteristics.

Consider a cell that has a total capacity of B bbu. Two types of calls share the bandwidth of the cell: new calls and handoff calls. Traffic arriving at the cell is partitioned into K separate classes based on bandwidth requirements. The adaptive bandwidth model and QoS parameters are explained in the following sections.

3.1. Adaptive Bandwidth Model

A multimedia call can have its bandwidth dynamically changed depending on the network load situation during its lifetime. Similar to [6, 11, 12], we adopt the layered coding approach where the bandwidth of a call can take a set of discrete values. Moreover, multiple classes of adaptive multimedia are taken into consideration. Each class-i connection requires a discrete bandwidth value, $b_{i,j}$, where $b_{i,j}$ belongs to the set $B_i = \{b_{i,1}, b_{i,2}, ..., b_{i,j}, ..., b_{i,Ki}\}$ for $i = 1, 2, ..., K$, and $b_{i,j} < b_{i,j+1}$ for $j = 1, 2, ..., K_i 1$. Here, we express discrete bandwidth values in terms of basic bandwidth unit. Thus, $b_{i,j}$ is an integer number of basic bandwidth units. The minimum and maximum values that a class-i connection can take are $b_{i,1}$ and b_{i,K_i}, respectively. Also, K_i denotes the number of possible values of bandwidth that a class-i calls can be allocated. The requested bandwidth of a connection of class-i is denoted as $b_{i,request}$, where $b_{i,request} \in B_i$. We assume that all the connections in the same class use the same requested bandwidth. Hence, the requested bandwidth for each class is predetermined.

3.2. QoS Parameters

In this section, we describe the QoS parameters that will be used to evaluate the performance of our adaptive multimedia framework. In addition to the new call blocking probability (NCBP) and the handoff call dropping probability (HCDP), we introduce bandwidth utilization, degradation ratio (DR) and QoS performance indicator. We define the bandwidth utilization as the ratio of the bandwidth used by completely serviced calls to the total bandwidth capacity. DR is defined in [8] for multiple classes of adaptive multimedia in order to effectively characterize the bandwidth degradation for different classes. Thus, DRi represents the degradation ratio of class-i and is equal to the average ratio of the number of degraded calls of class-i to the number of ongoing calls of class-i.

One important QoS performance indicator from the service providers point of view is the percentage of QoS satisfaction for each class-i (%QoS$_i$). %QoS$_i$ refers to the ratio of the number of class-i calls for which QoS is met to the total number of calls served in a given time interval T. This indicator helps the service provider to see how satisfied the mobile users are in relation to the requested QoS. Based on the statistics of the QoS performance indicator, the provider will be able to design an efficient and effective system that provides better QoS guarantees to mobile users and at the same time maximizes system revenues.

4. Adaptive Multimedia Framework

Our adaptive multimedia framework consists of three main components: a threshold-based bandwidth reservation policy, a call admission control algorithm, and a bandwidth adaptation algorithm.

4.1. Threshold-Based Bandwidth Reservation Policy

The main principle of our bandwidth reservation policy is based on reserving bandwidth for aggregate handoff calls, thus giving them priority over new calls and providing them with lower handoff call dropping probability. In addition, the policy prioritizes between different classes of handoff calls according to their QoS constraints by assigning a series of bandwidth thresholds t_0, t_1, ..., t_K, such that $t_0 \leq ... \leq t_i \leq t_{i+1} \leq ... \leq t_K$, where, t_0 denotes the maximum number of total basic bandwidth units that can be allocated to new calls, and t_i, $1 \leq i \leq K$, denotes the maximum number of total basic bandwidth units that can be allocated to class-i handoff calls. Note that, if the different handoff calls were allowed to completely share the bandwidth, calls with lower bandwidth requirements will have a higher priority to allocate bandwidth than those with higher bandwidth requirements. Threshold values are tuned to satisfy the handoff dropping probabilities for each individual traffic class. Note that $\sum_{i=1}^{K} t_i$ may be greater than B, which also allows further sharing of bandwidth. Figure 2 shows the accessible bandwidth regions for a two-class system.

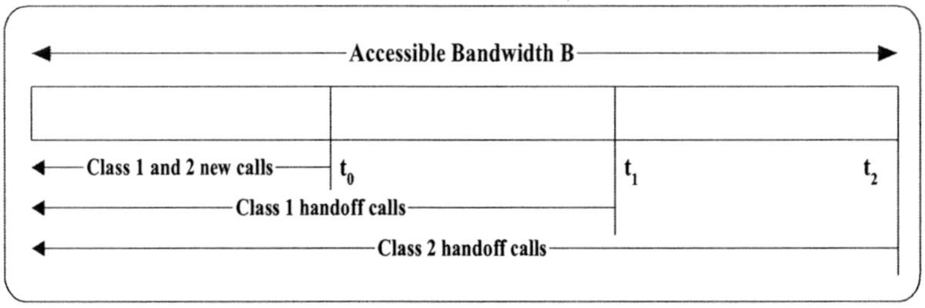

Figure 2. Accessible bandwidth for a two-class system.

4.2. Call Admission Control Algorithm

The objectives of any Call Admission Control (CAC) are to satisfy the QoS requirements and to utilize the system resources efficiently [13, 14, 15]. In this section, we provide a simple threshold-based CAC algorithm. The algorithm utilizes the threshold values that are assigned in the bandwidth reservation policy to make the admission/rejection decision for an arriving call request.

A newly arriving call of class-i is blocked if its requested bandwidth plus the current total bandwidth of ongoing new calls for all classes is greater than t0, or no more bandwidth is available in the given cell to accommodate the new call. On the other hand, we allow an incoming handoff call of class-i to be always momentarily accepted regardless of its bandwidth requirements. If necessary, bandwidth adaptation is performed. Final acceptance of a handoff call is constrained by the following condition: The bandwidth adaptation algorithm, described in the following section, can allocate enough bandwidth for the accepted request. If this is not satisfied then the handoff connection will be dropped.

4.3. Bandwidth Adaptation Algorithm

A Bandwidth Adaptation Algorithm (BAA) increases or decreases the assigned bandwidth of ongoing calls in the cell depending on the network load situation. BAA performs two main procedures: reduction and expansion. The reduction procedure is activated when an incoming handoff call arrives to an overloaded cell. On the other hand, the expansion procedure is activated when there is an outgoing handoff call or a call completion in the given cell.

In this work, our BAA seeks to satisfy the following objectives at any time instance:

- Minimizing the degradation ratio.
- Minimizing the number of calls of class-i with lower assigned bandwidth than requested.

We do not apply bandwidth adaptation for incoming new calls as our main objective is to minimize the handoff dropping probability. Therefore, the bandwidth adaptation reduction will only be performed for the incoming class-i handoff calls, $b_{i,request}$, according to the following cases:
Case 1: no more bandwidth is available in the cell to accommodate this connection

$$b_{i,request} + (\sum_{i=1}^{K}\sum_{j=1}^{m_i} b_{i,assigned_j} + \sum_{i=1}^{K}\sum_{j=1}^{n_i} b_{i,assigned_j}) > B$$

Case 2: no more bandwidth is available under class-i to accommodate this connection, i.e., the current total bandwidth of ongoing class-i handoff connections is greater than t_i, $1 \leq i \leq K$

$$b_{i,request} + \sum_{j=1}^{n_i} b_{i,assigned_j} > t_i, \forall i$$

where mi represents the number of class-i new calls and ni represents the number of class-i handoff calls that are present in the system at the time of the call request.

Based on these two cases the system will decide which reduction algorithm should be invoked as shown in Table 1. The main objective of both algorithms is to accommodate incoming handoff calls and to assign the maximum bandwidth ($b_{i,Ki}$) for such connections. As aforementioned, the application of BAA may result in the rejection of handoff calls if all tests fail.

When a call of class-i departs the cell (outgoing handoff call or a call completion), the available bandwidth of class-i, B_{A_i}, increases. The system will invoke the expansion procedure to increase the bandwidth for one or more calls of the degraded calls of class-i starting from class-1 and ending with class-K. The detailed description of the expansion algorithm is shown in Figure 4, where $b_{upgrade}$ denotes the required bandwidth to upgrade the bandwidth of the corresponding call and $b_{i,current}$ denotes the currently allocated bandwidth of the corresponding call.

```
SCRA(int b_{i,request}, int i) // i: class index
{
  if (BAi ≥ b_{i,request})
  {
    allocateBandwidth(b_{i,j}, b_{i,request}, b_{i,K_i});
    allocationIsDone = 1; // the call is accepted and allocated a bandwidth
  }
  else
  {
    reduceAll(b_{i,request});
    if (B_{A_i} ≥ b_{i,request})
    {
      allocateBandwidth(b_{i,j}, b_{i,request}, b_{i,K_i});
      allocationIsDone = 1;
    }
    else
    {
      reduceAll(b_{i,1})
      if(B_{A_i} ≥ b_{i,1})
      {
        allocateBandwidth(b_{i,j}, b_{i,1}, b_{i,request});
        allocationIsDone = 1;
      }
      else allocationIsDone = 0; // the call is dropped
    }
  }
return allocationIsDone;
}// end of SCRA

allocateBandwidth(int assignedValue, int minBW, int maxBW)
{
  allocate assignedValue to the incoming call such that ...
  (assignedValue ≥ B_{A_i}) and (minBW ≤ assignedValue ≤ maxBW)
}// end of allocateBandwidth function

reduceAll(int minLevel)
{
  order the calls decreasingly by bandwidth;
  while(B_{A_i} < b_{i,request} and exist a calls bandwidth > minLevel)
  {
    reduce the calls with more than minLevel to minLevel ...
    starting with the largest bandwidth;
    add the extra bandwidth to B_{A_i};
  }
}// end of reduceAll function
```

Figure 3. Multiple Classes Reduction Algorithm.

Table 1. Bandwidth-reduction procedure decisions.

Case 1	Case 2	Action
False	False	The call is accepted
False	True	Execute Single Class Reduction Algorithm (SCRA)
True	True	Execute Multiple Class Reduction Algorithm (MCRA)
True	True	Execute MCRA if SCRA is successfully executed

```
Expansion Algorithm()
{
  for (classIndex = 1; classIndex  $K$; classIndex ++)
  {
    while( exist degraded calls of class classIndex to upgrade)
    {
      $i$ = classIndex;
      order the most degraded calls increasingly by bandwidth ...
      starting with the most degraded call from above step;
        $b_{upgrade} = b_{i,request} b_{i,current}$;
        if ($B_{A_i} \geq b_{upgrade}$)
          allocateBandwidth($b_{i,j}$, $b_{i,request}$, $b_{i,K_i}$);
    }// end of while
  }// end of for
}// end of expansion algorithm
```

Figure 4. Bandwidth Expansion Algorithm.

5. Performance Evaluation

In order to evaluate the performance of our proposed adaptive multimedia framework, we developed a simulation model which is described in next section. The performance measures are connection-level QoS parameters, the new call blocking probability (NCBP) and the handoff call dropping probability (HCDP), the degradation ratio (DR), the bandwidth utilization and the QoS performance indicator. These performance measures are plotted as a function of the offered load (call arrival rate). We compare our adaptive multimedia framework with the non-adaptive case. The bandwidth of a call of class-i (i =1, 2, 3) in the non-adaptive multimedia networking paradigm is fixed throughout the simulation lifetime and is equal to the requested bandwidth value of class-i (i =1, 2, 3).

5.1. Simulation Model

The following describe the assumptions used in our simulation. The cellular network studied assumes a standard 7-cell configuration with identical mobility and traffic conditions (homogenous traffic environment). The diameter of each cell is 1 km (i.e., micro cellular environment). The base station resides at the center of each cell. Here, we simulate an environment in which there are three classes of adaptive multimedia services. The simulation parameters that consist of the discrete set of bandwidth values for each class and the traffic parameters are shown in Table 2. B is the total bandwidth capacity in each cell, T_{sim} is the simulation time, and T is the measurement time period.

Table 2. Simulation parameters

Class index	Bandwidth set		$b_{i,request}$		$1/h$ (sec)		$1/\mu_i$ (sec)
1	$\{1, 3, 5, 7\}$		5		100		500
2	$\{2, 4, 6, 8\}$		4		100		700
3	$\{8, 16, 24, 32\}$		16		100		900
Other parameters							
B	T_{sim} (sec)	T (sec)	t_0	t_1	t_2	t_3	Call arrival rate (λ call/sec)
100	1000	4	50	100	100	100	$\lambda_{nc_1} = \lambda_{nc_2} = \lambda_{nc_3} = \lambda$

Call requests are generated according to a Poisson process with rate λ (calls/second) in each cell. Note that $\lambda = \lambda_{nc_1} = \lambda_{nc_2}$ and the handoff call arrival rate of class-i is assumed to be proportional to the new call arrival rate of class-i by $\lambda_{h_i} = \frac{h}{\mu_{b_i}} \lambda_{nc_i}$.

The lifetime of a class-i call (i.e. call holding time) is exponentially distributed with mean $\mu_{b_i}^{-1}$ seconds and it is selected when the call is first admitted into the network. For the mobility model, we consider three parameters, the initial position of a mobile, its direction and its speed. A newly generated call can appear anywhere in the cell with an equal probability. When a new call is initiated, a mobile is assigned a random initial position derived from a uniform probability distribution function over the cell area. As for handoff calls, the initial position is determined when the handoff event is scheduled as described below. A mobile is assigned a random direction upon entering a cell. The distribution of the direction reflects the correlation between the different cells. Different mobility scenarios can be modeled by proper selection of the distribution. For example, in order to model the morning rush hour mobility pattern, the distribution is selected to be biased to the direction leading to the downtown area. While for the evening rush hour, the distribution is biased towards the direction that leads to the suburbs. A constant randomly selected speed is assigned to a mobile when it enters a cell either at call initiation or after handoff. The speed is obtained from a uniform probability distribution function ranging between $V_{min} =$ 10 km/hr and $V_{max} =$ 60 km/hr.

From the three parameters described above, along with the radius of the cell R, and the network topology, the simulation tool calculates the mobiles residence time in the cell. It also determines the cell to which the mobile will handoff to if a handoff is taking place, as well as the initial position of the mobile in the next cell.

In [16], an analytical model is developed for the above system, and it was found that the average residence time for a new call, t_{nc}, is given by:

$$t_{nc} = \frac{8RE[1/V]}{3\pi},$$

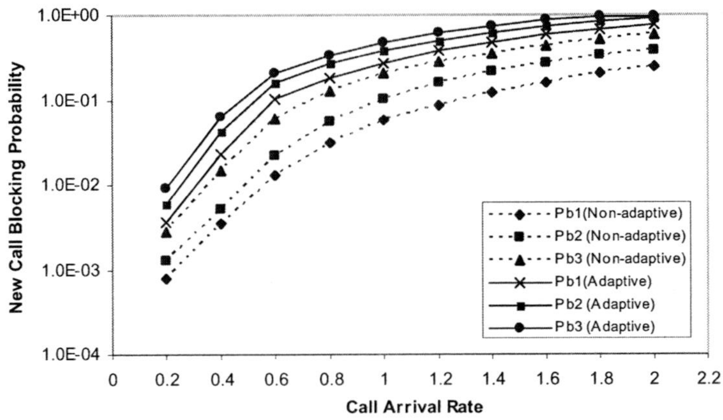

Figure 5. Effect of varying the new call arrival rate on NCBP for non-adaptive/adaptive multimedia.

While the average residence time of a handoff call, t_h, is given by:

$$t_h = \frac{\pi R}{2E[V]}$$

where R in the radius of the cell and V is the average speed of a mobile in the cell. Therefore, the handoff rate of new calls, h_{nc} equals $1/t_{nc}$ and the handoff rate of handoff calls, h, equals $1/t_h$.

5.2. Numerical Results

The effect of varying the new call arrival rate on the connection-level QoS parameter NCBP for both frameworks, adaptive multimedia and non-adaptive multimedia, is illustrated in Figure 5. The NCBP for both frameworks increases as the new call arrival rate increases. However, the NCBP in the case of the non-adaptive multimedia is lower than that of adaptive multimedia. This is because there are already many ongoing calls in each cell in the adaptive multimedia paradigm as there is almost no handoff dropping (HCDP near zero). Note that among the evaluated classes, class one has the lowest NCBP values while class three has the highest NCBP values. This is due to the fact that the bandwidth requirement decreases as the class index decreases, giving a better chance to accept calls of the smallest indexed class.

Figure 6 illustrates the performance of the proposed adaptive multimedia framework and the non-adaptive case in terms of the HCDP. It is apparent that by applying the CAC and BAA algorithms in the manner described above, the HCDP of the adaptive multimedia framework is minimized (lower than 0.00048) and, therefore, it outperforms the non-adaptive case. However, as the traffic load increases, the HCDP increases. Note that while the CAC scheme will accept the handoff calls all the time, when the traffic load becomes heavier the threshold-bandwidth reservation policy can still reject some handoff arrivals due

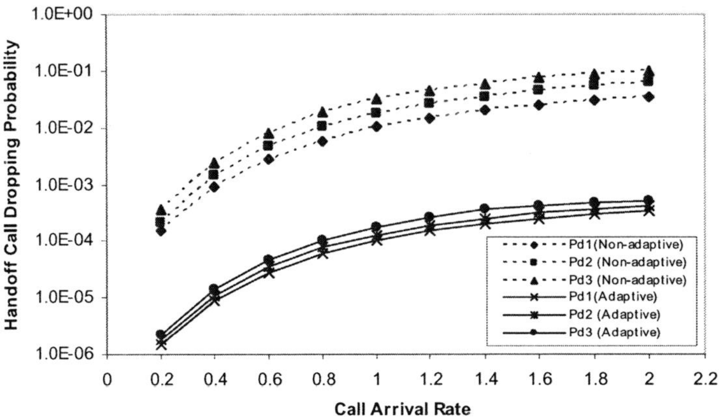

Figure 6. Effect of varying the new call arrival rate on HCDP for non-adaptive/adaptive multimedia.

to the restrictive threshold occupancy. In the non-adaptive case, an incoming handoff connection of class-i is immediately dropped if the current total bandwidth of ongoing class- i connections is greater than t_i, $1 \leq i \leq K$, or no more bandwidth is available in the cell to accommodate this connection. This results in increasing the HCDP class- i as the new call arrival rate increases. Furthermore, the results in Figure 6 show that the HCDP increases as the class index increases for both cases. This is due to the complete sharing between the handoff connections ($t_1 = t_2 = t_3 = 100$). Hence, prioritization among different classes is achieved.

Figure 7 shows the degradation ratio (DR) versus the offered load (call arrival rate) for all classes. As expected, DR increases for all the three evaluated classes as the call arrival rate increases. However, the degradation ratio of the different classes can be clearly distinguished, where class one has the lowest DR values while class three has the highest DR values, i.e., $DR_1 \leq DR_2 \leq DR_3$. This behavior can be explained as follows: the requested bandwidth of class-i increases as the class index increases and since all the three classes are sharing the same bandwidth threshold, $t_1 = t_2 = t_3 = 100$, there are more calls to share the degradation under class-i as the new call arrival rate increases

Figure 8 shows the bandwidth utilization versus the offered load (call arrival rate) for adaptive multimedia framework as opposed to a non-adaptive multimedia framework. Clearly, the bandwidth utilization of the adaptive multimedia framework outperforms that of the non-adaptive multimedia framework. As the offered load increases, the advantage is more evident. This follows since a highly restrictive threshold-bandwidth reservation policy (non-adaptive case) can result in high call dropping before their completion. Whereas for the adaptive multimedia framework, the usage of the bandwidth adaptation algorithm allows the system to offer services whenever there is sufficient amount of bandwidths by intelligently adjusting bandwidth reservation. This results in a near zero HCDP. Therefore, more calls are able to complete their services and, consequently, better bandwidth utilization is obtained, meeting the service provider objectives.

Figure 9 shows the plot for percentage of class-i QoS satisfaction ($\%\mathrm{QoS}_i$) as the call

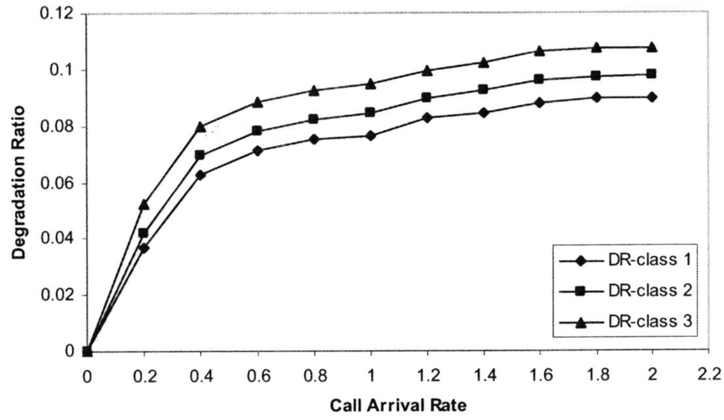

Figure 7. Effect of varying the new call arrival rate on degradation ratio for adaptive multimedia.

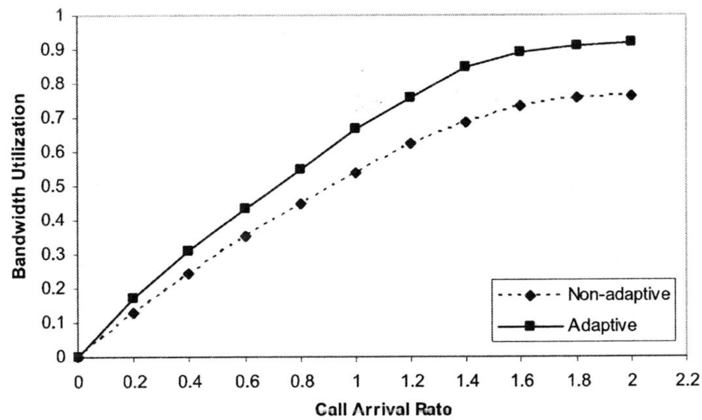

Figure 8. Effect of varying the new call arrival rate on bandwidth utilization for non-adaptive/adaptive multimedia.

arrival rate varies. We observe that at lower loads, all three classes achieve 100% QoS satisfaction since all calls get either the requested bandwidth or more. At higher load, however, the effect of increasing the call arrival rate on the $\%QoS_i$ becomes obvious. As expected, we notice that the $\%QoS_i$ decreases as the call arrival rate increases for all classes where class-1 experiences less $\%QoS_i$ decrements while class-3 has the highest $\%QoS_i$ decrements. This difference can be explained by the fact that higher bandwidth users experience more degradation ratio (see Figure 7), which means more users are assigned bandwidth less than requested which translates in decreasing of $\%QoS_i$.

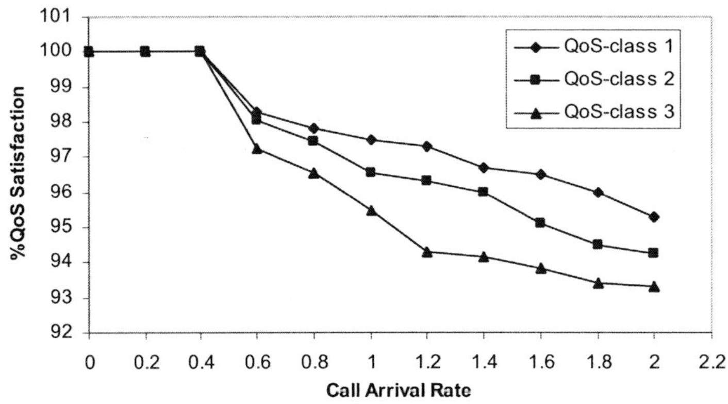

Figure 9. Effect of varying the new call arrival rate on QoS performance indicator.

6. Summary

An adaptive multimedia paradigm can play an important role to mitigate the highly fluctuating bandwidth resources in the wireless cellular networks. In this chapter, a novel adaptive multimedia framework called QoS-Adaptive Multimedia Wireless Access (QoS-AMWA) for the wireless cellular network has been presented. The proposed QoS-AMWA system considers the different priorities of the traffic classes and assures a guaranteed connection-level QoS for heterogeneous traffic. Three related components comprise the main building blocks of the framework: (1) a threshold-based bandwidth reservation policy, (2) a threshold-based call admission control algorithm, and (3) a bandwidth adaptation algorithm. As the mobile user may visit different cells during a connections lifetime, the system components work together to provide the mobile user a means of seamless handoff communication through the wireless network. Thus, the proposed QoS-AMWA system maintains the QoS of ongoing connections in the network, and at the same time minimizes the new call blocking probability and the handoff call dropping probability.

Simulation results show that the handoff call dropping probability is always lower than the new call blocking probability. Moreover, the handoff calls with lower bandwidth requirements always have lower dropping probability than those with higher bandwidth requirements. The performance of the adaptive multimedia framework is compared to that of the existing non-adaptive case. The results show that although the new call blocking probability of the adaptive multimedia framework is greater than that of the non- adaptive multimedia framework, the overall performance of the adaptive multimedia networking is very attractive in that the handoff call dropping probability is near zero (negligible) and that the bandwidth utilization increases as the offered traffic load increases. Observing the reported results, we perceive that this work is a very powerful and novel contribution in wireless cellular network design as it can be used to support high bandwidth multimedia services in a manner that enhances the overall system performance. In addition, since our framework is cell-centric, it can be implemented on a cell-by-cell basis. The framework hence features a low complexity that is desirable in practical deployment.

References

[1] N. Nasser and H. Hassanein, *Dynamic Threshold-based Call Admission Framework for Prioritized Multimedia Traffic in Wireless Cellular Networks*, in Proceedings of the IEEE Global Telecommunications Conference (Globecom), Dallas, Texas, Novermber 2004, pp. 644–649.

[2] N. Nasser and H. Hassanein, *Robust Dynamic Call Admission Control Framework for Prioritized Multimedia Traffic in Wireless Cellular Networks*, in International Journal of High Performance Computing and Networking (IJHPCN), 2005, to appear.

[3] C. Chou and K. Shin, *Analysis of Combined Adaptive Bandwidth Allocation and Admission Control in Wireless Networks*, in Proceedings of IEEE Infocom, New York City, New York, June 2002, pp. 676–684.

[4] S. Kim, T. Kwon, and Y. Choi, *Call Admission Control for Prioritized Adaptive Multimedia Services in Wireless/Mobile Networks*, in Proceedings of the IEEE Vehicular Technology Conference (VTC), Tokyo, Japan, May 2000, pp. 1536–1540.

[5] T. Kwon, Y. Choi, C. Bisdikian, and M. Naghshineh , *QoS Provisioning in Wireless/Mobile Multimedia Networks Using an Adaptive Framework*, in Kluwer Wireless Networks, vol. 9, no. 1, 2003, pp. 51–59.

[6] T. Kwon, Y. Choi, C. Bisdikian, and M. Naghshineh , *Call Admission Control for Adaptive Multimedia in Wireless/Mobile Networks*, in Proceedings of the ACM Workshop on Wireless Mobile Multimedia (WoWMoM), Dallas, Texas, October 1998, pp. 111-116.

[7] Y. Xiao and C. Chen, *Improving Degradation and Fairness for Mobile Adaptive Multimedia Wireless Networks*, in Proceedings of the IEEE International Conference on Computer Communications and Networks (ICCCN), Phoenix, Arizona, October 2001, pp. 598–601.

[8] Y. Xiao, C. L. P. Chen, and B. Wang, *Bandwidth Degradation QoS Provisioning for Adaptive Multimedia in Wireless/Mobile Networks*, in Computer Communications, vol. 25, no. 13, August 2002, pp. 1153-1161.

[9] Y. Xiao, C. Chen and Y. Wang, *Fair Bandwidth Allocation for Multi-Class of Adaptive Multimedia Services in Wireless/Mobile Networks*, in Proceedings of the IEEE Vehicular Technology Conference (VTC) Rhodes, Greece, May 2001, pp. 2081–2085.

[10] G. Kesidis, J. Warlrand and C. Change, *Effective Bandwidth for Multiclass Markov Fluides and Other ATM Sources*, in IEEE Transactions on Networks, vol. 1, no. 4, August 1993, pp. 424–428.

[11] M. Naghshineh and M. Willbeek-LeMair, *End-to-End QoS Provisioning in Multimedia Wireless/Mobile Networks Using an Adaptive Framework*, in IEEE Communications Magazine, vol. 35, no. 11, November 1997, pp. 72–81.

[12] A. Talukdar, B. Badrinath, and A. Acharya, *Rate Adaptation Schemes in Networks with Mobile Hosts*, in Proceedings of the ACM/IEEE MobiCom Conference, Dallas, Texas, October 1998, pp. 169–180.

[13] C. Ho and C. Lea, *Improving Call Admission Policies in Wireless Networks*, in ACM Wireless Networks, vol. 5, no. 4, 1999, pp. 257–265.

[14] T. Kwon, S. Das, I. Park, and Y. Choi, *Bandwidth Adaptation Algorithms with Multi-objectives for Adaptive Multimedia Services in Wireless/mobile Networks*, in Proceedings of the ACM Workshop on Wireless Mobile Multimedia (WoWMoM), Seattle, August 1999, pp. 51–58.

[15] Y. Xiao, C. Chen, and Y. Wang, *Optimal Admission Control for Multi-Class of Wireless Adaptive Multimedia Services*, in IEICE Transactions on Communications, vol. E84-B, no. 4, April 2001., pp. 795–804.

[16] K. Yeung and S. Nanda, *Channel Management in Microcell/Macrocell Cellular Radio Systems*, in IEEE Transactions on Vehicular Technology, vol. 45, no. 4, November 1996, pp. 601-612.

In: Adaptation Techniques in Wireless Multimedia Networks ISBN 1-59454-883-8
Editors: Wei Li and Yang Xiao, pp. 91-108

Chapter 6

ADAPTIVE TECHNIQUES FOR ROBUST REAL-TIME VIDEO COMMUNICATIONS*

Liang Cheng*[†] *and Magda El Zarki
Bren School of Information and Computer Science,
University of California, Irvine

Abstract

It is challenging to offer real-time video services over current networks that lack quality of service(QoS) provision. The real-time video services require strong error recovery capability, stringent synchronization control, and low end-to-end latency. In this chapter, we will present a discussion of selected issues that are related to a robust real-time video application. Expecting no special treatment from the underlying network, we focus on the application layer. Specifically, we will investigate two problems: 1) how to control the transmission error and eliminate its impact and 2) how to maintain low end-to-end delay and reasonable intra-stream synchronization.

Keywords: video coding, synchronization, low-latency, error resilience

1. Introduction

With the increased availability of next generation wireless networks, sophisticated wireless services with rich content multimedia will become increasingly pervasive. Two-way videophones are certainly becoming more popular with the advent of 3G cellular networks. The success of Internet-based video streaming applications will be expanded to millions of mobile users throughout the world. In addition to their public appeal as a new communication method, real-time wireless video services are particularly important for crisis situations. For instance, mobile tele-diagnosis devices may reduce the delay caused by geographical distance, allowing doctors to monitor patients at remote sites and to place timely medical orders.

---

*This research was in part funded by NSF and a gift from Conexant, Newport Beach, CA through the auspices of the Center for Pervasive Communications and Computing (CPCC) at UC, Irvine.

†E-mail address: liang.cheng@ieee.org

Real-time communication refers to the communication that takes place without lengthy delays - also known as synchronous communication. Real-time video applications can be roughly categorized into conversational applications (such as video conferencing, videophony, and tele-presence) and streaming applications (such as web-television and VOD). Conversational applications have more stringent delay requirements than streaming applications. Normally, full interactivity requires the end-to-end latency to be around 0.4 second to ensure the mutual awareness of tele-presence; regular video streaming services may relax the delay bound to around 1-3 seconds to fulfill VCR-like commands, such as "play" and "stop".

Wireless links add another complication to real-time video applications. It is well-known that wireless links are error prone, band limited, and time varying. Many unpredictable environmental factors influence the reliability of wireless links. Different modulation and MAC mechanisms could co-exist on a single path between the two communication parties. Dramatically varied transmission performance could result due to different control mechanisms deployed to combat the temporal outages on wireless links. It is virtually impossible to provide error free transmission over wireless networks in a punctual manner. In summary, the timing requirements and the lack of quality of service (QoS) on wireless links, in conjunction with the high quality expectation of wireless users on the video service, make the robust real-time video application a fairly challenging task.

Among numerous requirements of any general wireless application, real-time video services must fulfill at least two prerequisites in order to enable the "error robustness" and "real-time". First, despite the fact that transmission errors are inevitable over the wireless link, the impact of transmission errors on the video fidelity must be eliminated punctually. Second, the temporal fidelity, i.e., the temporal ordering among the Media Data Units (MDUs) in the video stream need to be preserved and the end-to-end delay must be controlled, such that the user may experience a smooth and interactive low latency video playout.

In section 2., we will discuss the transmission error and its impact on the predictive video. We will also investigate the solutions that may mitigate and/or stop the error spreading. In section 3., we will analyze the end-to-end transmission delay and then introduce a number of adaptive techniques, by which we aim to smooth the source video rate control and to compensate for the transmission jitter with the delay constraint.

2. Error Control

The Internet in general is time varying and band limited [1]; a wireless network in addition has high bit error rates due to fading [2]. The stringent timing requirement makes the error free data reception impossible. Transmission errors, including random bit errors and packet erasure errors, continuously cause end-to-end infidelity across the network. Although channel characteristics usually conform to some stochastic models in the long run, the transient occurrence of error is highly random and unpredictable. Depending on compression methods, the location of the loss, and punctuality requirements, the impacts of the transmission errors range from negligible to objectionable.

Essentially, random bit errors and packet erasure errors are due to different reasons. Random bit errors are normally associated with wireless link interference and fading, whereas packet erasure errors are known to be the result of channel congestion. Random bit errors can be partially recovered using channel coding. The residual bit errors have a similar impact on the compressed video data as packet erasure errors because almost all modern video communication standards consider the minimum decodable unit as a packet. A bit error, if not recovered, may convert to a packet loss, i.e., the whole packet that contains the error bit is dropped.

2.1. Error Propagation

The damage caused by packet erasure is usually exacerbated by the error propagation. The prediction loop at the compression layer, which is well devised to increase the compression efficiency, may spread errors temporally and spatially. By inserting a resynchronization point at the beginning of each packet [14], we can limit the spatial error propagation within the boundary of packets. However, temporal error propagation is difficult to suppress, because temporal errors spread via the motion compensation loop, affecting successive frames and remaining visible for an uncertain long period of time. In addition, if the motion vector (MV) is non-zero, the area affected by the error may expand, aggravating the video fidelity distortion in the subsequent frames, as is illustrated in Figure 1.

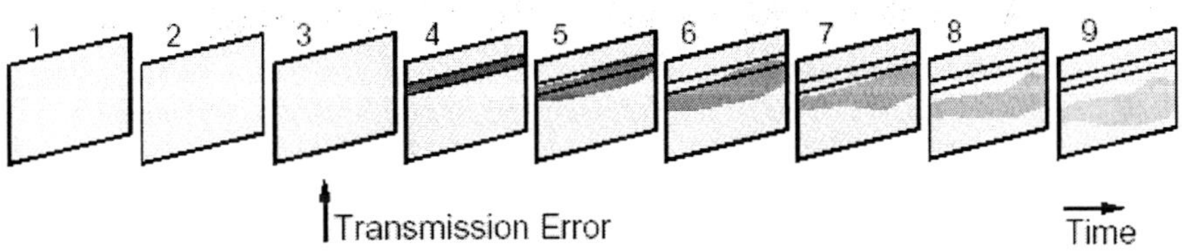

Figure 1. Effect of error propagation [4].

Now we explain the effect shown in Figure 1. In Figure 2, we illustrate the error spreading process over one frame interval. We use MB to denote macroblock. We denote $MB^T_{i,j}$ as the MB at the ith row and the jth column at time T; $MV^T_{i,j}$ is the motion vector of $MB^T_{i,j}$. In Figure 2, we assume there is one lost packet at time $T-1$, which results in the error at $MB^{T-1}_{i,j}$. By utilizing the temporally and spatially redundant information, we can partially conceal the lost $MB^{T-1}_{i,j}$. (For more information on error concealment, refer to [15]). The residual error inevitably exists and potentially propagates to its neighboring MBs through the motion compensation process. For instance, in Figure 2, at time T, each of the nine MBs (including the lost MB and its eight neighboring MBs) are decoded with reference to their corresponding predictors at the previous frame. Some of these referenced area overlap with the lost MB at time $T-1$. The overlapping part may spread the error when the predictor is used for motion compensation at the decoder. For example, as seen in Figure 2, the error in $MB^{T-1}_{i,j}$ is spread to its six neighboring MBs at time T via this process. In practice, The temporal error propagation will continue if no control mechanism (i.e., force-intra-updating the affected MBs) is applied.

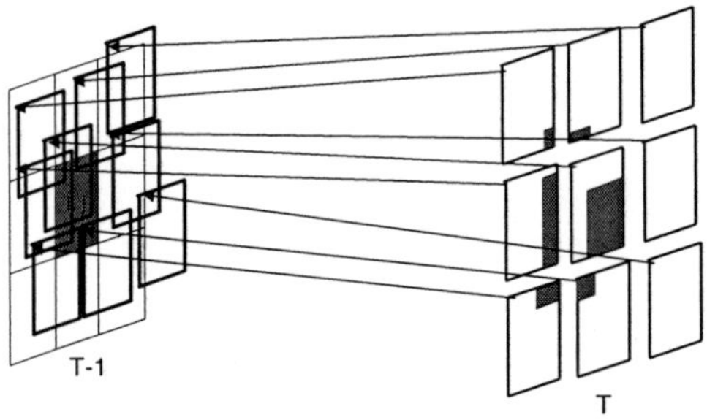

Figure 2. Temporal error propagation. (Note that we mark the original concealed MB and the affected area by "dark" color in the figure.)

From the illustration in Figure 2, we know that 1) the degree of the error distortion in the affected areas is determined by the initial error concealment algorithm and 2) the range of error propagation is determined by the motion vectors that indicate the location of the predictor window in the reference frame.

2.2. State-of-the-Art Error Robust Techniques

Error robust techniques are generally composed of 1) channel error control [3] at the transmission layer 2) error concealment at the decoder and 3) error resilient source coding [15] at the encoder. Depending on the interaction between sender and receiver, error robust techniques can also be categorized into non-feedback techniques and feedback based techniques, both at the compression layer and at the transmission layer, as seen in 3.

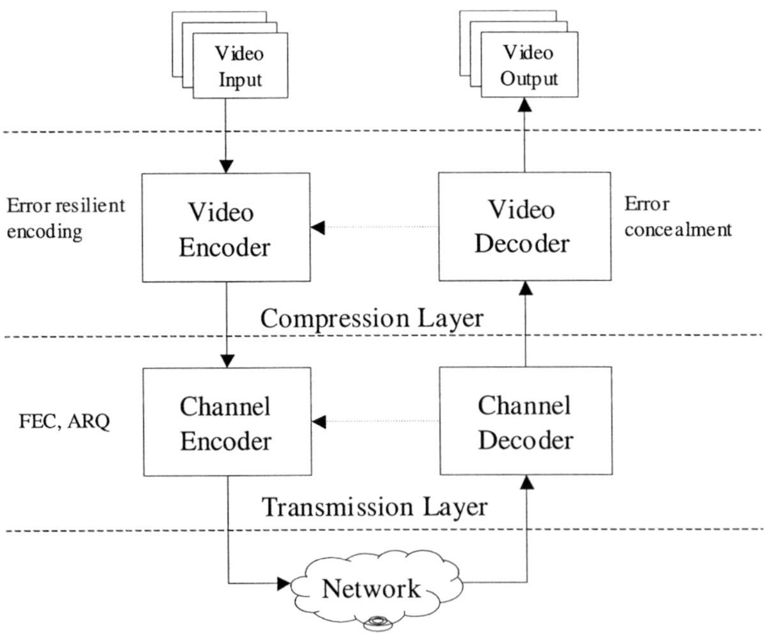

Figure 3. Error control schemes

2.2.1. Channel Error Control

Channel error control schemes provide protection capabilities to the encoded video data against the channel errors. These schemes are video content agnostic and aimed at flipping the error bits directly.

Channel error control coding normally includes Forward Error Coding (FEC) [16] and Automatic Repet reQuest (ARQ) [17][18][20][19]. FEC is based on Reed-Solomon coding. The size of the Reed-Solomon codeword is $N = K + R$, in which the number of check bytes R and codeword size N are determined by the system requirements. FEC can be applied in an open-loop mode: no interaction is required between the encoder and decoder. The encoder adds parity bits; the decoder detects and corrects the errors. Hence, FEC based techniques are preferred for real-time video applications because they maintain a constant throughput and a bounded delay. The disadvantage of FEC is that redundant parity bits may waste a certain amount of bandwidth. By adjusting the number of the redundancy bits, FEC can achieve unequal protection capability, which is best suited for prioritized data streams [16].

Different from FEC, ARQ techniques are closed-loop methods. ARQ requests that packets received in error be retransmitted. The combination of FEC and ARQ, called *hybrid ARQ*, has been developed to combine the advantages of both schemes [17][20]. Depending on the response of the encoder to the error feedback, hybrid ARQ can be further divided into type-I [19] and type-II [17]. Type-I hybrid ARQ schemes include parity bits for both error detection and error correction in every transmitted packet. If the number of erroneous bits in a received packet is within the error correction capability of the code, the errors are corrected and the receiver accepts the decoded message. Otherwise, the packet is rejected, a retransmission is requested and the original packet is sent again. For type-II hybrid ARQ schemes, the erroneous packet is kept for future use rather than discarded. Only redundancy bits are transmitted when they are needed. Normally ARQ cannot be used in real-time applications, since retransmission may violate delay requirements. However, if round-trip delay is short with respect to the maximum allowable application delay, a limited number of retransmissions may be viable, and thus the delay-constrained ARQ methods [21] are proposed for real-time error control.

2.2.2. Error Concealment

When the stream arrives at the decoder, there are still residual errors that are un-recovered. Decoder error concealment is the last resort to combat channel loss. The result of concealment directly determines the perceptual quality fidelity. In general, spatial, spectral or temporal redundancies of received video data are utilized to perform the error concealment [15]. More advanced error concealment strategies, such as hybrid switching of different types of error concealment, are described in [22][23][24]. Besides these heuristic schemes that assume the smoothness or continuity of neighboring pixel values, more complicated yet more effective loss prediction using context-based model is proposed in [25].

Error concealment neither introduces extra delay nor lowers the compression ratio. All practical video decoders have error concealment algorithms built in. Error concealment can more or less reconstruct the lost data; it, however, cannot completely recover the decoded video from the error. Temporal error propagation still exist after the error concealment, as seen in Figure 2. Hence, error concealment may partially eliminate the perceptual impact of errors but can by no means clean the errors.

2.2.3. Source Error Protection

Source error protection strategy is aware of the video syntax/semantics and is meant to protect the important video stream and recover the usable bits if necessary. It generally includes scalable video coding (i.e., layered coding) [26][27][28], multiple description coding (MDC) [29], and reversal variable length coding (RVLC) [30][31].

Scalable video coding produces a hierarchy of bitstreams, where different parts of an encoded stream have unequal contributions to the overall quality. Compared to the single-layer video coding, scalable video coding better fits the network with inherent quality of service (QoS) classification. Different layers can be transmitted separately such that more important layer can reach the receiver with higher priority [26].

MDC has the ability to produce descriptions that may be equally important. Each description can be decoded independently to give a usable reconstruction of the original signal. By combining two or more descriptions, the decoded video quality can be improved. MDC does not require prioritized transmission and increases the chance of receiving at least a video of usable quality. Thus, MDC is better suited for a network with no hierarchical support, such as the Internet. The disadvantage of both MDC and layered coding is their low compression efficiency compared to single layer coding.

RVLC, as the name suggests, can be decoded in two directions and can therefore be used by a decoder to enhance robustness in the presence of transmission bit errors. RVLC involves little or no efficiency loss relative to the corresponding non-reversible variable length codes [30]. Most of the video standards, like MPEG-4 [39] and H.263+ [40], have adopted the RVLC as the error resilience tools [14]. One known problem to the bit-level RVLC, however, is that the random bit errors may confuse the decoder - the RVLC decoder might not be able to distinguish some error bits if these bits have correct syntax. Compared to the packet-level data dropping and concealing scheme, decoding the syntax decodable error data sometimes may cause much more objectionable quality.

2.2.4. Source Error Elimination

None of the mentioned anti-error schemes are capable of completely preventing the occurrence of errors. As we discussed earlier in Section 2.1., the residual errors may spread to other MBs both spatially and temporally. In particular, temporal error propagation may lead to the infidelity of the whole system for an uncertain length of time. Hence, eliminating the impact of the temporal error propagation is of significance.

The insertion of an intra-coded frame (I-frame) for a specified interval is the conventional scheme to stop the temporal error propagation [41]. The I-frame does not use temporal prediction techniques and thus it is self-contained. For this reason, the I-frame is considered as the key frame, providing both the function of "fidelity refreshing sync point" in the compression domain and the function of "random access" for play-back. In MPEG-2 [38], the I-frame and its associating group of picture (GOP) are mandatory for high quality digital TV broadcasting services. One GOP consists of one I-frame and a number of predictive frames (P-frames) and bi-directional frames (B-frames). However, I-frame needs much more bits to represent than the P-frame, which causes delay and low quality for the low bit rate video application.

Recently, several papers propose schemes to mitigate the impact of error propagation without using I-frames except for the very first frame in the stream. The delay of the first I-frame is combined with system signaling time. Based on some criteria and assumptions, they selectively intra-code a number of MBs in the P-frame to mitigate errors and their propagation effect due to the inter-frame prediction. For example, adaptive intra refreshing (AIR), which is standardized in MPEG-4 [39], periodically refreshes only those MBs that have differences from the corresponding MBs in the previous frame higher than a specified threshold. In [7], the intra-MBs are randomly distributed in the frame; the number of the intra-MBs is determined to minimize the overall mean squared error (MSE). More advanced schemes are proposed in [8](ROPE) and [9]. The proposed algorithms select the intra/inter mode for each MB to achieve end-to-end rate-distortion optimization based on *a priori* knowledge of the channel error statistics and the decoder's error concealment scheme.

Despite the fact that these schemes [39][7][8][9] are capable of mitigating the error propagation, they all lack the basic function that the I-frame is meant to provide: frame-level refreshing capability. As a result, residual errors might propagate inside the motion compensation loop for a uncertain long period of time, which seriously affects the fidelity of the decoded video. Furthermore, the error concealment schemes may be practically irreproducible [25] at the encoder, which makes the quality optimizations proposed in [8] and [9] difficult to achieve.

When a feedback channel is available on the application layer (i.e., source coding level), the decoder may inform the encoder of the location of the error MBs. By keeping the synchronization between the encoder and the decoder, error propagation can be stopped effectively. In paper [41], the author introduced three methods to mitigate the transmission errors using feedback. They are error tracking(ET), reference picture selection(RPS) and error confinement(EC). If no error is reported, all of these three schemes use inter-mode to achieve the highest coding efficiency. For ET, when the encoder knows the location of the error MBs, it stop the inter-frame error propagation by selecting the intra mode for all those reported error MBs and affected MBs thereof. For RPS, both encoder and decoder store a number of consecutively reconstructed pictures prior to the current encoding picture. The encoder selects a reference picture from these stored pictures such that the picture with error is not used for motion estimation/compensation by other pictures. For EC, no force intra-updating is applied. Instead, the areas that are determined to be in error are subtracted from the motion estimation searching range.

For these three schemes, the accuracy and the promptness of the feedback are assumed; otherwise, the lost MBs, if not being reported, will continuously affect the decoded video without being noticed by the encoder. In practice, however, the uplink channel, like the downlink channel, is not error immune. Thus, the encoder might take it for granted that it is tracking the decoded video but in fact it is not. For example, Figure 5-a shows the decoding PSNR loss for an assumed round-trip delay of 3 frames when the ET method [41] is used. The encoder fails to recover the picture quality when a feedback reporting the loss at frame 4 is lost, even though the next packet loss (at frame 23) are tracked and corrected. The encoder may be aware of the feedback loss (e.g., by checking the feedback index number); it, however, loses the distortion details of the decoding video sequence, which may hinder the encoder from making the optimal error resilient decision. The retransmission mechanism, notably the TCP protocol [10][11], can be used to guarantee the reception of the feedback. Still, the feedback may be useless when it is received, because it may exceed the tolerable feedback latency, i.e., the assumed maximum round-trip delay, which is determined by the encoder's computing capability and memory capacity.

In [33][32], we proposed a scheme named "progressive group of picture" (PGOP), by which we aim to stop rather than mitigate the temporal error propagation without using I-frame. By periodically refreshing a uniform segment of the video frame and a selected number of MBs that are refreshed but affected at the time of next segment being refreshed, PGOP is able to achieve a frame-level fidelity refresh in a timely manner without using feedback [33]. Like the I-frame, PGOP is designed for the universal encoder and can fit into the heterogeneous environments: the refresh process does not rely on the statistics of the network transmission nor on the decoder's implementation. Once a PGOP is received, any errors existing inside the motion compensation loop will be refreshed. We exemplify PGOP in Figure 4. As illustrated in Figure 4-a, at time $(T+2)$, two MBs in the first column that have just been refreshed at time $(T+1)$ are still able to use the unrefreshed area for motion compensation. Even though this unrefreshed area to be referenced is being updated at time $(T+2)$, the existing errors on this area at time $(T+1)$ will propagate to these two MBs. To solve this problem, we analyze the motion vectors (which give an indication of the error propagation as is discussed in Section 2.1.) and augment the regular column-by-column refresh process by intra-updating the MBs that are affected by the residual error propagation across the MB column that is being refreshed. We call this additional intra-updating process *stride back*, as seen in Figure 4-b, by which we are able to prevent the error from spreading back across the refreshing column and the "dirty" area (where the error might exist) can be progressively reduced until the whole frame is completely refreshed.

In [32], we apply the PGOP scheme in a quality feedback framework using the ITS quality feature extraction technique [12]. Once a quality degradation is detected at the encoder, one PGOP is encoded and sent out to stop the temporal error propagation. The quality adaptive PGOP can survive even when the quality feedback is delayed or lost. In Figure 5-b, we notice that our proposed scheme enables the whole system to completely recover from errors even if the first quality feedback is not received. Hence, even though its refresh period may be long when compared to the error tracking scheme, the proposed scheme demonstrates a more robust error refresh. Of course, the combination of PGOP and ET can be studied and is expected to demonstrate both system robustness and prompt error refresh.

3. Real-Time Transmission

3.1. End-to-End Delay

The low-latency video communication on wireless systems must fulfill the stringent timing requirements in order to enable the "awareness of real-time existence". The "real-time" can be further explained as 1) low end-to-end delay and 2) temporal fidelity.

The end-to-end delay T can be decomposed into

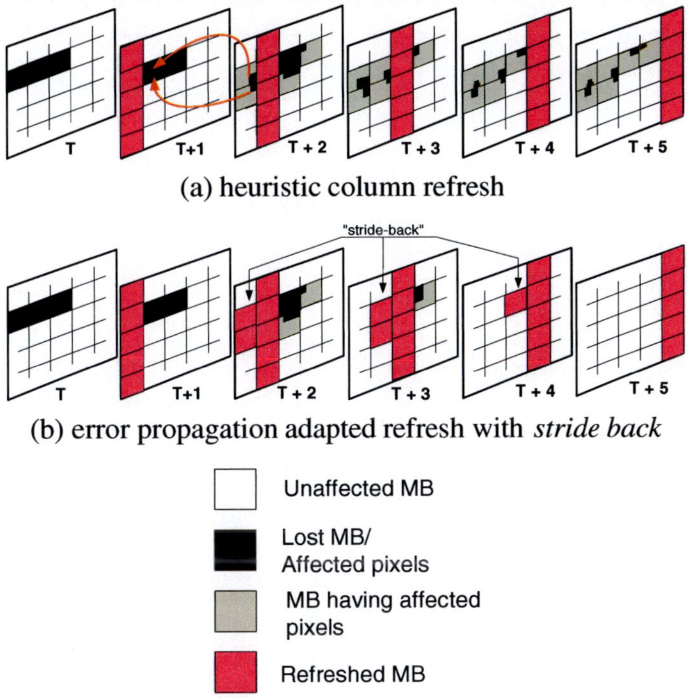

(a) heuristic column refresh

(b) error propagation adapted refresh with *stride back*

Figure 4. Illustration of PGOP with *stride back*.

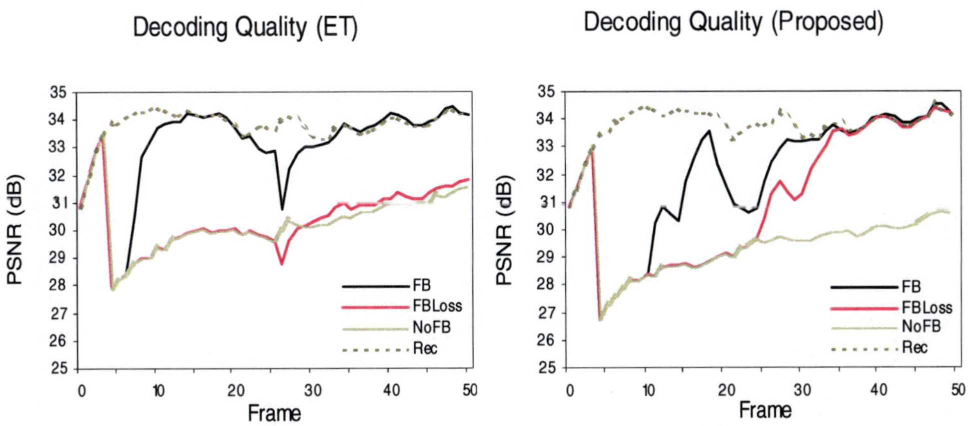

(a) Error Tracking

(b) Quality Feedback Adapted PGOP

Figure 5. Influence of the lost feedback [34]. FB: two packet losses are received correctly; NoFB: no packet losses are received; Rec: encoding quality; FBLoss: the first feedback is lost; packet length: 1600 bits; sequence: *Foreman*; bit rate: 64 kbps; frame rate: 10 fps; format: QCIF.

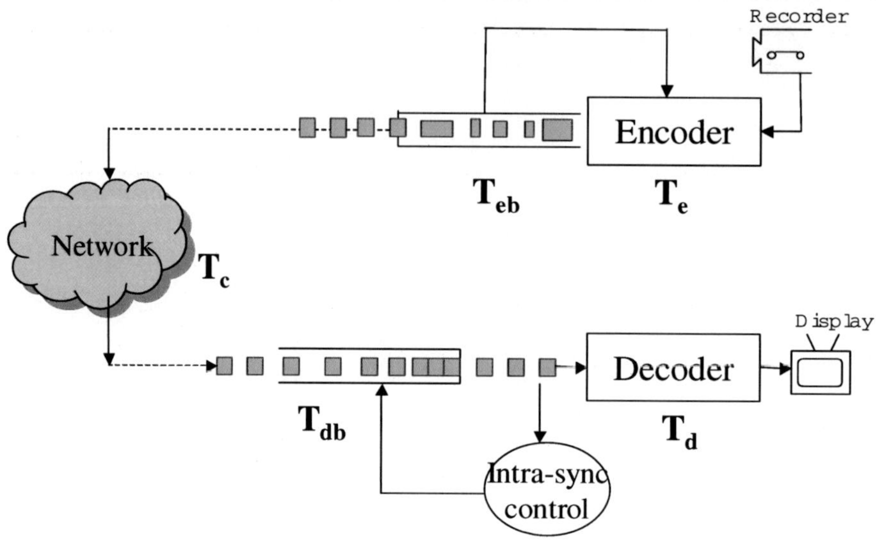

Figure 6. End-to-end delay.

$$T = T_e + T_{eb} + T_c + T_{db} + T_d, \quad (1)$$

where T_e,T_{eb},T_c,T_{db}, and T_d are, respectively, the encoding delay, the encoder buffer delay, the transmission delay, the decoder buffer delay, and the decoding delay, as is shown in Figure 6.

Essentially, T_e and T_d are determined by the computing capability of the hardware and the algorithm complexity. For most of the interactive video communication, MPEG-4[39] or H.263[40] is used. For simplicity without losing generality, we assume that T_e and T_d are small enough for every video frame.

The transmission delay T_c involves the intermediate network infrastructure. Most of our networks (Internet and wireless network) use router to store and forward data packets. Hence, the intermediate router algorithms and other underlying re-transmission mechanism (such as ARQ) determine T_c. These underlying functions, however, cannot be accessed from the application layer. A general assumptions in our discussion is that T_c in average has already been confined to a satisfied limit despite of some types of variation for a short period of time.

The last two end-to-end delay elements, T_{eb} and T_{db}, will be discussed in the following sections. The basic rules are to 1) use T_{eb} to absorb the bit rate fluctuation caused by the encoding process such that the time used to transmit each picture is similar and 2) control T_{db} to compensate the end-to-end delay variation caused by all of the rest delay elements, i.e., T_e, T_{eb}, T_c, T_{db}.

3.2. Rate Control

The purpose of conducting rate control is to fit the encoded bit stream into the fixed bandwidth network. Two distinct rate control modes are discussed in [43]. They are reactive mode and preventive mode. The reactive scheme denotes that the encoder iterates different encoding modes if a violation of its encoder buffer constraint is detected. The reactive mode rate control can only be used in streaming applications where the video stream is encoded in advance. The preventive scheme denotes that the encoder takes action only if the encoder buffer occupancy reaches a certain threshold, which is suitable for the real-time applications, though the preventive mode rate control cannot guarantee that the encoder buffer not overflow or underflow. Throughout our discussion, the preventive mode rate control is assumed. For more details of the reactive mode rate control, see [44][45][46].

Encoder buffer is the key element to fulfill the rate control function. At the encoder, the encoding process is adjusted by checking the status of the encoder buffer, i.e., decreasing the bit rate when the encoder buffer is near full and increasing the bit rate when the encoder buffer is near empty. If bits accumulated in the encoder buffer are far beyond the encoder buffer capacity, the next frame will be skipped from the encoding loop to quickly flush the bits out of the buffer. The classic prevalent rate control algorithms include MPEG-2 test model 5 (TM5) [47] and MPEG-4 verification model 5 (VM5) [48].

The encoder buffer delay (T_{eb}) is produced by bits accumulated in the encoder buffer. On one hand, we need to reduce the buffer size to decrease the buffer delay and to lower the memory overhead of the device; on the other hand, a small buffer tends to cause buffer overflow or underflow. The size of the encoder buffer is largely determined by the rate control algorithm. Depending on the frequency of checking the encoder buffer status, the rate control algorithms can be divided into two types: frame-level rate control and MB-level rate control. The frame-level rate control conducts coarse rate control. For the frame-level rate control, the encoder buffer's status is checked once each frame. The frame-level rate control adds light complexity overhead, but it needs relatively larger encoder buffer capacity than the MB-level rate control, because the frame-level rate control cannot regulate the bit rate very effectively, which often results in the buffer violation - buffer overflow/underflow. The advantage of the frame-level rate control is that it may maintain the consistency of the video quality in frames that are not skipped. The MB-level rate control is fine rate control. The encoder buffer's status is checked once each MB, which results in more effective bit rate regulation with rarer frame skipping, but higher complexity overhead. In certain circumstances, when the video content changes dramatically or the intra-frame is used, MB-level rate control may cause visible video quality degradation for the frames that would need more bits to represent. A variety of algorithms have been proposed to improve the accuracy of the rate control and/or better optimize the quality and the bit rate. For frame-level rate control improvements, refer to [51][52]; and for MB-level rate control improvements, refer to [49][50].

Despite these improvements, neither frame-level rate control nor MB-level rate control is able to effectively regulate the bits used to represent the I-frame without noticeable quality drop. For most of the video coding standards, group of picture (GOP) is used to refresh the impact of the transmission error on the predictive coding loop - motion compensation. In [33], we demonstrate that our proposed PGOP could possess the same error refresh capability as the I-frame. Moreover, because PGOP pre-allocates the high-bit-budget intra-MBs into a sequence of P-frames, PGOP significantly reduces the bit rate fluctuation under both frame-level rate control and MB-level rate control. As a result, the encoder buffer size that is used to absorb the bit rate fluctuation can be decreased, which leads to less T_{eb}.

Given the same encoder buffer size and the same error refresh capability, PGOP outperforms GOP by achieving less frame skipping for the frame-level rate control and avoiding the video quality fluctuation for the MB-level rate control. Figure 7 demonstrates the buffer-level variations and the encoding quality for both GOP and PGOP when using the MB-level rate control mode. It shows: 1) the MB rate control effectively prevents the buffer from overflowing for both PGOP and GOP; 2) the encoding spatial quality degradation occurs every time an I-frame is coded for the GOP scheme; and 3) with PGOP, the MB-level rate control is able to preserve a consistent spatial quality. Figure 8 shows the buffer variation and the encoding quality of GOP and PGOP with frame-level rate control. It can be seen that: 1) the buffer occupancy of the frame-level rate control shows more fluctuation than that of the MB-level rate control in general; 2) the buffer-level surge occurs periodically and thereby a considerable amount of frames are skipped with the GOP scheme; 3) the PGOP scheme facilitates the bit rate regulation and thus greatly reduces the fluctuation of the buffer level and the occurrence of buffer overflow.

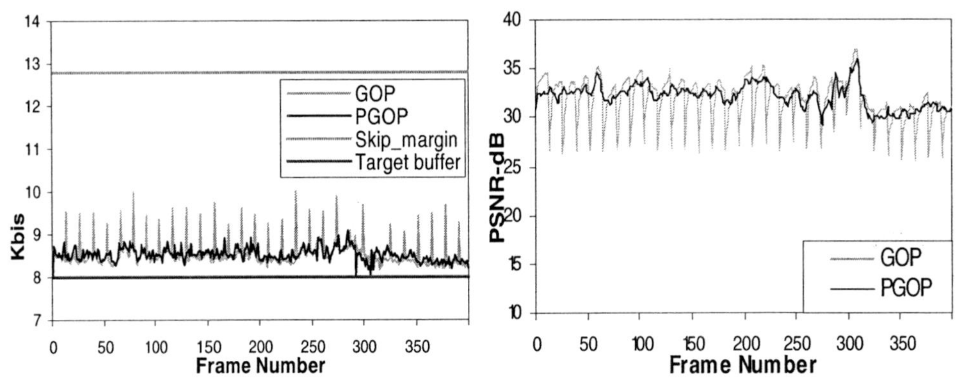

(a) Buffer-level variation (b) Encoding quality

Figure 7. MPEG-4 MB-based Q2 rate control. Buffer size = 16 kbits; Encoding buffer delay = 0.25s, sequence: *Foreman*.

3.3. Intra-stream Synchronization

As seen in Figure 6, after the encoder rate control, the video stream enters the network. The intermediate network mechanisms (e.g., retransmission mechanism ARQ, router's buffering, etc) tend to incur variable T_c, which in turn causes end-to-end delay fluctuation, as seen in Figure 9.

A straightforward solution to this problem is to use a size-predetermined buffer (T_{db}) to compensate for the end-to-end delay jitter. However, because the network behavior is unpredictable, it is not always feasible to find an optimal T_{db} that satisfies the requirements of the application. A small T_{db} might incur excessive MDUs skipping as illustrated by "static playout timeline (2)" in Figure 9. On the other hand, a large T_{db} can reduce MDUs skipping by extending the playout deadline. But, as illustrated by the "static playout time-

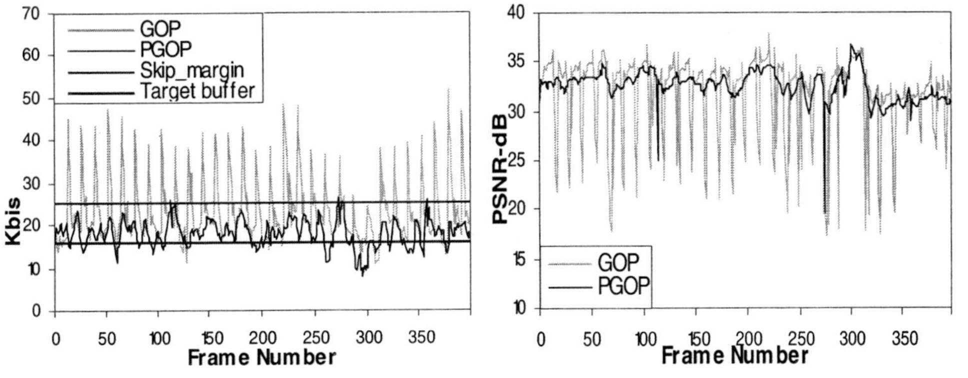

(a) Buffer-level variation

(b) Encoding quality

Figure 8. MPEG-4 frame-level Q2 rate control (a) Buffer-level vairation. (b) Encoding quality. Buffer size = 32 kbits, Encoding buffer delay = 0.5s, sequence: *Foreman*.

line (1)" in Figure 9, the large T_{db} leads to a larger end-to-end delay, which will deteriorate the perceived "real-time" tele-presence. As such, a buffering scheme that is capable of dynamically striking a balance between the delay and the MDUs skipping is considered ideal [5][6], as shown by "ideal playout timeline" in Figure 9.

In [35][36], a solution is proposed to employ an adaptive delay and synchronization control scheme, as seen in Figure 10. The essence of this scheme is that it is receiver-based and application-level QoS oriented. This proposed scheme 1) uses distributed clocks with a virtual timing mechanism, 2) monitors the synchronization status in real-time, and 3) piece-wisely adjusts the buffering delay (i.e., T_{db}), to compensate for the delay jitters. It increases the buffering delay to compensate for the delay jitters while the intra-stream synchronization error has exceeded the threshold, and decreases the buffering delay automatically whenever the synchronization performance permits.

4. Conclusions

Due to the tremendous complexity in administration and the nature of wireless communications, provision of hard-guarantees for services will not happen in the foreseeable future. As such, providing robust real-time video services on wireless systems becomes fairly challenging. The unpredictable behavior of the end-to-end transport channel leads to the fidelity loss of the video communication: not only data losses occur during the transmission but the original temporal relationships among MDUs are also destroyed.

With little assumption for the underlying layers, we discussed a set of application-level solutions to achieve the robust and real-time video service. Specifically, we described the transmission error, its propagation and a spectrum of error control strategies, including channel error correction, error protection and error elimination. Based on these techniques, the encoded video stream can survive even the worst channel loss situation. To achieve the

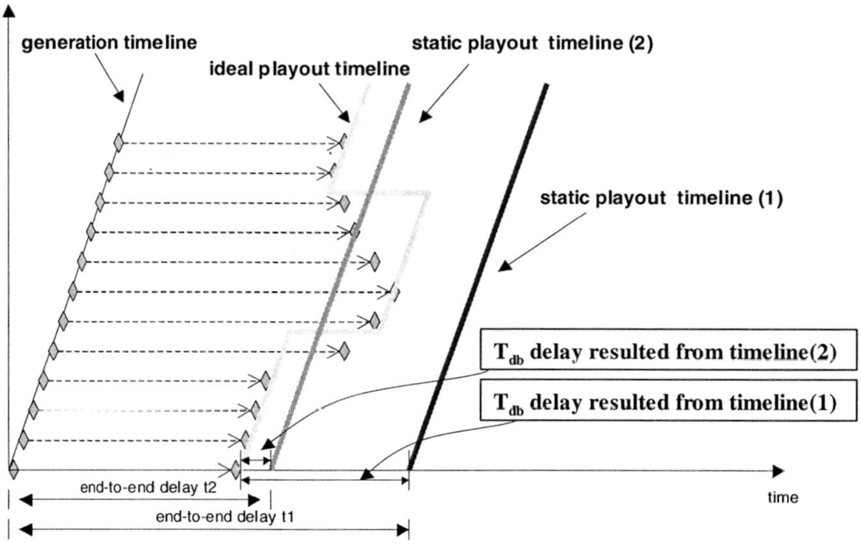

Figure 9. Illustration of Delay Equalization [36].

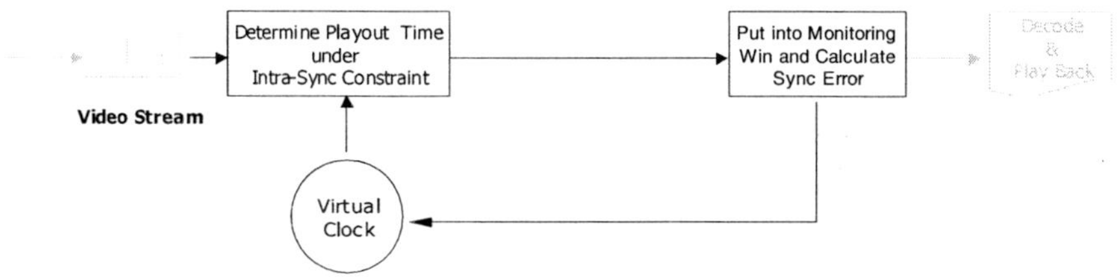

Figure 10. Synchronization control scheme

punctual and synchronous video play back, we explored the end-to-end delay, and explained how to control the encoding buffer and the decoding buffer. In particular, we proposed a scheme - PGOP - by which we reduced the encoding bit rate fluctuation and greatly ease the rate control process. PGOP is also proposed to replace the I-frame: PGOP is able to achieve the equivalent error refresh capability as GOP and to avoid the low coding efficiency problem innate to the I-frame. At the decoder side, we discussed an adaptive decoding buffer control strategy, by which we can both maintain a low end-to-end delay and avoid the unnecessary video data dropping.

In summary, by selecting the appropriate video error resilient coding schemes and the adaptive delay and synchronization control, we are able to preserve the video data fidelity and to balance the synchronization requirement and latency preference for the real-time video applications, despite of those uncertainty within the network transmission infrastructure.

Acknowledgments

The authors thank Mr. Haining Liu for his valuable help and suggestions.

References

[1] J. Boyce and R. Gaglianello, "Packet Loss Effects on MPEG Video Sent Over the Public Internet," *in ACM Multimedia 98 - Electronic Proceedings*, September, 1998.

[2] A. Konrad, B. Y. Zhao, A. D. Joseph, R. Ludwig, "A Markov-Based Channel Model Algorithm for Wireless Network," *in Proceedings of Fourth ACM International Workshop on Modeling, Analysis and Simulation of Wireless and Mobile Systems (MSWiM 2001)*, 2001.

[3] H. Liu, H. Ma, M. El Zarki and S. Gupta, "Error control schemes for networks: An overview," *in Mobile Networks and Applications*, vol. 2 pp. 167-182, 1997.

[4] B. Girod and N. Farber, "Wireless Video," chapter in "Compressed video over networks" by M.-T. Sun and A. R. Reibman, publisher Marcel Dekker, ISBN 0824794230, 2001.

[5] H. Liu and M. El Zarki, "A synchronization control scheme for real-time streaming multimedia applications," *in Proc. of Packet Video 2003*, France, 2003.

[6] H. Liu and M. E. Zarki, "Delay and synchronization control middleware to support realtime multimedia services over wireless PCS networks," *in IEEE Journal of Selected Area in Communications*, vol. 17, no. 9, pp. 16601672, 1999.

[7] N. Farber, K. Stuhlmuller, and B. Girod,"Analysis of error Propagation in hybrid video coding with application to error resilience, " *in Proceeding of ICIP 1999*, Oct., 1999.

[8] R. Zhang, S.L. Regunathan, and K. Rose, "Video Coding with Optimal Inter/Intra Mode Switching for Packet Loss Resilience, " *in IEEE Journal of Selected Area in Communications*, vol. 18, pp.966-976, June, 2000.

[9] D. Wu, Y. T. Hou, B. Li, W. Zhu, Y.-Q. Zhang and H. J. Chao, "An End-to-End Approach for Optimal Mode Selection in Internet Video Communication: Theory and Application," *in IEEE Journal on Selected Areas in Communications*, pp. 977-995, vol. 18, no. 6, June, 2000.

[10] V. Cerf and R. Kahn, "A Protocol for Packet Network Intercommunication," *in IEEE Transactions on Communications*, pp. 637-648, vol. COM-22, no. 5, May, 1974.

[11] H. Schulzrinne, S. Casner, R. Frederderick, and V. Jacobson, "Transmission Control Protocol (TCP)," *in IETF RFC 793*, September, 1981.

[12] S. Wolf and M. H. Pinson, "Spatial-temporal distortion metrics for in-service quality monitoring of any digital video system," *in Proceeding of SPIE International Symposium on Voice, Video, and Data Communications*, September, 1999.

[13] G. Cote, S. Shirani and F. Kossentini, "Optimal Mode Selection and Synchronization for Robust Video Communications over Error-Prone Networks, " *in IEEE Journal of Selected Area in Communications*, vol. 18, pp.952-965, June, 2000.

[14] I. Moccagatta, S. Soudagar, J. Liang, and H. Chen, "Error-Resilient Coding in JPEG-2000 and MPEG-4," *in IEEE Journal on Selected Areas in Communications*, vol: 18, pp. 899-914, June, 2000.

[15] Y. Wang and Q.-F. Zhu, "Error control and concealment for video communication: A review, " *in Proceedings of IEEE*, vol. 86, pp. 974-997, May, 1998.

[16] E. Ayanoglu, P. Pancha, A. Reibman, and S. Talwar, "Forward Error Control for MPEG-2 Video Transport in a Wireless ATM LAN," *in Proceedings of International Conference on Image Processing*, pp. 833-836, September, 1996.

[17] S. Kallel and D. Haccoun, "Generalized type II hybird ARQ scheme using punctured convolutional coding," *in IEEE Transactions on Communications*, pp. 1938-1946, vol 38, no. 11, November, 1990.

[18] S. Lin, D. Costello, M. Miller, "Automatic-Repeat-Request Error-Control Schemes," *in IEEE Communications Magazine*, pp. 5-17, vol. 22, no. 12, December, 1984.

[19] L. K. Rasmussen and S. B. Wicker, "Trellis-Coded, Type-I Hybrid-ARQ Protocols Based on CRC Error-Detecting Codes," *in IEEE Transactions on Communications*, pp. 2569-2575, vol. 43, no. 10, October, 1995.

[20] J. hagenauer, "Rate-compatible punctured convolutional codes (RCPC codes) and their applications," *in IEEE Transactions on Communications*, pp. 389-400, vol. 36, no. 4, April, 1988.

[21] Hang Liu and M. El Zarki, "Performance of H.263 Video Transmission over Wireless Channels Using Hybrid ARQ," *in IEEE Journal of Selected Areas in Communications*, vol. 15, no. 9, December, 1997.

[22] S. Valente, C. Dufour, F. Groliere and D. Snook, "An efficient error concealment implementation for MPEG-4," *in IEEE Transactions on Consumer Electronics*, pp. 568-578, vol. 47, no. 3, June, 2001.

[23] F. Delicado, P. Cuenca, A. Garrido, L. Orozco-Barbosa, and F. Quiles, "On the capabilities of error concealment in MPEG-2 communications over wireless ATM," *in IEEE International Conference on Multimedia and Expo*, pp. 1443-1446, vol. 3, July, 2000.

[24] S. Cei and P. Cosman, "Comparison of error concealment strategies for MPEG video," *in Wireless Communications and Networking Conference*, pp. 329-333, vol. 1, September, 2000.

[25] T. P.-C. Chen and T. Chen, "Second-generation error concealment for video transport over error-prone channels," *in Wireless Communications and Mobile Computing Special Issue on Multimedia over Mobile IP*, pp. 607-624, vol. DOI: 10.1002/wcm.83, October, 2002.

[26] W. Luo and M. El Zarki, "Quality Control for VBR Video over ATM Networks," *in IEEE Transaction on Selected Areas in Communications*, pp. 1029-1039, vol. 15, August, 1997.

[27] W.-P. Li, "Overview of fine granularity scalability in MPEG-4 video standard," *in IEEE Transaction on Circuits and Systems for Video Technology*, pp. 974-997, vol. 11, no. 3, March, 2001.

[28] H. Radha, M. van der Schaar, Y. Chen, "The MPEG-4 Fine-Grained scalable video coding Method for Multimedia Streaming over IP," *in IEEE Transaction on Multimedia*, March, 2001.

[29] V. K. Goyal, "Multiple description coding: Compression meets the network," *in IEEE Signal Processing Magazine*, pp. 74-93, Semptember, 2001.

[30] J. Wen and J.D. Villasenor, "Reversible variable length codes for efficient and robust image and video coding", *in Proceeding of Data Compression Conference*, pp. 471-480, April, 1998.

[31] Y. Takishima, M. Wada, and H. Murakami, ""Reversible Variable Length Codes," *in IEEE Transactions on Communications*, pp. 158-162, vol. 43, no. 234, February, 1995.

[32] L. Cheng and M. El Zarki, "Perceptural Quality Feedback Based Progressive Frame-Level Refreshing for Robust Video Communications," *in Proceeding of IEEE Wireless Communications and Networking Conference 2004*, Atlanta, GA, March, 2004.

[33] L. Cheng and M. El Zarki, "PGOP: An Error Resilient Technique For Low Bit Rate and Low Latency Video Communications," *in Picture Coding Symposium 2004*, San Fransisco, CA, December, 2004.

[34] L. Cheng, "Real-Time Mobile Video Communications," *Univerisity of California at Irvine*, Ph.D. thesis, 2005.

[35] H. Liu, L. Cheng, and M. El Zarki, "Towards Robust AV Conferencing on Next-Generation Networks," *in Proceeding of Multimedia Computing and Networking 2005*, San Jose, CA, January, 2005.

[36] H. Liu and M. El Zarki, "On the Adaptive Delay and Synchronization Control for Video Conferencing Over the Internet," *in International Conference on Networking (ICN'04)*, France, 2004.

[37] Q.F. Zhu and L. Kerofsky, "Joint Source Coding, Transport Processing and Error Concealment for H.323-Based Packet Video," *in Proceeding of SPIE VCIP'99*, vol. 3635, San Jose, CA, Jan., 1999.

[38] ISO/IEC JTC 1/SC 29 13818-2, "Information technology - generic coding of moving pictures and associated audio (MPEG-2): Video," 1994.

[39] ISO/IEC JTC1/SC29/WG11, *in MPEG-4 video verification model 10.0*, Feb. 1998.

[40] ITU-T, "Video codec for low bitrate communication, Version 2," *in ITU-T Recommendation H.263*, January, 1998.

[41] B. Girod and N. Farber, "Feedback-Based error control for mobile video transmission," *in Proceeding of IEEE, Special Issue on Video For Mobile Multimedia* , vol. 97, no. 10, pp. 1707-1723, October, 1999.

[42] Y. Ishibashi, S. Tasaka, H. Ogawa, "A comparison of media synchronization quality among reactive control schemes," *in Proc. of INFOCOM 2001*, Anchorage, Alaska, April, 2001.

[43] P. Nunes, F. Pereira, "MPEG-4 compliant video encoding: analysis and rate control strategies," *Proc Asilomar Conf. on Signals, Systems, and Computers* , October, 2000.

[44] R. Egawa, A. Aydin Alatan, and A.N. Akansu, "Compressed domain MPEG-2 video editing with VBV requirement," *in Image Processing, 2000. Proceedings. 2000 International Conference on*, pp. 1016 - 1019, vol. 1, September, 2000.

[45] S. Bozoki and R.L. Lagendijku, "Scene adaptive rate control in a distributed parallel MPEG video encoder," *in Image Processing, 1997. Proceedings., International Conference on*, pp. 780 - 783, vol. 1, September, 2000.

[46] Y. Yokoyama and S. Nogaki, "A rate control method with preanalysis for real-time MPEG-2 video coding," *in Image Processing, 1997. Proceedings., International Conference on*, pp. 514 - 517, vol. 3, October, 2001.

[47] ISO-IEC/JTC1/SC92/WG11, "Test Model 5," 1993.

[48] T. Chiang and Y.-Q. Zhang, "A New Rate Control Scheme using Quadratic Rate Distortion Model," *in IEEE Transaction on Circuit System and Video Technology*, pp. 247-250, vol. 7, no. 1, February, 1997.

[49] J.-W. Lee, D.-H. Kim, and Y.-S. Ho, "A Rate Control Algorithm Using Macroblock Types," *in IEEE Speech and Image Technologies for Computing and Telecommunications*, pp. 761-764, 1997.

[50] S. Bilato, G. Calvagno, G.A. Mian, R. Rinaldo, "Accurate Bit-Rate and Quality Control for the MPEG Video Coder," *in Proceedings of Image Processing*, vol. 3, pp. 571-574, 1997.

[51] J. Ribas-Corbera and S.-M. Lei, "A Frame-Layer Bit Allocation for H.263+," *in IEEE Trans. on Circuit and System for Video Technology*, vol. 10, no. 7, Oct. 2000.

[52] S. Aramvith, I.-M. Pao, and M.-T. Sun, "A Rate-Control Scheme for Video Transport over Wireless Channel," *IEEE Trans. on Circuit and Systems for Video Technology*, vol. 11, no. 5, pp.569 - 580, May 2001.

In: Adaptation Techniques in Wireless Multimedia Networks ISBN 1-59454-883-8
Editors: Wei Li and Yang Xiao, pp. 109-130

Chapter 7

DCT-Based Audio Decomposition: Bandwidth Adaptation for Real-Time Audio Streams

***Kevin Brown**[*]**and Leann Christianson**[†]*
Department of Mathematics and Computer Science, California State University, East Bay, 25800 Carlos Bee Blvd. Hayward, CA 94542, USA

Abstract

Audio and video-based applications typically generate data streams which use constant amounts of bandwidth (on the average). If the network cannot provide the required capacity, the resulting losses and delays significantly reduce the quality of playback at the receiver. Unfortunately, mobile wireless environments tend to exhibit high loss rates, widely varying bandwidth, and frequent disconnections. Real-time applications, in particular, cannot rely on retransmissions in order to correct for losses as the added delay is prohibitive. A possible solution is to provide support for bandwidth adaptation so that the application can match its output level to the network's capacity, and in so doing, reduce the loss rate and delay experienced at the receiver. In this chapter, we propose a DCT based audio decomposition (DAD) scheme for real-time general audio streams. This scheme allows audio quality to be traded off against reductions in bandwidth usage and is appropriate for both unicast and multicast environments. We compare the performance of DAD, using objective and subjective measures, against two other schemes which also support adaptation to restricted bandwidth.

Keywords: coding, adaptive, wireless, multicast

AMS: 68-02, 68M10, 68M12, 68P30

1. Introduction

Audio and video-based applications typically generate data streams which use constant amounts of bandwidth (on the average). This data must be delivered to the receiver within

[*]E-mail address: kevin.brown@csueastbay.edu
[†]E-mail address: leann.christianson@csueastbay.edu

strictly enforced limits on maximum delay and jitter. Interactive or *real-time* applications, such as audio conferencing or shared application control require even more stringent bounds on delay. As such, traditional error control solutions (*e.g.*, Automatic Repeat Request) may not be used due to the prohibitive delay induced. Unrecovered losses or packets experiencing excess delay will result in dropouts in audio streams, possible loss of synchronization between video and audio streams, and difficulties in sharing application control.

The mobile wireless environment unfortunately provides a harsh environment for real-time applications. This is due to the high loss rates, widely varying capacity, and frequent disconnections which are typically experienced by receivers using wireless technologies. Most wireless systems provide very weak or no guarantees on the maximum delay and minimum capacity that a receiver may expect. As a receiver encounters obstacles or signal interference, or as traffic levels from other users rise and fall, the receiver will see their own service levels vary.

In this chapter, we propose a solution to the problem of supporting real-time audio applications deployed in a wireless environment, using either unicast or multicast techniques for delivery. Applications which lend themselves to multicast delivery in the wireless environment include CD-quality digital radio broadcasts, multicasts of live concerts, and delivery of audio tracks to accompany video-on-demand services. Multicast may be used in these situations to efficiently deliver the same audio stream to a number of receivers which can effectively scale from a few receivers to many thousands. Unicast delivery would be appropriate to provide audio conferencing service, tourist information regarding historical sites, and audio to accompany advertisements aimed at mobile consumers.

Audio applications pose a specific challenge as they typically produce data streams at *fixed* data rates (*e.g.*, 1.4 Mbps for uncompressed stereo CD-quality audio or 192 Kbps for stereo MP3 streams). When available bandwidth drops below the level needed to transport the stream, packet losses, bit errors, and delays that degrade the quality of the received signal may occur. Lost or delayed packets and bit errors cause dropouts and static which lead to an audio stream which sounds choppy and may be unintelligible to the user. The human ear has been shown to be very sensitive to discontinuities in speech and other types of general audio signals.

A possible solution to this lack of bandwidth is to reduce the quality of the audio stream so that continuous playback is possible within the restrictions of the available bandwidth. A lower quality audio stream will require less bandwidth for transmission. It is not reasonable, however, to simply deliver the lowest quality (and hence lowest bandwidth) audio stream at all times in an attempt to reduce packet losses and delays. Application users should expect to receive high quality audio when bandwidth is available and lower quality audio when bandwidth is restricted. Hence, the application must *adapt* its output to the available bandwidth with *continuity* of the transmission being the primary goal. When considering the multicast environment, however, a second issue arises, that of defining "available bandwidth". Some receivers may be connected via high-speed wireless LANs (*e.g.*, 11 Mbps) and others via low-speed CDPD (19 Kbps). Hence, a high data rate would overwhelm the CDPD users, but a low data rate would not take advantage of the bandwidth available to the wireless LAN users. As such, adaptation must be done on a per-receiver basis.

In the unicast environment, adaptation may be done at either the source or the receiver. If done at the source, the source can tailor its output to the specific bandwidth available to the one receiver. If adaptation is to be done at the receiver, there must be some mechanism to ensure that the bandwidth of the data stream is reduced before it appears at the receiver. The reduction cannot be done at the receiver itself because the damage is done by the time the data stream arrives at the receiver. The stream has already undergone random losses due to the restricted bandwidth. Hence, adaptation can be controlled by the receiver, but must be effected by the source or an upstream neighbor of the receiver. Note that this is in contrast to the progressive image techniques in which the entire data stream is sent but the receiver may stop processing the stream at any point. In the multicast environment, due to the difficulty of choosing a single data rate for a group of receivers, adaptation is seldom done at the source. A known strategy to handle adaptation at the receiver requires the source to encode and transmit signals at a selection of data rates and allow each receiver to choose the rate which "best" fits her available bandwidth. Intermediate nodes would be required to ensure that only the chosen streams were delivered to the receiver.

In either case, choosing a stream which is transmitted at a lower rate would imply a lower quality stream, where the loss of quality is dependent upon the method used to reduce the data rate. A stream whose data rate matches the currently available bandwidth, however, would be much less likely to suffer dropped packets and delays. The choice here is between a reduction in quality which can be specified and controlled by the receiver and a reduction in quality which is randomly effected by the network as it drops and delays packets which cannot be delivered due to bandwidth restrictions. Clearly, controlled reduction is preferable as the lowest priority information may be eliminated rather than a random selection of packets.

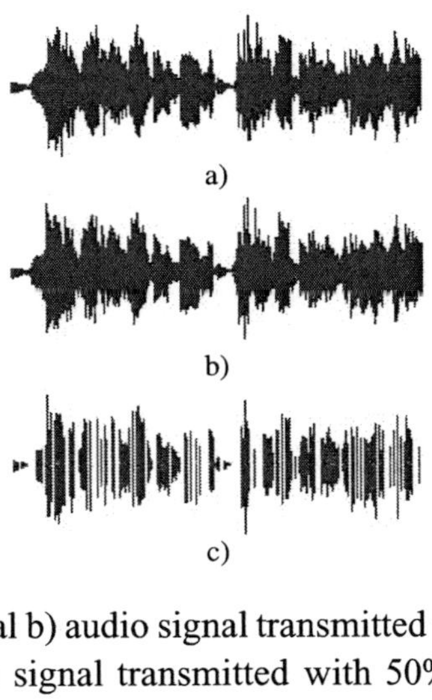

Figure 1. a) original audio signal b) audio signal transmitted with 50% bandwidth reduction via DAD mechanism c) audio signal transmitted with 50% bandwidth reduction due to random packet loss

Figure 1 shows the difference between these mechanisms for a mono CD-quality audio stream. The top signal is the original audio signal, which was sampled at 44,100 Hz using 16 bit samples. It requires 705.6 Kbps of allocated bandwidth for transmission without loss. The middle and bottom signals both require only half that amount (352 Kbps), but the reduction in bandwidth is achieved in two different ways. The bottom signal is the result of the original signal suffering a 50% loss of data packets during transmission. The lost packets cause large gaps, or silence periods, in the signal. The result is poor quality music that is choppy and often annoying or unintelligible to the user. The middle signal is the result of recoding the original signal and lowering the signal resolution by discarding the least important 50% of signal information. Although the middle and bottom signals require the same amount of bandwidth, the effect of the restricted bandwidth on the two signals is very different. We will show in Section 5. that listeners prefer the lower resolution sample to the higher resolution sample with loss. The key, again, is that the reduction in bandwidth is controllable rather than random and that playback remains continuous.

The encoding scheme that we propose in this chapter is called DCT-based Audio Decomposition (DAD). It provides support for bandwidth adaptation in either the multicast or unicast environment. In either case, audio samples are transformed using the Discrete Cosine Transform (DCT), and then separated into groups based on their importance to the signal. Reception of a base group provides minimal audio resolution, and each additional group increases the signal resolution (at the cost of increased bandwidth usage). In the unicast scheme, the size of the base group is matched to the available receiver bandwidth and only the base group is transmitted. When the receiver chooses to reallocate existing bandwidth or to compensate for changes in available bandwidth, she notifies the audio source to change the size of the base group to match the desired data rate. In the multicast scheme, the group sizes are fixed and each group of DCT coefficients is transmitted as a separate multicast stream. The receiver chooses to subscribe to as many groups as she can feasibly receive. When bandwidth adaptation is desired, the receiver may decide to subscribe to an additional group (to increase quality) or cancel a subscription to an existing group (to decrease bandwidth usage).

The rest of the chapter is organized as follows. Section 2. contains background and related work, and Section 3. presents our design. Section 4. presents objective results and Section 5. presents qualitative results from listening tests. Section 6. describes the implementation of the DAD unicast and multicast tools and Section 7. concludes the chapter.

2. Background and Related Work

DAD is designed to provide audio adaptation in the wireless environment. If we first consider the wireless environment, we can identify several characteristics which make it different from traditional wired networks.

- Low bandwidth
- Varying bandwidth
- Frequent disconnections
- High error and loss rate

Wireless technologies which support end-user mobility tend to provide relatively low bandwidth. Wireless LANs (*e.g.*, 802.11g) provide as much as 54 Mbps of shared bandwidth but provide limited mobility. Microcellular solutions (*e.g.*, Ricochet system) can provide as much as 128 Kbps of shared bandwidth but only in isolated service areas. Cellular phones typically can provide only 9.6 Kbps of reserved bandwidth or less although newer 2.5G phones may provide 384 Kbps or more. In any case, these bandwidth levels are much lower than the bandwidth available on wired networks.

Microcellular and wireless LAN technologies share the available bandwidth among all the wireless users within range of an access point, which can lead to drastic and frequent changes in the bandwidth allocated to a particular mobile host due to the movements of other hosts into and out of the microcell. Signal interference and multipath fading can lead to variation in available bandwidth in cellular systems as well. Mobile users may also experience frequent and lengthy disconnections due to movement behind obstacles or during handoff when moving into a new cell. Finally, disconnections, interference and fading all contribute to much higher rates of packet corruption and subsequent loss than seen on the fixed network. All of these characteristics argue for adaptation in real-time applications.

2.1. The Multicast Backbone

IP multicast is an efficient way to deliver the same data stream to multiple recipients. Rather than sending the same packet over a link multiple times (once for each recipient), IP multicast relies on intelligent nodes, called multicast routers, to duplicate packets when necessary. In other words, a single copy of a packet is transmitted until it reaches the point where two destinations lie in different directions. At that point, the packet must be duplicated and a copy sent in each direction. As additional functionality is required at the router, not all routers provide support for routing multicast packets. The set of routers which do support multicasting make up the multicast backbone or mbone.

Many conferencing tools have been designed to take advantage of the mbone, including vic and ivs (video), vat and rat (audio), and sdr (session directory). Most of these applications depend upon the Real-Time Protocol (RTP)[1], which is used to support continuous media data transport in the Internet. RTP is made up of two components: a real-time data transfer protocol (RTP) and a control protocol (RTCP). The data transfer protocol prepends an RTP header which includes timing information necessary for playback at the receiver. The Real Time Control Protocol (RTCP) performs Quality of Service monitoring, intermedia synchronization, and participant identification.

2.2. Audio Coding

Three phases of audio coding may be identified, sampling, quantization, and encoding. In the sampling phase, instantaneous measurements of the audio signal are taken. In the quantization phase, sample values are rounded to the nearest of a finite set of reference values. In the encoding phase, the quantized samples are mapped to an efficient representation for transport (or storage). Sampling must be done at a rate which is twice as high as the highest frequency component to be represented. As such, wideband audio (*e.g.*, music) is sampled at a much higher rate than telephone quality speech. CD quality audio uses 41,100 samples/sec where each sample is quantized to 65,536 possible reference values. These samples are then mapped to a 16 bit representation. A single uncompressed CD quality audio channel requires 705.6 Kbps while stereo audio requires 1.411 Mbps.

As part of the encoding phase, quantized audio samples are commonly compressed to reduce the bandwidth used by the data stream. There are many algorithms and standards for compressing speech and general audio, using lossy compression methods. These include PCM and GSM for voice, and MPEG I, layer 3 for general audio.

2.3. Related Work

As described in Section 1, bandwidth adaptation by the source is not appropriate for the multicast environment due to the heterogeneous requirements and capabilities of the receivers. Hence, we concentrate here on adaptive solutions which can be used in either the multicast or unicast environments. Most of these solutions fall into one of two categories, either transcoding or hierarchical coding, both explained below.

Several transcoding schemes for video and audio streams have been proposed. These schemes are usually oriented towards the multicast environment although they can easily be modified to be used in the unicast environment as well. When using transcoders to provide bandwidth adaptation, a proxy or gateway which is located close to the receiver intercepts the original data stream, reconstructs it, and recodes it to a lower data rate (and quality). The rtpgw[2] provides a transcoding service for H.261-based video streams which operates on a per-packet basis. Audio streams, however, were not addressed. In [3], a scheme called Self Organized Transcoding (SOT) was proposed to support multicast audio transmissions. The transcoding service was provided by the Robust Audio Tool (rat), and SOT provided the control mechanism. SOT forms groups of receivers who have similar bandwidth requirements. A representative of the group contacts a transcoder and requests service. The transcoder then receives feedback regarding the current loss pattern that the representative is experiencing. The transcoder may then adapt to current network conditions by increasing or decreasing the data rate.

The transcoding solution in general has several problems. The transcoding process increases end-to-end delay, which can cause significant problems for real-time applications. Multiple transcodings will increase this delay even more. If transcoding is done on a per-receiver basis, a large amount of processing power must be available within the network to do the transcoding and significant messaging must be done to locate and control the transcoders. If transcoding is done on a group basis, groups of receivers with similar characteristics must be formed (as in SOT). If these groups are not geographically disjoint, then receivers from multiple groups may be located on the same subnet, requiring multiple data streams of differing bandwidths to be delivered to the subnet, actually wasting bandwidth. This problem is especially troublesome in the wireless environment where receivers in the same cell may be allocated different amounts of bandwidth due to reception characteristics, battery power, or the CPU power of the mobile device.

Hierarchical encoding has been used successfully in many image and video applications and more recently has been applied to audio applications. Hierarchical encoding allows each user to balance level of quality with bandwidth usage or download time. Data transmission is stopped when the desired level has been received. Hierarchical image encoders often use transformation techniques such as the Discrete Cosine Transform (DCT) or wavelets to transform the digitized samples into a new representation. Larger magnitude transform coefficients represent average or coarse characteristics while smaller coefficients add detail. Hierarchical encoding involves organizing the transform coefficients based on overall importance to reconstruction quality. The coefficients which contribute most to reconstruction, generally those representing average characteristics, are transmitted first with detail coefficients following.

The MPEG I layer 3 audio standard [4] uses a modified DCT to aid in the identification of the coefficients that contribute most to the quality of the audio signal. In particular, it feeds the DCT coefficients into a psychoacoustic model and then maps the results to the signal output of a subband filter. Subband samples that match the DCT coefficients identified as most important are encoded for transmission. The MPEG audio standard, however, is a fixed rate coder, is not hierarchical, and cannot dynamically adapt to available bandwidth.

In [5], Wang proposes a method, Layered Unequal-Sized Packetization (LUSP), to divide an MP3 audio stream into a base layer and several enhancement layers based upon the frequency ranges represented by the individual DCT coefficients. Seven unequal sized layers are proposed which could then be used to support adaptation to bandwidth-restricted destinations. This decomposition scheme is similar to one we propose below (first n) in that the coefficients are assigned to groups based on their position in the DCT frame rather than their magnitude. Also described is a sophisticated interleaving scheme, Multi-Stage Interleaving (MSI), which is used to counteract the effects of loss, and which does not require that the coefficients be decompressed before error concealment is done. No mechanism for providing sender or receiver-based control of the number of layers delivered is described, however. In addition, the author states that the scheme will increase delay by a few hundred milliseconds to a few seconds making it unacceptable for real-time applications.

The MPEG-4 natural audio standard [6] uses the Advanced Audio Coding (AAC) method for encoding high quality general audio signals. This technique is similar to MP3 in that it uses a subband filter, modified DCT, and psychoacoustic model. The standard introduces a new bitstream format, however, which uses hierarchical coding to allow adaptation during transmission or at the receiver. A base layer of encoded audio samples is created along with two to three residual enhancement layers. These layers are encoded with Bit Slice Arithmetic Coding (BSAC). BSAC is a bit plane encoding technique in which groups of samples are encoded, with the most significant bit of each sample placed early in the frame and successive bits placed in successive frames. This allows the user to choose the audio rate she desires and to receive bit plane frames equal to that rate. This type of hierarchical coding, however, allows for only a maximum of two to three coding layers [7] which does not provide enough levels of bandwidth usage to match the wide range of capacities seen in wireless networks. In [8] and [9] variations of this approach have been studied.

In [10], the authors propose an embedded source coding algorithm called ZeroChain Quantization (ZC) which is based upon Embedded Zerotree Quantization, a wavelet compression algorithm. As ZC is fully embedded, when the transform is applied to the audio stream, the most important coefficients are generated first and may be assigned to the base layer. Coefficients which are less important to signal reconstruction are generated later and assigned to enhancement layers. Our best n scheme, described below, takes a similar approach. ZC can provide finer-grained bandwidth scalability in the unicast environment as a prefix of any length of the coefficient stream may be transmitted and decoded. The authors also describe an error recovery scheme called Internet Modulation in which they append progressively lower resolution copies of each packet to subsequent packets.

A layered approach to delivering video streams which supports adaptation controlled by the receiver is proposed in [11]. The authors describe a hybrid DCT/wavelet based video coder which is combined with a layered multicast scheme, Receiver-driven Layered Multicast (RLM). The video coder is aimed at supporting heterogeneous networks and stripes data across the layered streams, which are then multicast, to provide the various bandwidth levels. A receiver measures its own reception statistics to detect insufficient bandwidth. If found, the receiver drops a stream subscription via IGMP to reduce its bandwidth needs. Detecting unused excess bandwidth, implying that a stream subscription should be added, is more problematic. The RLM algorithm specifies that receivers should periodically perform *join experiments*, where they add a stream and quickly decide if the adaptation has caused congestion. A key aspect of the protocol is that it must scale to large numbers of users. Hence join experiments use a backoff technique similar to TCP to ensure that they are not performed too often (which would cause congestion). Receivers also may use the results of the experiments of other nodes to further reduce the impact of experiments. This process is called *shared learning*. The authors have implemented the RLM protocol using a layered video encoder but have not addressed audio. Also, since the adaptation process is fairly slow to respond, and humans are more sensitive to the impact of losses in audio streams than video, the adaptation scheme will not be able to provide adequate quality in the wireless audio environment.

In [12], Turletti proposes a layered transmission scheme for audio, specifically speech. A temporal decomposition is used to divide the original data stream into layers. In one example, an 8 KHz stream was divided into three 2.7 KHz streams by assigning each sample from a group of three samples to a different stream. An adaptation mechanism was also proposed. Rather than requiring explicit or implicit communication between receivers as in RLM, bandwidth usage (and hence number of streams subscribed to) is based upon the amount of bandwidth an equivalent TCP connection would use. An individual receiver measures its current loss rate and round trip time to the source and uses these values to calculate the equivalent TCP bandwidth. It then subscribes to streams or drops subscriptions to match this bandwidth. The temporal decomposition scheme, however, does not differentiate between samples which are important to playback quality and those which are not. As a result, even when a fairly large proportion of groups are subscribed to, the quality of the signal may be low. In addition, when only a small percentage of the total number of groups is subscribed to, long strings of samples will be missing. Even using interpolation, this will lead to discontinuities in the audio stream which will lower playback quality and may be detected by the receiver. We believe that decreased per-sample resolution, which allows a continuous stream of samples to be received, is preferable. We will examine the performance of a subsampling encoder like Turletti's in Sections 4. and 5..

3. Design

We now turn to our own hierarchical DCT decomposition (DAD) scheme. There are three main components of DAD. These include a decomposition scheme, a mechanism for layered transport, and an adaptation model.

3.1. Decomposition Scheme

In a hierarchical coding scheme, samples are decomposed based on their importance to the signal. The most important values are sent in a base group and values which add levels of resolution are sent in auxiliary groups. In the unicast environment, the source may adjust the sizes of the groups to match the requests of the receiver. The most straightforward solution is to size the base group to exactly match the specified bandwidth and to transmit only the base group as the others are unnecessary. This will help to reduce complexity at the receiver, which is typically a low-powered portable device. In the multicast environment, group sizes are typically fixed and all groups are transmitted as adaptation is left to the receiver. The sound data may be divided into groups exponentially or linearly. In either case though, the base group should correspond to some minimal level of resolution that requires only a small amount of bandwidth to transport so that all possible receivers may feasibly accept the associated stream. Using a linear decomposition, each auxiliary stream adds some multiple of the base group's resolution at a linear cost in bandwidth. This approach is likely inappropriate for supporting CD-quality audio as a large number of individual streams would be required. It is appropriate, however, when the original signal bandwidth is fairly low. For a 128 Kbps audio stream and a base group and auxiliary group size of 16 Kbps, only eight groups are needed to fully represent the signal (see Figure 2b). It may also be appropriate to use linear decomposition if it is known that all receivers differ in bandwidth allocation by only a small amount. The auxiliary streams could then be tailored to provide the best fit to this set of receivers. This may be possible in a conferencing environment where the set of receivers is controlled or if it is known that all receivers are connected via a specific connection type (*e.g.*, CDPD).

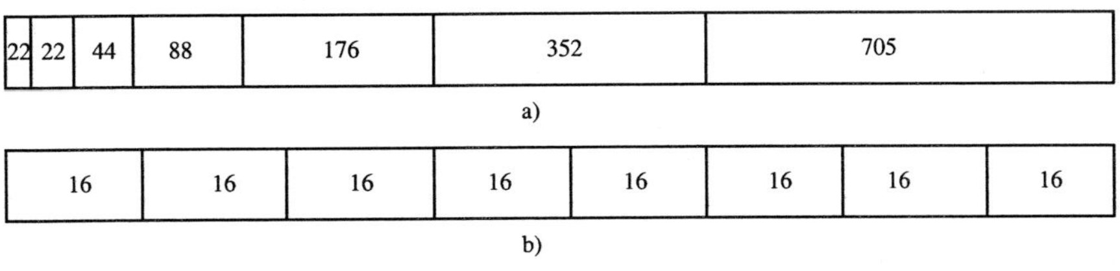

Figure 2. a) 1.411 Mbps audio stream decomposed into 7 substreams using exponential decomposition b) 128 Kbps stream decomposed into 8 substreams using linear decomposition

Using an exponential decomposition, each auxiliary group added would double the resolution of the reconstructed stream at a cost of doubling the required bandwidth. This scheme is appropriate when the original audio stream requires high bandwidth (*e.g.*, stereo CD-quality audio) and when receivers may be connected by a mix of high (*e.g.*, Ethernet) and low-speed (*e.g.*, wireless) interfaces. Using the exponential scheme will limit the number of streams required in order to fully represent the signal. For instance, for a stereo CD-quality stream running at 1.411 Mbps, and using a base group of 22 Kbps, only seven streams are required to represent the entire original signal (see Figure 2a). We have chosen to use an exponential decomposition in the DAD project.

The DCT-based Audio Decomposition (DAD) encoding scheme works in this way (see Figure 3):

1. The source audio stream is broken into frames. The frame size is chosen to ensure that packetization delay is not excessive.

2. The Discrete Cosine Transform (DCT) is applied to the frame. The DCT transforms the samples allowing the coefficients which represent the most signal energy to be easily identified.

3. (a) In the unicast case, the source selects a set of coefficients and transmits them to the receiver along with an indication of which coefficients from the original DCT frame are present. The size of the set is based upon feedback from the receiver regarding desired bandwidth usage.

   (b) In the multicast case, each DCT coefficient is assigned to one of a set of hierarchical streams. The streams are multicast independently along with an indication of which frame coefficients are present.

4. At the receiver, in either case, the coefficients are collected, the reverse transform is applied, and the audio is played out.

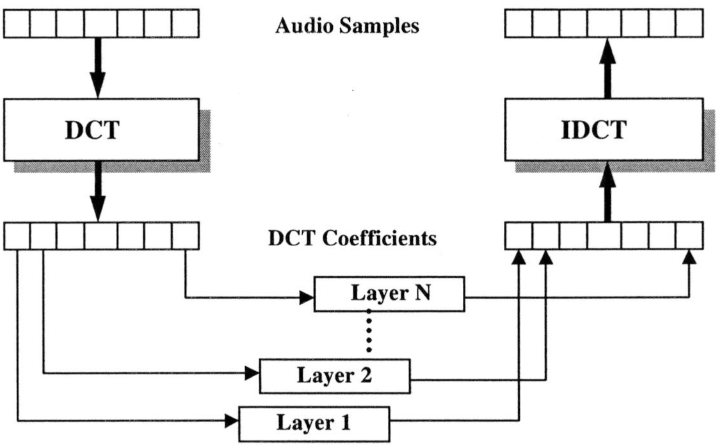

Figure 3. Flow chart of DAD mechanism

We have experimented with various methods of assigning the DCT coefficients to a data stream. In one, which we call *first n*, the samples are sequentially assigned, starting at the beginning of the frame, first to the base stream, then to the next higher level stream, and so on. The order of the coefficients determines their assignment to a group as opposed to their values. Since low frequency components map to the beginning of a DCT frame, however, and typically most of the energy of a signal is contained in the low frequency components, it is likely that the first n scheme will assign the most important coefficients to the base group as desired. The first n scheme requires no additional transmission overhead to specify the positions of the DCT coefficients in the reconstructed frame and requires only minimal processing at the source and destination to select the coefficients from the original frame and reconstitute the frame respectively. Both of these factors are important when dealing with low bandwidth wireless links and the slow CPUs typical of handheld wireless devices.

In a second scheme, called *best n*, the coefficients with the most energy (largest magnitude) are located and assigned to the base stream. Coefficients with less energy are assigned to higher order streams. In this case, the most important coefficients are sure to be assigned to the base stream, but at the cost of additional overhead. Some indication of the relative positions of the transmitted coefficients, (*e.g.*, a bitmap), must be included in the data stream, and it is important to ensure that the position information does not represent a significant portion of the data stream. Consider that a 1,024 sample frame of 16 bit samples would require 2,048 bytes (uncompressed) and the accompanying bitmap would require 1,024/8 = 128 bytes or 6.25% of the data size. That ratio might be considered acceptable, however, the frame size stays the same even as the number of coefficients sent becomes smaller. If a receiver could accept only 10% of the coefficients using available bandwidth, the size of the collapsed frame would be only 2,048*0.10 = 204.8 bytes and the bitmap would represent 62.5% of the size of the data. Clearly, this is unacceptable. It may be observed that a sparse frame also leads to a sparse bitmap. Hence traditional compression algorithms such as run-length encoding should be very successful in reducing the size of the bitmap and this scheme may well be practical. A second issue is the added complexity which is required in order to find the best n coefficients. Finding the n coefficients takes O(n) time using the first n scheme and no more than O(nlogn) using the best n scheme depending on the algorithm used. As coding is done at the source, however, which is likely a relatively powerful machine on the wired network, the added processing should not prove to be an issue. Reconstitution of the frame at the receiver is only somewhat more complex than the first n scheme.

Regardless of the mechanism used to select samples or coefficients to be transmitted, it is clear that compression techniques must be applied to reduce the bandwidth required for transmission. We do not address the type of compression to be done after the fact, but rather focus on the problem of selecting the most appropriate set of samples or coefficients to undergo compression. In Sections 4. and 5., we will compare the DAD scheme to other possible schemes for choosing the most appropriate set of samples.

3.2. Layered Transport

In the multicast scenario, traditional IP multicast is used to transmit the base and auxiliary streams on distinct multicast addresses. A receiver must subscribe to the multicast group associated with each address in order to receive the data stream. This subscription is achieved via the Internet Group Management Protocol (IGMP)[13]. An IGMP join message is sent to the local multicast router which then begins forwarding the requested stream. IGMP is well suited to serve mobile hosts as the spontaneous appearance or disappearance of a mobile host (due to movement into or out of a cell) may be easily handled. A multicast router forwards a data stream down a link if there are one or more receivers currently registered to receive the stream in that direction. Hence a mobile which moves into a new cell may be able to immediately receive a subscribed data stream if another receiver has already registered for that stream. If not, the mobile must register itself to begin the data flow, which implies delay and a possible gap in the audio stream. In areas with high user density, the former case should be the common case. In areas with low user density, the technique suggested in [14] may be used to decrease the probability of temporary disconnection. Singh suggests that the same stream be multicast to a set of cells surrounding the mobile's current cell. Then, when the mobile moves into one of these cells, it immediately begins receiving data.

Regarding movement out of a cell, a mobile may not have the opportunity to drop its multicast subscriptions before moving into a new cell. As part of their operation, however, multicast routers periodically query their registered receivers to make sure that they are still interested in the streams which are being forwarded. Therefore the mobile's departure will be detected automatically and the multicast router can stop forwarding the stream (if no other receivers are present).

In the unicast scenario, only the base layer is transmitted, making it a degenerate case of the multicast scenario. In addition, since direct communication between the source and receiver is practicable (as there is no scaling issue), there is no need to involve multicast routers. The source transmits a unicast stream to the receiver and mobility may be handled using standard Mobile-IP.

3.3. Adaptation Model

When considering an adaptation model, the first question that must be answered is: At what granularity should adaptation be done, in terms of the duration of the change? Attempts to adapt to very short-term bandwidth variations (on the order of milliseconds) may be ineffective or even detrimental to connection quality. This is due to the likelihood that the adaptation mechanism will not be able to keep up with the rate of change in the allocated bandwidth, either because the adaptation mechanism itself has a lag time or because the feedback required to make the adaptation decision is not available in time. There are, however, many opportunities to capitalize on coarse-grained bandwidth adjustments. Coarse-grained adaptation attempts to match application bandwidth usage to changes in available bandwidth which last seconds or minutes, rather than milliseconds. Consider the following scenarios:

An Internet telephony application user is connected to the Internet via a micro-cellular wireless data network. In such a network, wireless devices common to a microcell must share the bandwidth available there. As such, user movements in and out of the cell, as well as user decisions to launch or terminate applications will cause the number of active connections within the cell to vary. As the number of connections varies, the bandwidth available for each will also vary. The bandwidth changes will occur at intervals of several seconds or longer, however, as they are the result of the actions of a small group of humans.

Overlay networks [15] in the wireless environment allow a wireless client to switch from one type of wireless network to another as availability dictates. A wireless device might be equipped with several transceivers, one for an in-door high-speed wireless LAN, one for a lower speed CDPD cellular network, and one for a very low speed satellite connection. As the user moves from one network to another, (which might be required when they leave a building or leave the coverage area of a cellular system), the amount of bandwidth available to their connections will change. These changes will likely occur in intervals whose size is on the order of minutes.

Consider that, much like current voice networks, it is likely that in the future, connections routed over the Internet will incur a cost based on the time of day, network load, requested bandwidth, or roaming area. Users may then desire the ability to specify the data rates to be used for general audio and speech depending on the cost incurred. For example, a high quality, high rate connection would be desirable for a CD-track download while a lower quality, lower rate connection would be acceptable for an audio conference. A user might also prefer CD-quality radio when located in her own service area, but accept lower quality when roaming charges apply. These factors will require adaptation for timeframes on the order of minutes or longer.

Finally, when we consider a conferencing environment, users commonly employ multiple conferencing elements, *e.g.*, video, audio, shared whiteboard, shared application, etc. As these elements are launched or dismissed, even if the total bandwidth available to a client stays the same, the bandwidth available to any one application will vary. Bandwidth changes here will again occur on the order of seconds or minutes.

Given then that there are many opportunities to provide adaptation at a coarse granularity, the overall goal of the adaptation mechanism must be determined. Goals might include:

- Maximizing the amount of bandwidth used by the receiver within the constraints of the bandwidth available
- Minimizing the cost of the connection
- Reducing the delay experienced by a mission-critical application by decreasing the bandwidth provided to other applications
- Providing a lower bound on the bandwidth provided to a particular application at the expense of others

The adaptation mechanisms proposed in [12] (described above in Section 2.3.) and [11] both aim at yet a different goal, that of making a continuous media application act like a "good neighbor." They either explicitly probe the network and back off if congestion is detected or attempt to emulate the behavior of a TCP connection, which has a similar backoff strategy. Both mechanisms, however, will use the maximum bandwidth possible when congestion is not encountered and may provide inadequate bandwidth when it is. What is missing in this adaptation decision is *user preference*. A user may well wish to use less than the maximum bandwidth available due to cost constraints, may wish to trade off audio bandwidth in favor of another application (*e.g.*, FTP or application-sharing), or may not choose to use an application at all if certain minimum bandwidth requirements cannot be met. We have proposed that profiles be created that contain user preferences such as the priority of one application over another or the maximum amount of bandwidth that a user is willing to pay for on an application basis. Some examples include:

- QoS(1) = preferred bandwidth for voice
- QoS(2) = preferred bandwidth for video
- QoS(3) = preferred bandwidth for whiteboard
- QoS(4) = preferred bandwidth for application sharing

- QoS(5) = minimum acceptable bandwidth for video
- QoS(6) = maximum cost for wide-band audio
- QoS(7) = maximum cost for narrow-band audio
- QoS(8) = preference of voice versus whiteboard

Even given a set of preferences, users will likely override automatic adaptation changes based on their current needs. In order to anticipate these needs or to populate the quality of service preferences initially, we have suggested the use of a case-based memory to drive the adaptation, see [16]. Over time, a user will tend to use the same set of applications in similar ways. A case-based memory can be used to capture the decisions that the user makes regarding cost and bandwidth tradeoffs between concurrent applications when they engage in a conference. After several training sessions, the case-based memory system could identify scenarios similar to ones seen previously and, using the user preferences and past behavior as a guide, make the adaptation decisions automatically. Figure 4 shows the components of the model including the conferencing applications, the quality of service filter (preferences), the case profiles, and the interaction with entities on the network itself, both to initiate the adaptation changes and to learn about current network conditions.

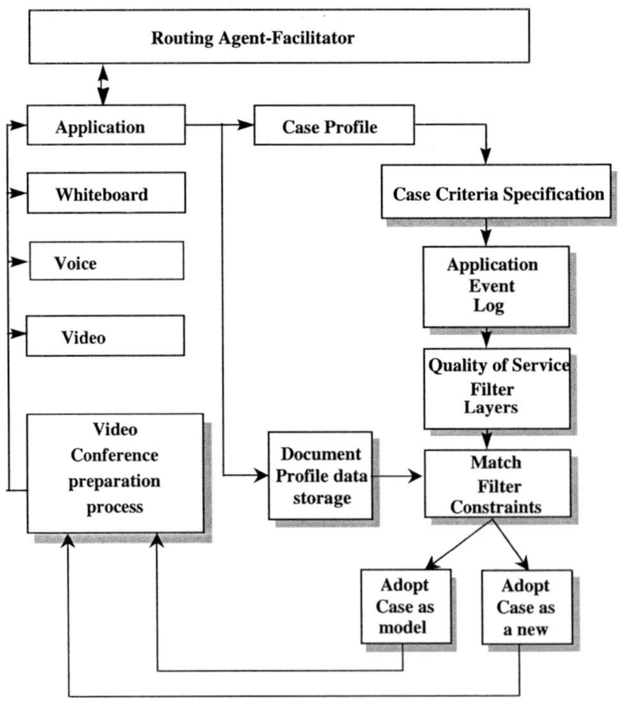

Figure 4. Data model for case memory integration

A case memory could be useful in other scenarios as well. Consider that a mechanism which can learn from past history could make reasonable predictions regarding the duration and severity of future bandwidth fluctuations. It could help to eliminate oscillations (cycles of increasing and decreasing bandwidth usage) and possibly initiate adaptation more quickly when indicators point towards more bandwidth changes.

When significant events which affect available bandwidth occur then, such as additional users joining a network and competing for bandwidth or the launch of a new application, the receiver may choose to adapt. In the unicast environment, the receiver instructs the source to modify its level of encoding so that the bandwidth of the transmitted data stream matches the bandwidth available to the receiver. In the multicast environment, the user communicates with her local multicast router. The receiver can react to significant bandwidth variation events without involving the source by locally subscribing to or unsubscribing from multicast groups. Either adaptation mechanism can be automatically driven by the case memory system or manually driven by giving the user override control.

4. Simulation

We constructed several programs to simulate the behavior of the DAD encoding scheme on sample audio streams. Our goal was to compare the results of the DCT-based schemes against other possible schemes for selecting received samples when bandwidth was restricted. As such, we examined two additional schemes, one in which samples were discarded as a result of packet loss, and another in which subsampling was used to select the samples to be transmitted. The subsampling approach is described above in Section 2.3. and packet loss is described below.

The packet loss scheme consists of subjecting the audio stream to random loss to effect the necessary bandwidth reduction. This is meant to mimic the results of restricted bandwidth on an audio stream if no adaptation were done. In this scheme, the full CD-quality audio stream is packetized and transmitted. When bandwidth is insufficient to carry the stream, entire packets are discarded in order to match the input rate to the available bandwidth. Packet size was set to 512 bytes which, assuming even a modest amount of compression, maps to at least 512 16-bit samples. Note that small packets are a best-case scenario when dealing with loss as the resulting runs of lost samples will be of minimal length. Packet loss was randomly and uniformly distributed. This is again a best-case scenario, as bursty losses would again increase the length of runs of lost samples. As shown in this section and Section 5., even best case results for the packet loss scheme produce unacceptable results, hence it is not necessary to examine average-case scenarios.

When packets are lost, it may be appropriate to attempt to correct for the lost samples. We simulated two different schemes for dealing with lost samples. In one, the lost samples were replaced with 0 values. In a second, the lost samples were replaced with the last successfully received sample. More elaborate schemes also exist, but due to the minimal processing power of many mobile devices, they may not be appropriate. Similarly, adding Forward Error Correction (FEC) may also be inappropriate due to the additional bandwidth required on the already bandwidth constrained wireless link.

The subsampling approach described in [12] was also simulated. While initially proposed for speech streams, we apply this scheme to CD-quality audio. We divided the original audio stream into 10 substreams, assigning the samples from the frame to groups in such a way as to minimize runs of lost samples when few streams were being received. Again, missing samples could be replaced with zero values or the last received sample. Like the DAD approach, the subsampling rate can be precisely specified by the receiver although multicast complexity may limit the number of streams in that scenario.

Finally, the two versions of the DAD encoding were simulated using either *first n* or *best n* schemes for selecting the received coefficients. The DCT transforms 16-bit audio samples into floating point coefficients that cannot be represented in 16 bits. Hence coefficients were scaled and truncated to 16 bit integer values before being reconstituted. Note that this implies that a scaling factor must also be transmitted with each frame unless a fixed scaling factor is used. Lost coefficients were always replaced with 0 values. This is an appropriate substitution as the missing coefficients are typically small in magnitude. In addition, there is no clear relationship between one coefficient and the next making extrapolation from the previous coefficient difficult. Finally the reverse transform was applied and values were again truncated to integer values.

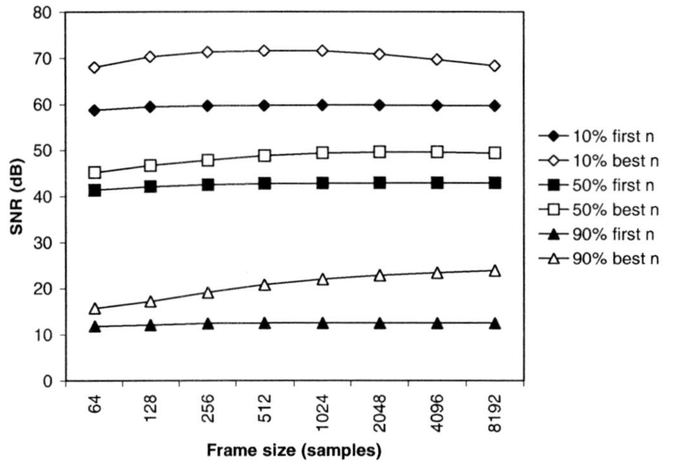

Figure 5. DCT frame size vs. SNR

A frame size had to be selected before our tests could be run. We took a CD quality sample (44,100 samples/sec, 16 bits/sample, mono) and subjected it to various levels of reduced bandwidth (*e.g.*, 10%, 50%, and 90%) using a range of frame sizes from 64 samples to 8,192 samples and calculated the signal to noise ratio (SNR) of the resulting audio clips. This was repeated for both the first n and best n schemes. Results are shown in Figure 5. In all cases except 90% best n, SNR leveled off after reaching a frame size near 1,024 samples. A frame that is too small may not capture local trends sufficiently, however, frames that are too large increase packetization delay. As SNR did not increase significantly for frame sizes over 1,024, we chose to use this size for our sample frames in the remaining tests.

Five CD-quality audio clips, each 5-7 seconds long, were subjected to various levels of bandwidth restriction using each of the three encoding mechanisms and no encoding (which led to packet loss). While the subsampling and DCT approaches to reducing bandwidth usage are deterministic, the packet loss approach requires randomness, which led to actual loss rates that varied slightly from the requested loss rates. In these cases, tests were repeated until the actual loss rate was within 1% of the requested loss rate. For the subsampling and random loss cases, missing samples were replaced with either zero (Subsample zero, Loss zero) or the last successfully received sample (Subsample last, Loss last). The signal to noise ratios (SNR) of the resulting clips compared to the original clips were calculated and used as a rough measure of quality. Informal qualitative observations regarding quality were made as well, prior to the formal listening test described in the next section. In Figure 6, results of the relative performance of the encoding mechanisms are shown for a Mozart piece. The four other pieces consisted of a cappella singing, pop music, rock music, and an opera piece, and produced similar results.

It can be seen that the performance of the packet loss mechanism is very poor, even for small restrictions on bandwidth such as 10%. Dropouts and noise were obvious during listening tests at all levels of restricted bandwidth and restrictions of 30% or greater led to reconstructed streams which were of unacceptable quality. Interestingly, when lost samples were replaced with the last sample rather than 0 values, the signal to noise ratios of the reconstructed signals went down. Listening tests showed that the loss correction added background noise that was more annoying than the dropouts. Note that the two lines for Loss zero and Subsample zero cannot be differentiated, as the results are nearly identical.

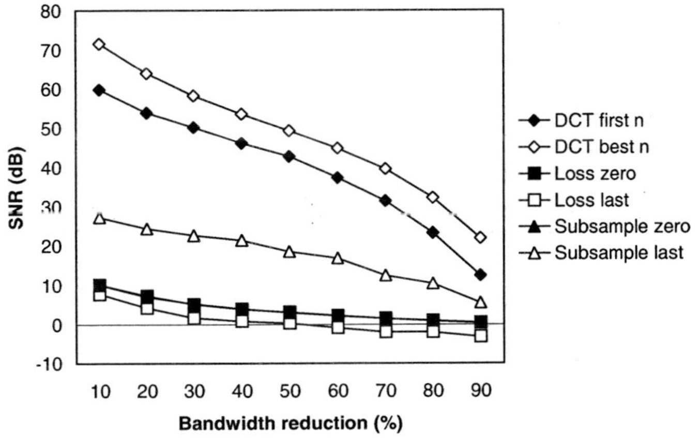

Figure 6. SNR performance of coders

The subsampling mechanism was more successful in retaining audio quality in the presence of restricted bandwidth. When the lost samples were replaced with zero values, the signal to noise ratios of the reconstructed signal were very low, almost identical to the packet loss results. On inspection of the reconstructed signals, it was clear that the frequent returns to zero were creating a persistent high frequency component which led to unacceptable reconstructed samples at any level of restricted bandwidth. When lost samples were instead replaced with the last successfully received sample, results were significantly improved. Signal to noise ratios rose and listening test participants described high quality reconstructed signals for low levels of restricted bandwidth. At 30% restriction, noise is apparent. At 60%, a high-pitched whine is added, and 80%, the whine makes listening very unpleasant. This scheme is, however, a great improvement over random packet loss.

The DAD mechanism provides another increase in the signal to noise ratios of the reconstructed signals. The first n scheme provides SNR of roughly twice that provided by the subsampling approach using last sample replacement over the entire range of restricted bandwidth tested. Subtle losses of quality in the reconstructed signal can be detected from 40% to 60% restricted bandwidth. At 70% restriction, loss of dynamic range begins which becomes more pronounced as restriction increases to 90%. At 90%, a low buzz is also detected and the signal becomes unacceptable.

The best n scheme increases SNR performance even further over the first n scheme adding 7-12 dB to the reconstructed signal across the range of restrictions. Subtle losses of quality were not apparent until 80% restriction was applied. Loss of dynamic range was apparent at 90% restriction, but the signal was still of reasonable quality.

5. Listening Tests

It has been regularly observed that objective measures such as the signal to noise ratio must be complemented with subjective measures when comparing audio coding schemes. Consequently, we also conducted listening tests to provide subjective evaluation of a subset of the coders. We included the DCT best n, DCT first n, subsampling with last sample replacement, and packet loss with zero replacement coders in the tests. Subsampling with zero replacement and packet loss with last sample replacement were clearly unacceptable solutions. We evaluated each of the four coders at 5 levels of bandwidth restriction, (10%, 30%, 50%, 70%, and 90%), on the set of 5 audio samples described in the section above. These combinations provide 100 test samples. A Degradation Category Ranking (DCR) test model was used, in which, for each test, the original sample is heard, then the sample after processing through the appropriate coder is heard. The listening test participants then ranked the second sample on a scale from 1-5, with meanings as shown in Table 1. Samples were placed in randomized order for each participant and total test time was roughly 30 minutes. 40 participants used headphones to listen to the samples.

Table 1. Rank definitions

Rank	Meaning
5	Transparent, no perceptible differences
4	Perceptible differences but not annoying
3	Perceptible and slightly annoying
2	Annoying
1	Very annoying and objectionable

Mean Opinion Scores (MOS) and 95% confidence intervals on the scores were calculated for each coder at each level of bandwidth restriction, as shown in Table 2. The scores support the SNR results, showing that the two DCT-based schemes provide better quality playback than the subsampling scheme, and that all three schemes provide much better quality than random packet loss. Subsampling appears to provide acceptable quality playback for lower levels of restriction, (30-50%) but breaks down at higher levels of restriction. The two DCT-based schemes both provide nearly transparent quality playback at restriction levels of up to 70%. The first n scheme quality drops off at 90% restriction but the best n scheme still provides acceptable quality at that level.

Table 2. Mean Opinion Scores

Coder/ Restriction %	MOS	95% Conf.	Coder/ Restriction %	MOS	95% Conf.
DCTbestn/ 10%	4.89	±0.11	Subsample/ 10%	4.64	±0.19
30%	4.88	±0.13	30%	4.43	±0.22
50%	4.84	±0.15	50%	3.96	±0.24
70%	4.87	±0.11	70%	2.95	±0.24
90%	3.71	±0.26	90%	2.17	±0.25
DCTfirstn/ 10%	4.84	±0.12	Packetloss/ 10%	2.64	±0.23
30%	4.90	±0.10	30%	1.69	±0.21
50%	4.78	±0.13	50%	1.28	±0.16
70%	4.28	±0.22	70%	1.12	±0.15
90%	2.93	±0.24	90%	1.05	±0.13

6. Implementation

We have implemented unicast and multicast versions of the DCT first n coders. Development was done on RedHat Linux 7.2 running on low-end PCs. Existing Linux sound tools were incorporated into the design in order to quickly produce a working prototype. The audio source and receiver both used an efficient and in-place version of the DCT algorithm, and Jrtplib, an open source C++ RTP library, was used to provide RTP and RTCP functionality. Source coding and destination decoding functionality was easily provided in real-time on 350 MHz Pentium processor-based machines. This is an encouraging result as it is important that the process not be CPU-intensive. Wireless devices tend to have slow processors in order to save power.

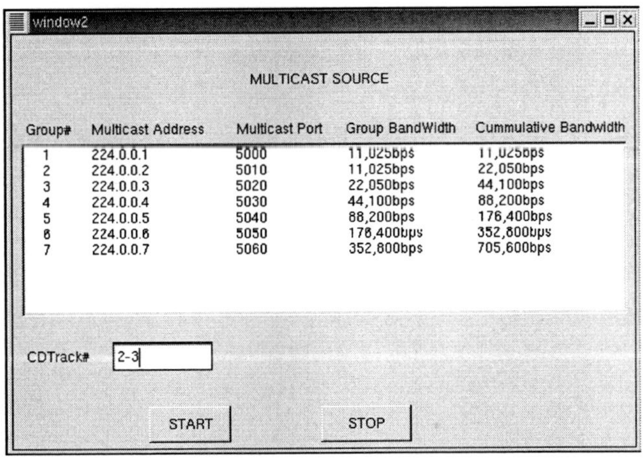

Figure 7. Screen shot of transmitter interface

The source makes use of the Linux cdda2wav tool in order to rip tracks off a compact disc. The sample stream is then rate-limited to 1.411 Mbps using the cstream utility and passed to the coder. The samples are placed in 1,024 sample frames, and the DCT is applied. Coefficients are decomposed into 7 multicast groups based on their position in the frame with consecutive runs of 16, 16, 32, 64, 128, 256, and 512 coefficients being placed into groups 1 through 7 respectively. Coefficients to be sent in the same packet are scaled and rounded to 16 bit integer values and the scaling factor is prepended to the packet. Packets are then sent on the multicast address associated with their group. In the unicast version of the source, an additional UDP port is monitored for commands from the receiver. These commands instruct the source to modify the number of coefficients placed in the base group. The GUI for both versions of the source provides the user the ability to choose tracks from the CD at will (see Figure 7).

The receiver accepts packets on multicast addresses associated with subscribed groups. Coefficients are reassembled into frames, missing coefficients are filled with 0 values, coefficients are scaled up using the scaling factor included in the frame, and the reverse DCT is applied. The reconstructed samples are then sent to the soundcard for playback using the Linux Open Sound System (OSS) library. The multicast version of the destination provides an interface to allow the user to add or drop multicast groups in order to match the bandwidth usage/quality desired (see Figure 8). IGMP messages are then sent to the nearest multicast router in order to subscribe to or unsubscribe from groups. In the unicast version, a message is sent to the source in a UDP packet including the data rate requested.

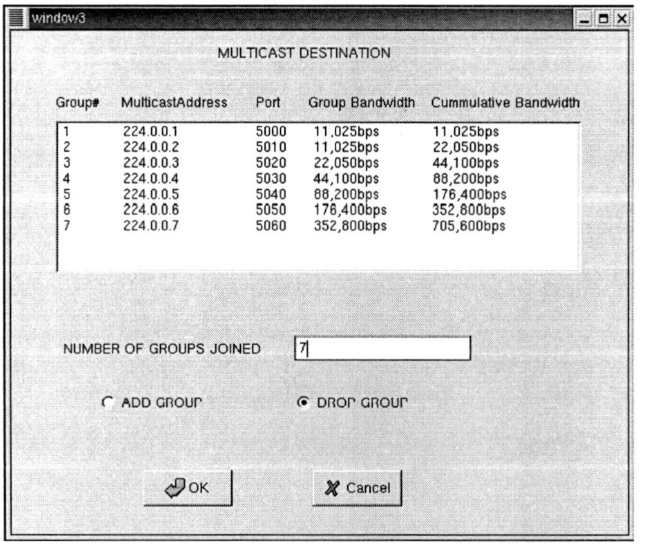

Figure 8. Screen shot of receiver interface

Both implementations of the first n coders performed according to design. In particular, the multicast version of the receiver did not induce detectable delay in the audio stream even when subscribed to all seven multicast streams. Since the multicast version is of greater complexity than the unicast version, both appear to provide acceptable support to real-time audio streams.

7. Conclusion

We have proposed two DCT-based audio encoding schemes which can be used to provide bandwidth adaptation to mobile devices in a multicast or unicast environment. Simulations and listening tests were conducted to show that the audio quality provided by the DAD encoding schemes under restricted bandwidth conditions is superior to the effects of random loss or the controlled adaptation of a subsampling encoder.

The DAD best n scheme provides the best quality reconstructed signal over the widest range of restricted bandwidths. It does, however, require additional processing and additional transmission bandwidth in order to include the positions of the selected coefficients. As a result, the first n scheme may be preferable, as its performance is comparable while not requiring the additional processing or bandwidth. The subsampling scheme requires even less processing than the DAD schemes but provides acceptable signal quality only for low levels of restriction. Similarly, listening test results show that the lack of an adaptive mechanism, leading to random packet loss, results in an unacceptable signal at any level of restriction tested. It is clear that source selection of samples to be transmitted leads to higher quality reconstructed signals than allowing the network to make those selections based on available bandwidth.

The DAD encoders provide a high quality reconstructed signal for bandwidth restriction levels of up to 80% and are low in complexity. As a result, they can be used to provide adaptable real-time audio streams to wireless and mobile devices.

References

[1] Schulzrinne, Casner, Frederick, and Jacobson, *RTP: a Transport Protocol for Real-Time Applications*, RFC 3550, July 2003.

[2] Amir, McCanne, and Zhang, *An Application-level Video Gateway*, ACM Multimedia'95, San Francisco, CA, November 1995.

[3] Kouvelas, Hardman, and Crowcraft, *Network Adaptive Continuous-Media Applications Through Self-Organized Transcoding*, Proceedings of Network and Operating Systems Support for Digital Audio and Video (NOSDAV'98), Cambridge, UK, July 1998.

[4] Brandenburg, *MP3 and AAC Explained*, Audio Engineering Society (AES) 17th International Conference on High Quality Audio Coding, Florence, Italy, September 1999.

[5] Wang, Huang, and Korhonen, *A Framework for Robust and Scalable Audio Streaming*, ACM Multimedia 2004, New York, New York, 2004.

[6] Brandenburg, Kunz, and Sugiyama, *MPEG-4 Natural Audio Coding*, Signal Processing: Image Communication, Vo. 15, No. 1, January 2000.

[7] Herre, et al., *The Integrated Filterbank Based Scalable MPEG-4 Audio Coder*, 105th Convention of the Audio Engineering Society, San Francisco, 1998.

[8] Dunn, *Efficient Audio Coding with Fine Grained Scalability*, 111th Audio Engineering Society Conference, New York, New York, September 2001.

[9] Zhou, et al., *Error Resilient Scalable Audio Coding (ERSAC) for Mobile Applications* , IEEE Workshop on Multimedia Signal Processing, October 2001.

[10] Leslie and Sandler, *Packet Loss Resilient, Scalable Audio Compression and Streaming for Wired and Wireless IP Networks*, 2nd International Conference on 3G Mobile Communication Technologies, pp. 119-123, March 2001.

[11] McCanne, Jacobson, and Vetterli, *Receiver-driven Layered Multicast*, ACM SIGCOMM'96, Stanford, CA, August 1996.

[12] Turletti, Parisis, and Bolot, *Experiments with a Layered Transmission Scheme over the Internet*, INRIA Research Report No. 3296, November 1997.

[13] Fenner, *Internet Group Management Protocol, Version 2 IGMP*, RFC 2236, November 1996.

[14] Seal and Singh, *Loss profiles: A quality of service measure in mobile computing*, Wireless Networks, Volume 2, Issue 1, pp. 45-61, 1996.

[15] Varshney and Jain, *Issues in Emerging 4G Wireless Networks*, IEEE Computer, Volume 34, No. 6, pp. 94-96, June 2001.

[16] Bordetsky, Brown, and Christianson, *Adaptive Management of QoS Requirements for Wireless Multimedia Communications*, Journal of Information Technology and Management, Volume 4, pp. 9-31, January 2003.

In: Adaptation Techniques in Wireless Multimedia Networks
Editors: Wei Li and Yang Xiao, pp. 131-190
ISBN: 1-59454-883-8

Chapter 8

MODELING ADAPTIVE SELF-ORGANIZING WIRELESS NETWORKS: METHODOLOGY AND EXPERIMENTS

James P. Davis, Kevin Walker, Achraf El Allali
Department of Computer Science and Engineering
University of South Carolina, Columbia, South Carolina 29208, USA
Michael L. Haggard
Signal Corps
United States Army, Arlington, VA 22201, USA
Cherrie Yuen
Blackbaud, Inc.
2000 Daniel Island Drive, Charleston, SC 29492, USA

Abstract

An important impediment in the adoption of mobile computing as a dominant paradigm in many industries has been the human "overhead" required to manage the configuration, tuning and optimization of mobile resources in wireless networks. The mere fact that mobile devices and resources are in motion—where each is denoted as a station on a link-layer wireless network—means that the topology is not fixed, but dynamic and subject to uncertainty regarding the ability to reach all stations at any given time. Furthermore, given that the stations communicate over a wireless medium means that the network channel is subject to random factors—noise, unpredictable congestion-causing traffic load, and adversarial threat—all *environmental pressures* affecting the *fitness* of the wireless network. Here, we denote fitness in terms of network reliability, robustness, security and integrity, performance, quality of service (QoS), and related metrics of network operation. The question we consider in this chapter is the following: What type of framework and associated design methods can we adopt that allow us to explore a range of adaptive behaviors allowing application-specific wireless networks to operate in a self-managed mode in the face of unpredictable traffic load and random (or malicious) disruption of channel integrity? We discuss a framework for exploring adaptive, self-organizing behaviors in wireless networks that involves a mapping between the problem domain using conceptual analysis models, and simulation domains using models representing both the network as a whole, and individual nodes in the network. It is this ability to model workflow-level, network-level and node-level behaviors in a co-design and co-simulation context that allows us to explore introduction of adaptive protocol changes into the nodes, so that the effects of self-organizing behaviors exhibited across the entire wireless network (as evidenced by appropriate fitness metrics) can be evaluated.

Keywords: 802.11 WLAN, adaptive systems, reconfigurable computing, network simulation, hardware co-simulation, intelligent agents

1. Introduction

1.1. Overview of the Research

Wireless LANs are becoming a means for communities of users to set up ad-hoc networks, to offer services, coordinate transactions or provide support for various business process operations. Operating in unregulated frequency bands of the radio spectrum, wireless LANs using the IEEE 802.11 protocol family are easy to set up and manage, particularly when the nodes are fairly stationary. However, when the nodes are moving relative to one another in an environment with unpredictable characteristics, there are security management, traffic congestion management, and resource management issues that must be dynamically addressed. If wireless networks using 802.11 WLANs are to remain robust under such conditions—i.e., not incurring considerable degradation in QoS and throughput or total disruption of service—they must be able to work without requiring considerable and constant management oversight.

In business domains where networks are formed around a stationary access point or server node, an individual or service provider can assume this management responsibility. For other applications of wireless network technology, it is simply not possible to have one or more individuals available to provide this service. Furthermore, if the topology of the network is changing, such that it is difficult for a human to respond quickly enough to prevailing environmental conditions, human remediation cannot be delivered effectively. What is required is for the network nodes, themselves, to monitor prevailing conditions and respond in such a manner that, collectively, they can self-manage and self-regulate themselves with minimal human oversight and without a centralized control node.

In the research described in this chapter, we apply principles of adaptive, self-organizing systems to this problem of distributed network management, exploring some initial solutions using these techniques. Wireless LAN nodes constitute a population of individuals that exhibit behaviors defined by a rigid protocol. What if this protocol were not so rigid, but adaptive along several dimensions, where each wireless station (namely, the behaviors embodied in the MAC layer node) were able to learn and adapt to changing conditions in its environment? What if these stations were able to share more than simply frame information, but also meta-level information governing their internal state and what they have learned from their local environmental "neighborhood"? Would it be possible to create behaviors in individual wireless nodes such that the behavior of the wireless network as a whole was capable of "intelligence"?

In this chapter, we present a methodology for exploring a range of adaptive behaviors in wireless networks, conducted in a problem domain where there are many stations moving relative to one another in unpredictable terrain. First, we present background of the problem domain. Next, we discuss modifications to the 802.11b MAC-layer protocol, such that we can extend the range of behaviors available to the MAC layer, and thus to the network as a whole. We also discuss how we subject this set of designs to experimental analysis. Our interest is seeing whether it is possible for individual MAC nodes to learn from their environment--consisting of conditions of the medium, information shared by surrounding nodes, and facts it is able to deduce from its own internal state. In order to explore this problem, we discuss it in context of a problem domain that has some interesting characteristics.

1.2. Problem Domain

Recent efforts by the Department of the Army to transform the land-based fighting force into a more deployable, agile, and versatile unit on the battlefield have led to a shift in the required technology footprint underpinning its operations [Shin99, Army01a, Noon02]. One of the components envisioned to support this change in the Army's deployment model is Land Warrior [Army01b], a collection of systems using an 802.11 wireless LAN structure as its backbone for battlefield communications [Maym02].

In the battlefield environment, the fighting force comprising the network is envisioned to constantly be in motion [Maym02], and the network must be able to adapt to changes affecting network quality. Some metrics—such as dropped nodes, dropped links and noise—give a direct indication of the quality of service (QoS) at any given time during the operation of the force in the battlefield. Adaptation and self-regulation of the network are especially important in this context, as the soldiers on the battlefield may not have the time, nor the skill-set, necessary to perform the ongoing WLAN management tasks to ensure robustness and QoS as the environment changes.

The Army has specified the Land Warrior system to assist the small infantry unit with two tasks crucial to winning an engagement [Army01]: pre-arranging the conditions on the field; and, striking the enemy with a decisive maneuver. For the first task, the small unit leaders must have real time access to all information about the upcoming encounter in order to develop the situation prior to contact with the enemy. A majority of the information is available from sources higher in the command structure, and can include information on terrain, obstacles, and composition and disposition of enemy and friendly forces. The primary requirement of the communications system for pre-arranging these conditions is rapid, efficient, and correct assimilation of information from multiple sources in different formats [Army01].

Just as the first task depends on information transfer and assimilation for a small unit to develop a plan allowing them to create the best conditions for mission success, the second task also depends on real-time information access to allow the unit to maintain superior situational awareness when striking the enemy. A majority of this information comes from the soldiers actively engaged in the field. The unit leader's access to the information developed while in enemy contact allows decisive maneuvering of subordinate units to positions maximizing their advantage in the field [Army01].

The communications system for Land Warrior is a wireless Local Area Network (WLAN) with the Mesh Radio providing the communications medium [Maym02]. Each soldier is a node in the network, which allows any soldier's communications system to route packets to any other soldier without the need to transmit to a central base station. Soldiers are initially grouped into squad networks. Two or three squads and the platoon leadership combine to create the platoon network. Accordingly, two to three platoons and the company leadership combine to create the company network cloud. This process continues up to battalion level and above.

The Army has, as an overall objective, to improve the infrastructure of the communications network through the development of self-healing systems, automatic fault detection and dynamic bandwidth adjustment capabilities to increase seamless connectivity. The infrastructure improvement is part of a larger effort to transform the force from one where information has previously been "stovepiped" in the organization in which it was created, to one where information is shared via a common network and transformed into knowledge the commander can use to make timely and accurate decisions [Noon02].

Given that the Land Warrior system is based on WLAN technology, this WLAN supporting an infantry fighting force must be able to support real-time information access required by all soldiers in the force at acceptable levels of service, throughput, and reliability. Traffic includes graphic intensive map overlays, voice transmissions, operations orders, soldier position information, soldier status, equipment status, and intelligence reports. All of this traffic must be correctly and efficiently delivered through the network despite topology changes caused by node movement, interference, congestion, or lost nodes [Noon02]. The Army is also interested in having the wireless station nodes being able to adapt to topology changes without human intervention (i.e. without requiring extensive network administrative function performed by a human) [Noon02].

The mobility and flexibility advantages of a wireless network are well suited to the needs of a mobile organization. First, the soldiers in the field benefit via increased mobility while remaining plugged into the network. Second, the flexibility of an ad-hoc wireless network allows the soldiers to quickly "task-organize", where a mission commander will organize elements from different subordinate units into a unit with the required resources to execute a particular mission. Flexibility of ad-hoc connectivity also supports final coordination meetings among subordinate unit leaders and their commander before a battle. The nature of data traffic and throughput requirements for the battlefield, in addition to the proximity of the fighting force members, makes wireless LAN topologies a choice being considered for linking members of a ground-based mobile units in ad-hoc, dynamic configurations [Army01b, Hall02].

Currently, the radios used in the Land Warrior system use the On Demand Multicast Routing Protocol (ODMRP) to account for loss of line of sight between radios. The ODMRP sits on top of the 802.11 protocol stack and routes packets to multiple stations. These packets can time-out or be discarded when they fail to reach their intended destination [Maym02]. The protocol builds a mesh of shortest paths between members in the group by having nodes periodically broadcast a "member advertising packet", which updates the route from the source node to all other members in the group. Even though this protocol is meant to reduce the overhead associated with similar *unicast* protocols, it still requires some overhead to build and maintain the routes between nodes [Lee00].

1.3. Organization of the Chapter

The chapter is organized as follows: first, we discuss the analysis methodology and its use in the Army's organizational domain, and present the domain model represented using the Unified Modeling Language (UML) notation [OMG03]. Next, we discuss transforming the domain model into a network simulation model and its implementation in *tcl* script for the *ns-2* network simulator [Fall00]. We then discuss evaluation of the model using battlefield scenario simulations, focusing on our results from simulation experiments—discussing network metrics, experimental data and behavior patterns—while comparing our work to that of other researchers, where appropriate. Finally, we discuss how this simulation model is integrated with simulation models of the MAC hardware to explore a range of possible agent-based, self-regulating behaviors being added to the MAC-layer for highly mobile and robust 802.11-based wireless networks in the field.

2. Background and Survey

2.1. Solution Attributes

In this section, we present the motivation for adaptive self-organization of wireless networks in terms of several important architectural characteristics of robust systems: reliability; availability; security and integrity; flexibility; and, maintainability. There are many architectural dimensions that we might consider; however, these tend to be predominant when developing systems using hardware and software techniques [Camp91, Barb96, Hos01, Geor02]. It should be noted that QoS could be described in terms of these attributes as well. In our work, we consider QoS solely from the standpoint of wireless network throughput of the prevailing types of traffic over the channel—which we consider from the MAC perspective, as opposed to other layers in the protocol stack [Slt01, Sms02].

2.1.1. Reliability

In considering this dimension, we are looking beyond MTBF and MTTR metrics, as such standards are well entrenched in production for military applications. Rather, we are interested in what types of system structure and behavior can be adopted to maximize reliability by minimizing exposed points of failure that might be susceptible to attack. In most networking systems, there is a mix of hardware and software components. There is a network processor or standard CPU, firmware or embedded software, among other components. Our approach is to use reconfigurable VLSI computing techniques to design wireless network nodes that reside fully in silicon. This affords opportunity to leverage the advantages of VLSI devices—compactness, performance, robustness, and high reliability under harsh environmental conditions—against the additional benefits of reconfigurable computing, namely, real-time, in-circuit reconfiguration of the circuit logic [Bak96, Sch02].

Reconfigurable, or adaptive, computing techniques have been used in networking applications, such as [Bell02, Bre02, Mms02]. Simply put, reconfigurable computing uses programmable logic devices as the medium for delivering embedded hardware applications. These devices support the download and programming (and subsequent reprogramming) of application-specific functions directly into the hardware, and are typically used in electronics systems along with microprocessors [Mit01], or as a replacement for microprocessors, to deliver high-performance hardened functionality into a small form factor [Sgv02].

The possibility of integrating the wireless protocol stack into a hardware package without requiring software components creates a package with higher resilience from failure, as long as the circuit design is sufficiently verified through hardware simulation. In addition, device vendors provide data on device failure based on logic fault testing of the field programmable circuit devices into which a design file is to be downloaded (the exact estimates for such packaging for reconfigurable computing devices is published based on the specific vendors' FPGA device components, such as can be obtained from Xilinx® and Actel®).

2.1.2. Availability

The availability of the wireless network can be discussed in terms of the availability of a given node in the network, or the availability of transmission capacity over which variable loads can be carried (i.e., the "network" itself). In ad-hoc networks, the capacity of the wireless medium can be affected by a number of factors [Jft01, Jic01, Kur03]: (1) number of nodes, and their placement relative to one another (which affects signal integrity and channel congestion), (2) the frequency bands, their bandwidths, and how many channels are available for use (affecting the protocols that can be used for interoperability), (3) the power budget required at the node to push traffic onto the medium, and (4) the distribution of traffic on the available wireless media.

Availability can likely be maximized through some of the following methods: (1) allowing individual nodes to be multi-functional, and allowing them to change function—as from an 802.11 station to an access point—in the face of changing environmental conditions such as traffic load, (2) allowing groups of stations to agree to switch to a different frequency band in the face of congestion or perceived threat, and supporting multiple PHY radio implementations in the same station, (3) allowing stations to unilaterally, or by agreement among groups of neighboring stations, to either cut or boost power to compensate for some environmental change—as might be the case if jamming, congestion or other disturbance were affecting the stability, throughput or QoS of the channel, and (4) again, allowing stations to alter their behavior—unilaterally or by agreement—during different periods of predicted or unanticipated load.

Another meaning of availability deals with the availability of "assets" of the network to authorized parties [Pfl97] through some authentication and access control mechanisms—where these same resources are not available to a party that should not have access (a "denial of service"), and where the system knows the difference. This is tightly woven with security.

Also, the notion of availability can apply to insuring that the network nodes share and use resources equitably, and that the resources are tolerant of faults arising from changes in the environment. Here, the critical resource is the wireless medium itself [Jft01], the available energy budgets of the network nodes [Rrh00, Jic01, Wub01, Nas02], or available buffer memory. Energy and memory resources with which to store and forward frames are of specific interested in supporting multi-hop routing [John96, Lee00, Sxt01, Toh01]. In an adaptive, self-organizing network, we might expect nodes to collaborate to insure that availability is maintained in such scenarios, as has been discussed for network management [Bern98, Whit98, Appl99, Baum99, Gava99, Puli99, Tint99], resource configuration [Hayz98, Gerb99], load balancing [Griv98], and traffic routing [Heu98, Mina99, Scho99] in telecommunication networks, and also in wireless sensor networks [Megu01, Aror03, Zhan02, Sing03].

2.1.3. Security and Integrity

The security and integrity of the network can be discussed in terms of physical security of the wireless devices themselves. It can also be discussed in terms of the security of the data transmitted over the medium. The integrity of the node implies that it is operating consistently with its programming and is not able to be tampered with, and that its assets can be modified only by those authorized to do so. Integrity of the data transmission is generally verified by the check sequence sent with frame data, according to CRC or other such scheme.

The security of the device implies that the packaging itself is tamper-resistant. The packaging might be fitted with sensors that could detect certain types of tampering, thus signaling the device's internal logic to take some specific action—such as alerting other nodes to the fact that it "believes" it is being tampered with. The security of the nodes themselves, the network, the data transmitted, and resources that are accessible through the network connection are all governed by basic notions of security, applied with the intention of providing controls to reduce vulnerability of the node, the data stream, the network and resources accessed through it.

There are a number of threats to the security, availability and integrity of the wireless network [Pfl97]:

(1) Interruption of service - where a node, transmission, resource or other asset is unavailable, lost or unusable. This can be affected by jamming, introducing errors into frames, or attacks on stations, access points, or other resources in an attempt to compromise the integrity of the resource or system. In most wireless networking contexts, interruption or disruption of service is caused by environmental factors such as noise or interference resulting in a "bursty" channel. In our problem domain, we also include the need for pattern detection among the random noise in order to detect and thwart an ordained malicious attack on the wireless network.

(2) Interception of an asset – where an unauthorized entity has gained access to the asset—a frame, resources of a node, etc.

(3) Modification of an asset – where an unauthorized party modifies an asset in an attempt to thwart or compromise the integrity of the system.

(4) Fabrication of counterfeit assets – where an unauthorized party creates assets—frames, signatures of actions, etc.—that are false, with the intent to mislead or misdirect.

Security is managed in a network environment through a number of different means [Pfl97, Kur03]: (1) identity-based authentication of messages, senders and receivers, and any other identifiable resources involved in transactions, (2) identity-based, or role-based, access control, (3) data encryption, in both the transmission medium and inside the stations themselves (to retain confidentiality in the face of direct attack on the device and/or its packaging), (4) integrity and non-repudiation through checksum schemes and signatures applied to the data stream and to the internal contents of the nodes (such as the logic and memory contents), so as to have high confidence that the recipient node is receiving something from the claimed sender, and (5) audit trails tracking a history of actions for *a posteriori* access checks.

All of these schemes are applied at layers above the Link Layer of a networking protocol stack. At issue is which measures can be applied at the MAC Layer level at which we are operating, so as to improve security, integrity and availability of the network assets [Kur03]. Today, the 802.11 protocol family has checksums and encryption applied to the frame data stream. However, we would want additional capabilities to make the MAC more aware of potential threats, able to assess likelihood of use for threats, and apply overlapping controls to limit exposure.

2.1.4. Flexibility and Maintainability

In considering flexibility, we are interested in the ability of the network and nodes to adapt to changes in the environment. Most systems handle change through direct modification by human systems developers. Protocol changes, software upgrades, and the like are part of this process. Maintainability, thus, includes specific ways in which we can make the network more resilient and flexible to planned change—such as by facilitating secure means to update the configuration of logic circuits as planned changes are instituted. In addition, we would want to make individual nodes, and the wireless network as a whole, more flexible in the face of unplanned and random environmental change. Defining a framework in which ideas for adaptive behavioral heuristics can be quickly encoded and explored via simulation has great value in this context.

2.2. An Overview of Adaptive Self-Regulating Systems

Adaptation is defined as any process operating on some structure where the structure is modified over time to improve its performance (of "fitness") in its environment. A structure's adaptations to its environment are "persistent properties" of the structure created as a result of its behaviors, where these behaviors act as a set of "structural modifiers" or "operators", applied according to some pattern (or "plan").

This notion of adaptation can be applied to systems, populations, individuals, or components thereof. The most commonly studied systems are insect colonies [Gor99, Joh01], cellular organic systems [Rum86, Joh01], or other systems involving large populations [Axel97]. Adaptive systems tend to have a common set of the following properties [Hol92, Axel97, Joh01]:

(1) They are very decentralized, generally without any "controller" function coordinating activities of all the population members (regardless of a role designated as "Queen");

(2) They are capable of using small adaptations (through simple learning maneuvers) in individual behavior to modify the collective behavior on a global scale in response to changes in their surroundings as perceived at the individual's level of operation in the population;

(3) In certain populations, individual members can change their behavior or function in the presence of indicators or signals triggering them to do so, based on what their neighbors are doing;

(4) In self-organized systems, the structural patterns exist at multiple levels of system abstraction, where the actions of a higher layer constitute the environmental changes of the lower layer, whereas behavioral changes at a lower layer percolate and propagate through the population of members at that layer to affect change in the layer above it; and,

(5) In certain systems, introducing random mutations in addition to environmental changes can cause new behaviors that may or may not be beneficial to the population; however, the reproductive effectiveness of the individual (or the propagation of the learned behavior through the population) based on the fitness imparted by its behaviors determines whether the trait will manifest itself in the wider population. It is not clear whether such assumptions of a "genetic" model are appropriate for considering adaptation and self-organization in wireless networks [Forr93]. This remains to be seen.

There are several different techniques and architectures that have been applied to the problem of adaptation in behavior of systems, including swarm intelligence [Dori96, Bon00], evolutionary algorithms, and intelligent agents [Bru91]. In the following sections, we summarize some of the properties of adaptive, self-organizing systems in terms of an agent-based paradigm, how they may relate to MAC functionality in a wireless network (using the 802.11 in particular), and how we can explore and exploit these characteristics in extensions to the 802.11 MAC.

2.3. An Agent-based Model for Adaptive Node Behavior

In the ASOWN environment, we are interested in creating a large population of individually acting wireless agents whose collective behavior improves the overall robustness of the wireless network landscape. This improvement in fitness may be defined in terms of performance, resource conservation and security threat avoidance. In consideration of this objective, it helps to examine the relevant characteristics of an agent-based model for system organization.

We can graphically explain the research landscape for ASOWN in terms of a software agent model, shown in Figure 1, depicting a general view of architecture—the structure and organization of functional components used to exhibit behavior in an environment—using the model of evolutionary biology as a guide [Den95]. Evolutionary models have long been the basis for much agent research [Mae95].

A software agent can be modeled on a biological organism, with salient characteristics extrapolated and mapped onto the domain of a software program. As with most higher level organisms, agents have the following: (1) internal needs and desires, (2) practiced means to satisfy those needs and desires using available resources, (3) means to sense what's going on in their surroundings, (4) means to evaluate perceptions for purposes of selecting appropriate response, (5) means to affect change in their environment relative to their needs and desires or relative to perceived environmental stimuli, and (6) means to evaluate the results of that change in order to move them closer to achieving a specific need or desire.

In nature, biological differences between organisms center on "behavioral machinery" and "architecture" tradeoffs evolution has made for coping with the complexities of their respective environments, where behavior results from applying the machinery. Higher-level organisms have more varied and complex means for sensing, evaluating and responding to their environments than do simpler organisms. Mechanisms such as memory and feedback enhance the level of response by giving the organism a means to build on prior experience of interacting in the environment. An internal representation of other organisms in its environment, and the means to communicate with others of its kind, provides opportunities for the organism to cooperate on satisfying innate needs (e.g., procreation to extend its genetic line). Richer biological constructs allow greater manifestation of "intelligent" behavior—such as through self-awareness, advanced learning strategies, advanced visualization techniques to "see" future outcomes (i.e., as a "plan") in its environment. Finally, it is often the case that organisms enter into symbiotic relationships with organisms of other kinds in order to achieve mutually beneficial objectives; strategies for cooperation and collaboration are important for survival and mastery of the environment.

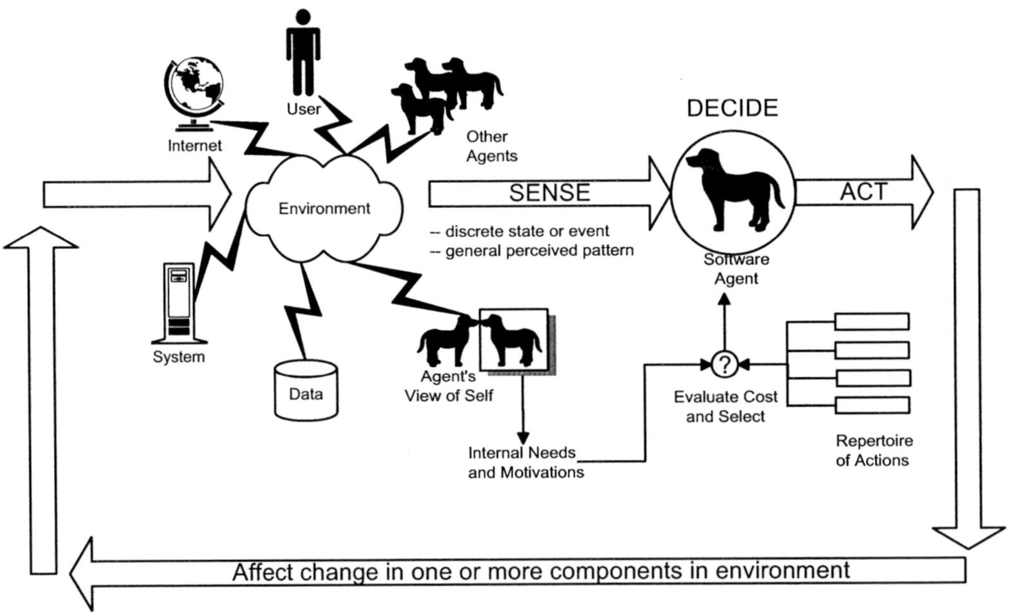

Figure 1. General Architecture for Intelligent Agents.

By abstracting these basic mechanisms in living organisms, we can discuss a simple and general architectural model for our wireless adaptive agents. First, agents must have some means to sense their environment. In our context, this notion of environment is the wireless network, and includes: (1) frame data received by a station from neighboring transmitting stations; (2) data collected regarding the channel; (3) goals, tasks and actions represented internally to the agents executing in the station, which might be stored as extensions to the Management Information Base (MIB) for the MAC layer [IEEE99]; and, (4) information about goals, needs, or state of other agents in the environment—either resident within the station itself or communicating via special frame types transmitted over the wireless channel.

The sensation of information from the environment can come in several forms: (1) discrete "events" from the environment (i.e., something that the agent is capable of sensing has been detected, such as the reception of a specific type of frame, from a specific source); (2) discrete "states" of other stations in the network; and, (3) general transmission, power, mobility and noise patterns detected as a result of statistical data collection, or historical and ongoing interaction with other stations in the network.

This last case presents an interesting scenario, namely, a station realizing that a particular sending station is the originator of an exceptionally high number of frames with integrity errors, as indicated by failure of the MAC's FCS decoder to generate a consistent CRC value for received frames. If such a condition were perceptible by the MAC layer, the MAC might develop a hypothesis that some entity is interfering with the integrity of the data transmission of this sending station (with some likelihood value). Alternately, it might conclude that the sending station's identity might be questioned suspect (with another likelihood value).

Once stimuli from the environment have been detected, identified and classified, the agent must decide what to do in response to each stimulus (either considered separately or collectively). In this regard, the agent might need to do one of the following: (1) react to externally-generated changes in its environment; (2) evaluate changes created by its own actions as a result of prior actions taken to further its own internal goals (what to do) and plans (how it is to do it); or, (3) evaluate its own performance on tasks in pursuit of its goals to see how to adapt its own behavior to improve performance.

In all cases, the agent must do the following: (1) look at its own state, comparing internal notions of goals and plans against perceived, identified and classified changes in its environment; (2) identify actions from its plans—or generate new unique plans--along with the "cost" or "utility" associated with taking each action, as part of formulating a set of "candidate" actions to consider in response to the environmental changes; (3) select one or more actions from the repertoire of possible responses, as a result of maximizing utility or minimizing cost, and schedule them for execution; and (4) execute the scheduled actions on the environment in the appropriate order.

This *sense-decide-act* process repeats, where the agent's effect on its environment is assessed as a result of sensing (or not sensing) change in that environment. When the agent is responding to external stimuli, it is engaging in "regulating" behavior—a primitive (and, still, most prevalent) form of behavior available in agents [Bru91].

When the agent is executing actions as part of tasks it executes in pursuit of its own goals, it is engaged in either "reasoning" behavior [Fra96] (inferring new information about the environment based on matching what it perceives in the environment against what it already knows) or "planning" behavior [Etz94] (creating or changing the sequencing of tasks to achieve its goals in response to new information), or both. When an agent has the ability to communicate and work together with other agents, it is engaged in "collaborative" behavior [Gen94]. Finally, when an agent is modifying its own strategies for interacting with its environment—in terms of its ability to respond, reason or plan effectively--for the expressed purpose of improving performance, it is engaged in "adaptive" behavior [Mae94] (i.e., it is capable of "learning" and applying new knowledge).

2.4. Adapting the 802.11 MAC to ASOWN

An important point to our research is to identify appropriate and suitable models and architectures for adaptive self-organizing behavior in the wireless network, then to make changes to the 802.11 MAC with the intent of building a modified version of the MAC protocol with which to perform experiments. Other research has focused on adding new layers to the protocol stack [Lee99, May02] so as not to affect the existing protocol and architecture of wireless networks. We believe that the types of changes in station and network behavior are only possible by modifying the core MAC protocol and architecture itself.

There are several reasons for this position: (1) there are "purist" reasons, in that a separate layer having access to internal state of the MAC violates the principles of separation between layers in a protocol stack, (2) there are security reasons for integrating the modifications directly into the MAC, as having such knowledge outside the layer would call into question the "boundary of trust" that exists between the MAC and entities outside it, and (3) there are performance reasons, in that the computational overhead of providing ASOWN behaviors will require more computational power, more memory, and other compute resources. These resources can be made available through a reconfigurable VLSI computing architecture, but the partitioning of functionality, along with the overhead to maintain security of its internal structure, adds a performance penalty.

In the table that follows, we present avenues for creating the ASOWN architecture, based on salient qualities of adaptive, self-organizing systems.

#	Adaptive Self-Organizing Systems Principle	MAC Layer Characteristics	Architecture Strategy
1.	Absence of a centralized "controller" for the entire population. The actions of individuals are scripted by simple hardwired plans, whose execution is triggered by a set of rules subject to perception of certain randomly occurring environmental cues [Joh01]. Individuals track a number of specific types of sensory inputs, levels of intensity of inputs, and count the frequency of certain events on those inputs, and correlate behavior to the inputs.	Wireless network protocols such as 802.11 operating in DCF mode have no central controller, but have simple hardwired functions embodied in the protocol definition. Unlike individuals in a colony population, wireless stations are not adaptive to changes in their environment to any appreciable degree—only to that degree based on the foresight of its designers. Nodes are mobile, and move into and out of ad-hoc configurations. Thus, the movement of nodes is somewhat "random". Nodes listen in on all frame traffic to which they can sense (so as to ascertain whether it is frame traffic being sent to	What if nodes had the ability to respond more flexibly to the random changes in their environment? The major areas where this might occur are in: (1) bandwidth and traffic congestion management, (2) security threat detection, internal resource optimization. What if the MAC monitored inputs, i.e., frame traffic, from its neighbors, and did more than simply throw it away once it extracted the NAV value? What if a node, like an ant, was sensitive to gradients, such as signal strength? What if a node kept track of occurrences and frequency or occurrences, for events in its iBSS/BSS/ESS? Could individual MAC nodes detect meaningful patterns in their environment? Could they

		them). They do minimal processing on a frame that is not identified as theirs (e.g., updating NAV Register to reset the wait time before attempting to access the medium).	affect the environment in an adaptive way, by using this new sensory information, to optimize their own performance or detect threats, from their limited point of view?
2.	The population is itself an adaptive system, whose behavior at a macro level is created by the summation of the individual behaviors of its individual members [Joh01]. There is systematic complexity, yet order is built out of local interactions [Hol92].	The functioning of the network (iBSS, BSS or ESS configuration) depends on conformance to protocol. WLAN nodes operate in an environment of limited capacity of its resources—bandwidth of the medium, energy availability (battery life), amount of buffer space, etc. Excessive congestion may cause feed-forward effects on bandwidth management—such as when excessive waiting traffic adversely impacts frame delivery, causing excessive retries, further clogging contention for the medium, affecting QoS.	The primary issue here is what happens when a very large population of nodes is present, thus taxing the available medium bandwidth? What if terrain conditions, or threats, cause rapid, dramatic change in their environment? What happens when there is rapid shifting of the population due to high mobility, and high-speed mobility? How does the effect of individual node movement affect the capability, capacity, reliability, security and maintainability of the entire network?
3.	Individual members of the population can change their behavior dynamically, to take on different roles, based on local changes in their environments. It has been documented that forager ants—on detecting certain concentration of other foragers, will change behavior to that of another role in the colony. Over time, roles played by an individual may change [Gor99].	In 802.11, the function of nodes (as stations, access points or bridges) is hardwired. The protocol defines a means to change to basic mode of operation (e.g., between DCF and PCF modes) to alleviate contention. A station operating in the distributed coordination function (DCF) may request that the AP set up point coordination function (PCF) so that each station must request, and send traffic, through the AP—in effect, setting it up as a "controller". This allows different traffic patterns to be set up, so the medium becomes and arbitrated resource. But, we	What if, based on monitoring traffic patterns in the frames received, and its own objectives, a node can change from a station into an AP? What if the conditions under which such a transition proved beneficial could be learned by the node and passed onto other nodes in the network? What if a node, by observing frame integrity and encryption checks (FCS and WEP frame fields), could detect a pattern of security threat or breach, and could set up tests to confirm its hypotheses, then communicate its findings to other nodes? What if it could devise new strategies to thwart attacks, jamming and other threats, and thus passing these insights to

		presume that, with a large number of stations in a BSS, this will cause severe traffic backlog, resulting in excessive timeouts, and dropped higher-layer packets.	neighboring nodes? Also, how might this trait affect multi-hop performance in an ad-hoc network?
4.	The nature of adaptive, self-organizing systems is they are recurring patterns at different levels of organization, and that the adaptation and learning occurs on different levels simultaneously, through the modification of some structures at each level by operators specific to that level [Hol92].	There is a hierarchy of abstractions in networking, involving the network as a whole (such as the Internet itself, or the platoon network in our domain), which consists of other networks linked by routers. In the wireless network component, the routers and nodes are stations. Its conceivable that we can think of the internal architecture of the MAC as a set of nodes that communicate among one another via another protocol (as is typical of the architecture of embedded systems).	Can this principle of layered abstractions, with adaptation occurring at micro and macro levels be harnessed to improve the fitness of the network at these different levels of abstraction? Can effects of local behavior changes in one level promote better performance in the wireless environment, or of the larger network as whole? Can the internal structure of the MAC (or other networking layers) be made adaptive and, to what degree so, through the use of reconfigurable computing techniques?
5.	Using behavioral "norms" to control interactions among agents, where, in certain social situations, individuals may choose to "defect" from these norms in order to achieve certain objectives [Axel97]. Considering norms in terms of precisely defined protocols, they can be *arbitrated*, *adjudicated* or *self-enforcing* [Pfl97].	MAC behavior constitutes a protocol, an orderly sequence of steps that all parties must follow to accomplish a task, where the regulated behavior is to the benefit of all [Pfl97]. The MAC protocol is fixed around certain agreed-upon conventions, such as CSMA/CA, to guarantee that the wireless medium is shared fairly. In addition, there is implicit prioritization of traffic based on using inter-frame delays [IEEE99].	What if, under certain circumstances, a given node could decide to "defect" from observing CSMA and CA so as to maximize its own access to the medium? What if the node's neighbors (i.e., other stations in the iBSS or BSS) were able to punish it under certain conditions? What if a station could assess its local conditions, for reasons of threat/security or resource management, and decide to "coerce" the protocol to satisfy some pressing objective?

Other work in this area focuses on using adaptive agents to optimize search problems. The ACO (ant colony optimization) [Bon00] is based on optimization to find shortest-path solutions for a broad class of problems. A derivative of this technique, referred to as ACR (ant colony routing) [Heu98] has even been applied to packet routing and load balancing in networks.

The difference between the approach being presented in this chapter and the prior work cited above is that we don't believe our problem is fitted to a shortest path optimization problem. We are basically dealing with a problem where there are a number of small, interacting sub-problems involving specialized adaptations of a wireless node to its changing environment. This solution can be characterized by a hierarchical collection of intelligent embedded agents, those that represent nodes in the wireless network, and those residing within a node. Basically, our macro-agent (i.e., a wireless station) is actually a collection of micro-agents that cooperate on tasks pertaining to threat detection ("patroller" agent), power and resource management ("foraging" agent), congestion management ("maintenance" agent), exception handling ("midden" agent) and MAC function management ("nest" agent). This taxonomy, although somewhat arbitrary at this point, follows the classification of ant roles observed in the field [Gor99].

Furthermore, just as there are multiple agents operating within the confines of a specific MAC unit, there are corresponding agents in each MAC unit. Thus, in a given high-density multi-layered network topology, there could be thousands of such agents interacting, collaborating, cooperating and competing with one another to achieve specific objectives associated with their own functioning—as dictated by the environmental landscape. We have conducted research on this type of agent architecture in prior research efforts in the healthcare domain [Dhb98]. The results from that effort provided us with guidance on how to initially proceed in defining an agent-oriented architecture for this domain. As such, the problem appears to be nothing like the shortest path optimization problem. However, there are likely to be formulations for cooperative learning, planning and diagnosis problem-solving activities to be performed by the agents—learning being fundamental to the adaptive character of the wireless network at both the individual node and for the network as a whole.

2.5. The Challenges of Wireless Platform Development

However, in this picture are several sets of assumptions involving our collective ability to develop and deliver wireless, always-on, ubiquitous, pervasive computing platforms and devices in support of wireless applications and those communities using them to manage process-oriented networks. First, there are challenges associated with the wireless infrastructure technologies, protocols and standards themselves [Goo00, Bja02, Che02, Isr02]. Second, there are challenges associated with the technologies employed to create wireless devices capable of meeting performance requirements in terms of data throughput and power consumption [Goo00, Jft01, Wub01]. Third, there are challenges associated with creating robust development environments supporting the creation of application-specific Internet-ready devices that are capable of meeting stated performance objectives [Tre96, [Dil00, Dzc02].

Within the space of technologies for building a wireless communications platforms and infrastructure, there are some specific issues and challenges that are germane to our solution, all of which pertain to the wireless device platform: (1) support of changing, and sometimes conflicting, communications standards; (2) provision for acceptable levels of performance in data rates, throughput and response time; and, (3) provision for effective management of energy consumption.

There are a number of protocol standards that are often perceived to be in competition with one another. In the wireless LAN arena, Bluetooth and 802.11b operate in the same frequency range, and are thus are considered contenders [Bra02]; however, some have reported on the ability for these protocols to coexist [Gvs01], and to provide differing capabilities based on their strengths [Bra02]. Different versions of the 802.11 protocol are available, mainly catering to different bandwidths, techniques for addressing signal interference, and how they handle area of coverage, and thus provide choices for platform developers, device makers and network managers alike [Isr02]. Also, as mentioned earlier, there are the protocol choices associated with wireless cellular and wireless LAN protocols, and the complexities associated with bridging these protocols and providing interoperability between them.

Each wireless protocol has been designed according to some set of performance criteria, and has some set of inherent limiting factors. Namely, the key issues affecting data throughput have to do with issues such as [Goo00, Ete01, Gsv01]: (1) frequency of operation (and the industry factors affecting this operating range) and the presence/absence of overlapping signal sources in the frequency channel that lead to signal interference; (2) coverage area over which a signal can be reliably transmitted and received, affected by topology (landscape, buildings, construction materials, etc.); (3) signal strength, affecting both the ability of a wireless source to interfere with other sources and its ability to conserve precious energy, and (4) transmission modulation techniques used to send a digital data stream as an RF signal in some allotted frequency channel at a predefined power level.

Within the framework of a given wireless system and its supported protocol(s), there are objective criteria that govern the types of platform solutions that can be developed. These include [Goo00, Ete01, Gsv01, Slt01] [Sms02] : (1) quality of service (QoS) requirements and satisfaction mechanisms, usually in the form of response time, latency, and end-to-end assurance, (2) interference management schemes, (3) power management schemes, (4) data traffic patterns (based on application profiles), and (5) capacity and congestion management. Some schemes attempt to address limitation through defining higher-level adaptations of a protocol to maximize one or more criteria. For example, one area has been the exploration of the feasibility of ad hoc, multi-hop transmission of data as a means to overcome performance limitations in noisy environments [Rrh00, Sxt01], and also as a means to limit power consumption through limiting signal strength [Jic00, Jft01]. Another example is the use of shorter "burst" messages as a means to communicate control information so as to minimize bandwidth consumption [Ete01].

Finally, the primary factor affecting the ubiquity of wireless devices is their power generation and energy consumption. The issue involves the fact that, although the innovation of more powerful and smaller batteries continues its pace [Nas02], a core part of the device power management challenge lies in the devices providing better means of conservation, budgeting and scheduling of power for its various functions. This is an area that has received an increasing amount of attention recently [Bak00, Jic00, Rrh00, Qhu01, Dzc02, Sgv02] as pertaining to both cellular and wireless LAN communications, including development in embedded systems, systems-on-a-chip (SoC) and reconfigurable computing platforms. The general consensus is that there are techniques that can be applied in the management algorithms to control the expenditure of power by such activities as regulating variable voltage supplies to the wireless device's circuits, for example. Thus, consideration of requirements of a platform for creating wireless solutions must take into account the issue of intelligent power utilization and energy management.

2.6. The Solution Space for Creating Wireless Platforms

To address these issues regarding wireless platform development, several approaches for building platform architectures are prevalent or are emerging. Embedded systems employing a real-time operating system (RTOS), Java, and other technologies running on a platform consisting of an embedded CPU core can be used to build wireless devices [Mit01, Hos01] and are touted as the most function-rich and flexible devices to support the kinds of collaborative applications users might want.

Another, emerging, approach involves the creation of wireless platforms using a predominance of VLSI custom logic and reconfigurable computing techniques in which to implement the core networking protocols and services [Bre02, Mms02, Dzc02]. This type of platform might, or might not, support an embedded operating system and application environment. Some such platforms are targeted for more ubiquitous applications, such as sensing and actuation in factory or medical environments [Rad00], and are not for applications associated with building social networks. Thus, whereas the former platform attempts to support all types of application functionality in a single package, the latter attempts to provide a base for building more application-specific devices targeted for certain types of wireless Internet interactions.

This second approach to wireless Internet platform development is based on two core technologies: (1) a collection of VLSI systems practices referred to as Systems-on-Chip (SOC) design [Die00] and (2) a set of architectures known as *custom computing machines* [Bak96] or simply, *reconfigurable computing* [Tre96].

Here, the objective is to appropriately partition the wireless device, the SoC, into hardware and software portions, and to realize these in the most efficient yet reusable configuration [Dil00]. To that end, a number of reusable components are being created and marketed in the industry, what is referred to as intellectual property (IP) development. It should be noted that this use of the term "IP" for intellectual property in the VLSI industry is not to be confused with the use of "IP" for Internet Protocol. Creating systems-on-chip for wireless platforms is becoming an important focus for this new approach, with commercial vendors and researchers looking to create solutions to the challenges to wireless platform development discussed in the previous paragraphs [Dzc02].

Adding to this SoC focus on techniques and IP for wireless platform design is the growing recognition that reconfigurable computing techniques can be used to deliver higher-performance, lower-power solutions without some of the inflexibility of some more standard VLSI design methods, many of which are employed in SoC design practices [Bak96, Tre96, Die00]. Results of using reconfigurable computing have recently been reported for applications involving IP (Internet Protocol) Router functionality [Bre02], network processor acceleration [Mms02], and in low-power energy management [Sgv02].

It is noted [Dil02] that Andy Grove, founder and former CEO of Intel, once said, "Silicon is just frozen software". This implies something many practitioners in the industry have understood for years, that what someone can develop in software can, with the appropriate level of VLSI device capacity, be implemented faster in hardware [Bak96, Dzc02]. The primary issue in considering a VLSI approach to wireless systems implementation—and one that attempts to maximize the use of reconfigurable custom logic for its functionality within an SoC design approach, is whether the wireless system being developed can run effectively at line speed and manage the different issues associated with protocol management, power conservation, and QoS maintenance, to name a few.

2.7. Importance of Node-level Behavior in the Wireless Network

An important point of facilitating the study of agent-based behaviors *in-situ* is the identification of appropriate and suitable models and architectures for adaptive self-organizing behavior in the wireless network, which guide us to a set of changes to the 802.11 MAC layer, with the intent of modeling a modified version of the MAC protocol in *ns-2* with which to perform experiments. Detailed discussion of the specific set of protocol changes is outside the scope of this chapter, and is being reported elsewhere. However, we focus on the methodology and experiments for two scenarios for adaptation that have been explored—namely: (1) localizing nodes into sub-nets by setting the transmission radius; and, (2) selecting the error coding scheme based on tradeoffs between code strength and cost in terms of channel utilization, and also given the prevailing conditions of the channel.

Other research has focused on adding new layers to the protocol stack [Lee99, Hall02] so as not to affect the existing protocol and architecture of wireless networks. We believe that evaluating the types of changes in both station and network behavior as a result of agency and adaptation is only possible by modifying the core MAC protocol and architecture itself, and incorporating such changes directly into the simulation environment. There are several reasons for this position: (1) a separate protocol layer having access to internal state of the MAC layer violates the principles of separation between layers in the stack; (2) there are security reasons for integrating the modifications directly into the MAC, as having such knowledge outside the layer would call into question the "boundary of trust" that exists between the MAC and entities outside it; and, (3) there are performance reasons, in that the computational overhead of providing adaptive networking behaviors will require more computational power, more memory and other computing resources. These resources can be made available through a reconfigurable VLSI computing architecture at a lower cost than through using more traditional embedded systems techniques [Bak96].

Furthermore, we are dealing with the interaction of wireless network nodes that communicate with nearest neighbors, which is the premise of operation behind the 802.11 MAC layer protocol [IEEE99]. In fact, given that all nodes receive the MAC layer frame traffic of all sending nodes that are within transmission range (i.e., "nearest neighbors"), extension of the existing MAC architecture to extract more information from received and discarded frames than currently dictated by the 802.11 standard is quite feasible. An example of such a modification would be to capture information on all neighbor nodes--their frame retries and dropped frame counts, for instance, information that is available in the frame structure parsed by the 802.11 MAC Layer Receiver module. To fully gauge the impact on node behavior as a result of system behavior, and vice-versa, we will need to explore how we can combine the information available at these two different levels of abstraction.

In extending MAC layer functionality of the node as an agent, we define its purpose as being to gather statistics relating to network quality and, if those statistics go above or below certain thresholds, change the MAC layer's behavior in some manner to respond to what it discovers, monitor its own performance as a result of the changes, and then broadcast the changes to its neighbors if they prove effective. Behavior transformations can involve changes in role and responsibility the node is playing in the network, and are based on two types of information: statistics relative to the node's operation itself; and, statistics gleaned from received packets (both packet destined for the node in question, as well as those not destined for it). Behavior changes could simply involve a change in some aspect of the transmission of data across the channel—such as for changes in the error-coding scheme—but would need to be coordinated among all neighboring wireless nodes in the network.

Information relative to the node's internal state includes packets received, packets forwarded, packets sent, packets dropped, ACK's received or not received, and queue size [Fall00]. In addition, the node can gather information such as the source node, destination node, packet's time to live, and byte size from each packet it receives. This information from the ns-2 simulation model correlates reasonably well with the actual data available for collection from received 802.11 frame traffic [IEEE99].

Some QoS requirements, such as minimum acceptable rate and max delay [Fall00] are difficult for a node to detect, except in terms of packets sent to its neighbors. For packets sent over multiple hops, it would be difficult for any one node between the source and destination to determine the global throughput or delay for that link. The source node may be able to determine the throughput on the entire link by dividing the number of data bytes sent by the time required for the entire atomic transaction. However, the problem with the link may be at one of the intermediate hops for which there is nothing the source node can do (i.e. no change the source can make to affect link quality at another hop).

Before discussing the specifics of evaluating adaptive behaviors in a mixed simulation environment, we first present the methodology for creating the system-level simulation model, based on the application of conceptual modeling techniques and mapping the resultant model into a lower level model suitable for parameter selection for simulation of the wireless network.

3. Discussion of the Wireless Ad-Hoc Network Model

We're discussing using simulation as a method for creating fighting force network models, to be used in evaluating a set of possible extensions to the 802.11 MAC layer, where we are exploring protocol extension via agent-oriented adaptive techniques to make the WLAN as a whole be more self-regulating in the face of dynamic environmental changes. Our research method involves combining conceptual modeling of the Army's battlefield domain with the model creation activities required to create a viable, robust and accurate simulation model of the networking infrastructure of the fighting force on the battlefield. Later, we discuss extensions to this simulation model to support integration of a network-level simulation model with a node-level model. However, first, we discuss how we devise a simulation model of the network from an understanding of the domain's conceptual model.

3.1. Conceptual Modeling of the Problem Domain

The first activity in our methodology calls for creating a conceptual model of the simulation domain—namely, that of the battlefield fighting force. Conceptual modeling is used in a number of software systems activities, including database design [Davi91], object-oriented software design [Rumb99] and agent-oriented software engineering [Sing99, Huhn01]. Conceptual models of a domain are useful in articulating the implicit structure and semantics of the domain in question, and serve as a "mediating" representation [John89] that can serve as a front-end to a number of systems engineering activities. In our research, we use the conceptual model creation activity as a means to drive the definition of the network simulation model framework.

To that end, we have adopted the Unified Modeling Language (UML) as the modeling notation for capturing the knowledge required for constructing an adequate simulation model. UML is a software engineering standard [OMG03] and has been used in a number of different modeling domains, such as those associated with conceptual modeling [Rumb99, Davi05]. It has been used in scores of application domains to describe the semantics of enterprises and the systems that operate in these enterprises, and has been used extensively in recent years for the modeling of agent-oriented software systems [Sing99, Baue02]. UML provides a multiple-view modeling notation and semantics, along with a corresponding set of analysis methods, for modeling complex domains in a disciplined fashion. It provides a means to capture the semantics of an arbitrary domain, represent its structural and behavioral semantics in a concise yet readable format, which can be both precise and correct (depending on the level of precision and accuracy desired in the modeling of an arbitrary domain of discourse). Although we use the graphical notations of the UML as a basis for presenting salient aspects of our wireless networking domain model in this chapter, detailed discussion of the UML notations—namely, the model views and their associated analysis methods—is beyond the scope of this chapter.

Our interest in using UML in the context of this research is based on the following needs: (1) to effectively capture the underlying relationships between the constituent units of the fighting force organization in a general way, so as to not tie the modeling of a network to one specific unit size; (2) to capture the conceptual entities, their relationships, properties and key domain constraints, that can be used to guide the process of synthesizing a communications model for use as input to simulation; (3) to provide an means to explore the abstraction of model properties—which properties should be captured and which should be ignored—as a basis for starting the exploration of modeling for simulation, and, (4) to provide an economical means of depicting the underlying domain semantics to those outside of the military who would need to have this basic understanding in order to fully comprehend the exploration of the solution to the research problem in context of the problem statement itself.

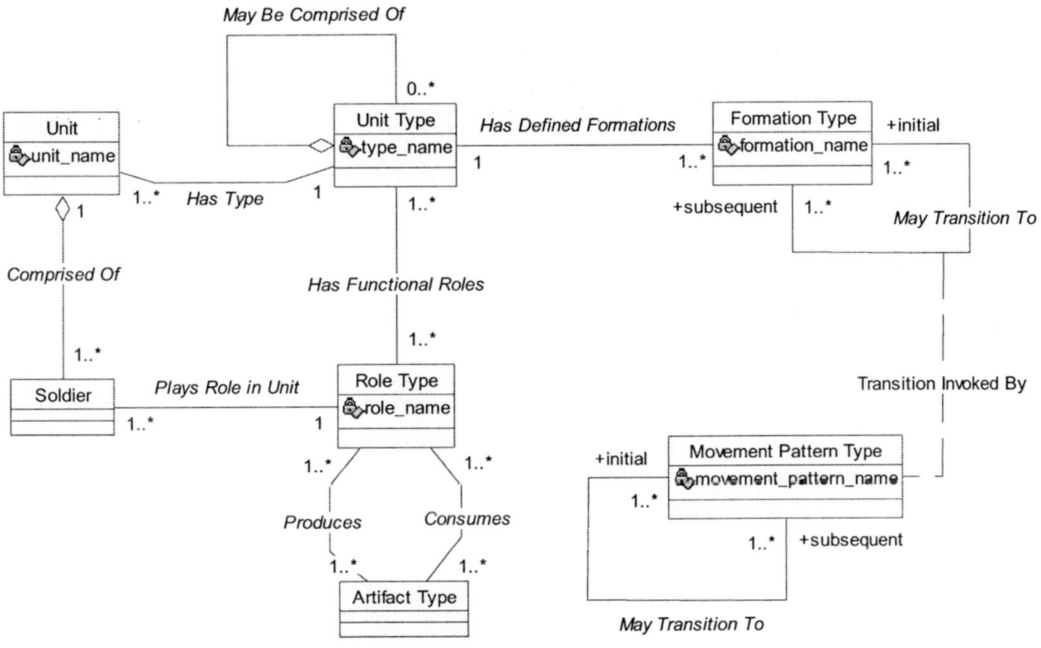

Figure 2. UML Class diagram for principal concepts in domain.

To capture salient aspects of the wireless networking environment, we used available documentation on the Army's Objective Force structure [Army01a], and carried out steps associated with conceptual modeling. One of the resultant structure models of the concepts and relationships in the fighting force domain is shown as a UML Class diagram in Figure 2. In this diagram, we see the principal concepts of interest in this domain—namely, concepts for the following: a *Unit*, comprised of some number of soldiers; *Unit Type*, whether the unit is a platoon, squad, fire team, etc., also showing the hierarchical relationship between composition of larger *Unit Types* from smaller constituent types; *Soldier*, which assumes a *Role Type* within the *Unit* of a given *Type*; *Role Type* is associated with sourcing and consuming *Artifacts* of different *Types*; and, finally, *Unit Types* are capable of being organized into specific *Formation Types*, where these formations of soldiers may transform from one into another while moving on the battlefield—as linked by a pre-defined *Movement Pattern Type* consisting of initial and subsequent *Formation Types*.

As shown in the figure, we can relate the role-playing of members of the fighting force (and their implied responsibilities) to the types of artifacts and, ultimately, to the relationship each soldier "role player" has to the artifacts in the inventory. In addition, we are able to establish relationships between the types of fighting force units with the types of formations defined for that unit type. Since formations are static, but precede and follow one another as a result of transitions made during the movement on the battlefield, we can define a "ternary" relationship between two allowable adjacent formation types and the movement pattern type in which the two formations are a part.

In other words, the fighting force adopts patterns of movement on the battlefield as a result of mission planning and as a response to conditions in the field. The Army, through its experience, has identified formations that maximize a set of objective criteria of the fighting force, and has defined a set of movement patterns that can easily be used to transition the fighting force between various formation types while moving on the battlefield. Just as with "hive-oriented" insects such as ants and bees [Gord99], the fighting force can be abstracted as an "organism" whose parts move in carefully choreographed "dances". However, unlike insects, military planners have carefully worked the effectiveness of these "dances" for years with intent of maximizing objectives in the field of battle. It is not an accumulation of random movements transmitted through inheritance and natural selection—although, the formations and patterns of movement can be disrupted by random and planned adversarial actions in the environment.

Given this strongly-typed model structure, we subject the properties of the model to scrutiny for the purpose of creating a quasi-formal mapping to an appropriately defined network model capturing those properties that are essential to exploring the impact of environmental effects on the robustness of the wireless LAN. To create our analysis model for devising 802.11 MAC-level WLAN simulation scenarios, we chose a U.S. Army light infantry platoon as the scope of the fighting force for analysis. The single infantry unit was selected because the Land Warrior system will be the primary means of communication for these units, scaling from fire team to division level. We chose the infantry *platoon* as the organizing unit because it is the primary means the Army uses for taking and controlling terrain [Shin99]. It also provides a fair representation of a typical organizational unit, which we can easily be scaled. The light infantry platoon is capable of performing the full range of required missions using a full complement of tactical formations and movement techniques. It also provides a volume of network traffic acceptable for exploring how the wireless network reacts under different dynamic modeling situations.

The abstract UML Use Case diagram depicted in Figure 3 consists of three role-based "actors" identified during conceptual analysis of the domain: *Command*, the *Fighting Force*, and the *Enemy Force*. This diagram depicts transactional exchanges involving these acting entities. They engage one another through the indicated Use Cases, abstract transactions constituting the inventory of scenario types being modeled for analyzing and exploring the wireless network under different operating conditions. We show later in this chapter how we decompose this model diagram into other diagrams that get to the heart of the network topology and the loading requirements.

The units of the fighting force carry out mission objectives through coordination of activities, modeled through the movement of domain "artifacts" among members of the force. For the most part these artifacts are standard documents defined for use in military processes; in the past, these documents had been physical entities, whereas now they are transmitted electronically. The fighting force executes its mission through coordinated movement of soldiers on the battlefield as they transition between stages of the mission, with document-oriented data flows augmented with GPS and video display data being streamed between members of the unit and the unit level headquarters. We next relate these two stages to the force unit's formations, patterns of movement and document artifacts exchanged as the mission progresses.

It is from this understanding of the domain that we are able to systematically move to a clear articulation of the network—namely, in terms of nodal movement and network traffic patterns and load—from which we easily transition to the model structure we adopt for use with the *ns*-2 simulator.

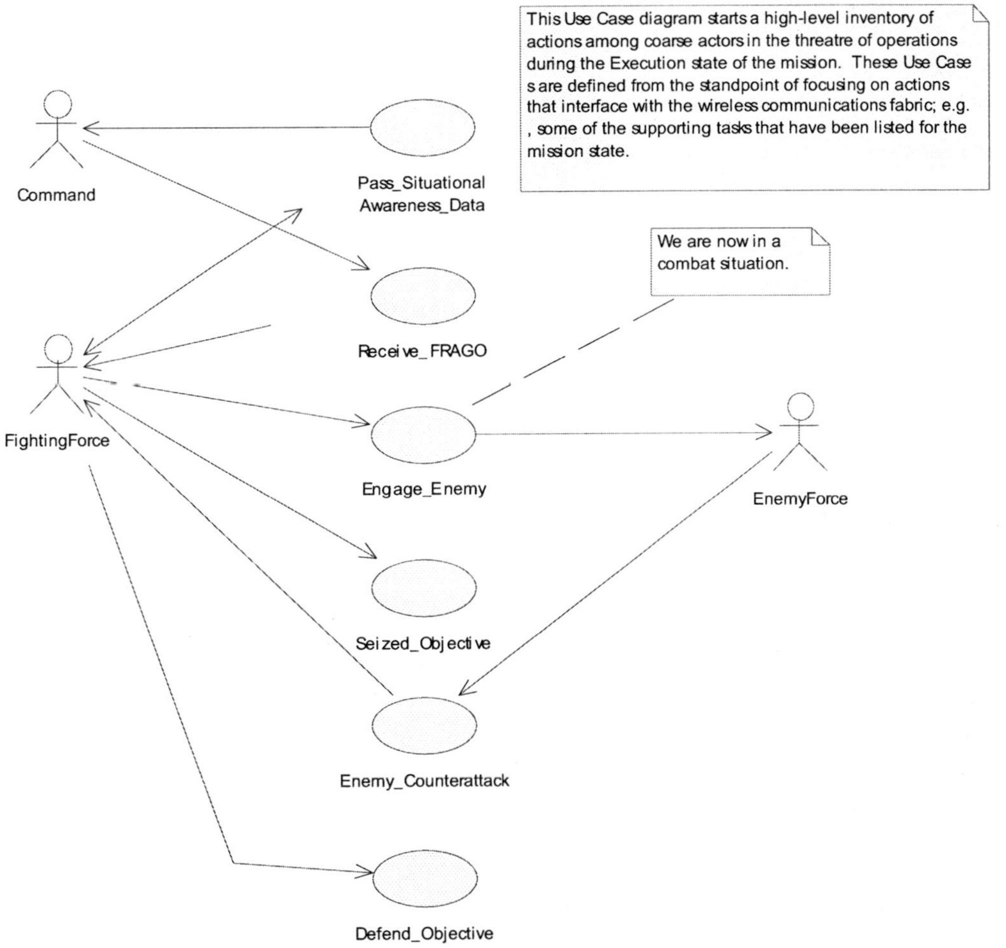

Figure 3. UML Use Case model of the fighting force domain.

The light infantry platoon is a hierarchical organization built upon teams of soldiers. It is primarily a three-level hierarchy, as shown in Figure 4. At the lowest level is the *fire team*, consisting of four soldiers—the Team Leader (TL), Grenadier (GNDR), Automatic Rifleman (AR), and Rifleman (RM). The GNDR, AR, and RM are soldiers responsible for maintaining different weapons, where their labels are derived. The TL maintains control of his team, controls their movement and rate of fire, and maintains accountability of all soldiers and equipment within the team. The fire team is the lowest maneuver element within the infantry and generally is the first unit to make contact with enemy forces. The light infantry *squad* consists of two fire teams and a squad leader. The squad leader trains the squad on combat tasks, manages the squad's logistical and administrative needs, directs maintenance on equipment and weapons, and maintains accountability of the soldiers and equipment in his squad. The light infantry *platoon* has three squads and a headquarters (HQ) team. The platoon leadership, the forward observer (FO), the radio/telephone operators, and two machine gun crews comprise the HQ team. A platoon leader (PL) and a platoon sergeant (PSG) combine to lead the platoon. The platoon leader (PL) commands the platoon and is primarily responsible for accomplishing the platoon's tactical mission. He develops and ensures execution of the tactical plan, moving around within the platoon to support mission objectives.

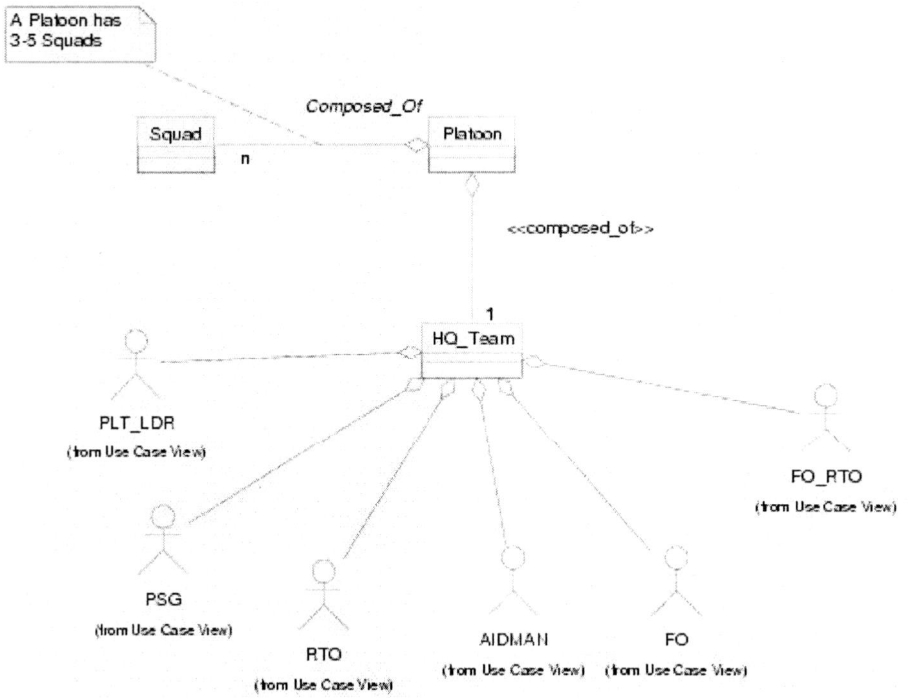

Figure 4. UML Class diagram of the fighting force hierarchy structure.

3.2. Modeling a Platoon's Movement on the Battlefield

Movement is one of three tactical operations undertaken by an infantry platoon, and is an essential component to the success of the platoon mission. Understanding how the platoon moves is essential to determining the movement's effect on the wireless network. During *Mission Planning*, the PL must decide on a formation and a movement technique. Movement techniques and formations differ in that formations are relatively fixed, while movement techniques are not. Formations describe the arrangement of squads, fire teams, and soldiers in relation to one another and are designed to allow the platoon to weigh its firepower in a particular direction. Movement techniques describe how the squads and fire teams are arranged in relation to each other during movement and allow the platoon to make contact with the enemy with the most efficient use of resources, allowing the rest of the platoon to maneuver to a more advantageous position.

The movement techniques employed changes during the course of the mission because the factors governing selection of a movement pattern change. The PL weighs the likelihood of enemy contact and how well the characteristics of control, dispersion, speed, and security for each technique meet the needs of the situation. Both squads and the platoon as a whole use the same movement techniques, but the meaning is slightly different at each level. For the squad, *traveling* requires approximately 20 meters between the lead and trail squad. For the platoon, it requires approximately 20 meters between each element within the platoon. With *Traveling Overwatch*, the squad increases the distance between the teams to 50 meters. In the platoon, the first squad will use *Traveling Overwatch* and increase its distance from the rest of the platoon to 50 meters. The trailing squads continue to use a basic *Traveling* technique.

Bounding Overwatch is a different movement pattern, used only when the leader believes an enemy is near and contact is expected, as depicted in Figure 5. The PL will position one team in an Overwatch position where they can fire at the enemy if needed. The other team is instructed to move to a position closer to the enemy. Once the second team is in position, it "overwatches" for the first team as they bound to a new position. The fighting force can watch for the enemy while still making forward progress. Depending upon the need for speed, squad-level movement can use either successive bounds or alternating bounds [Army01b]. For the platoon during execution of a *Bounding Overwatch* pattern, one squad will be in an *Overwatch* position, one will "bound" to a new position, and the final squad will be awaiting or receiving orders [Army01b].

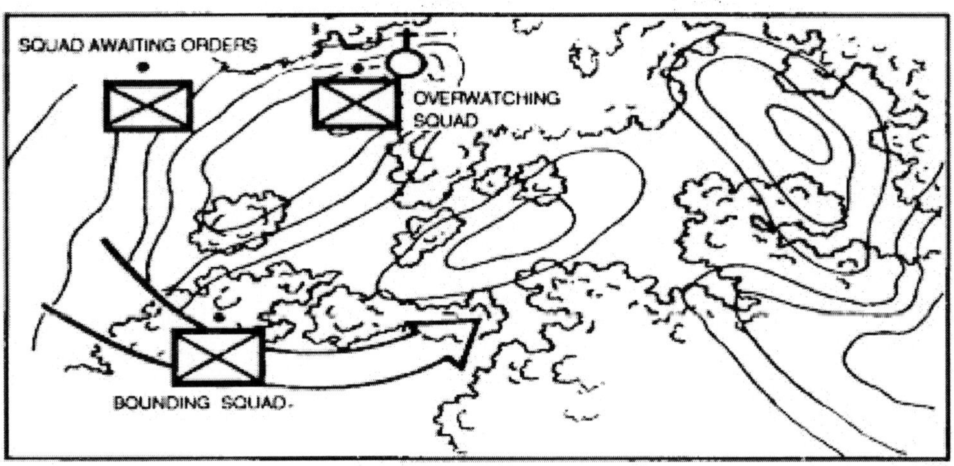

Figure 5. Platoon Bounding Overwatch example [Army01b].

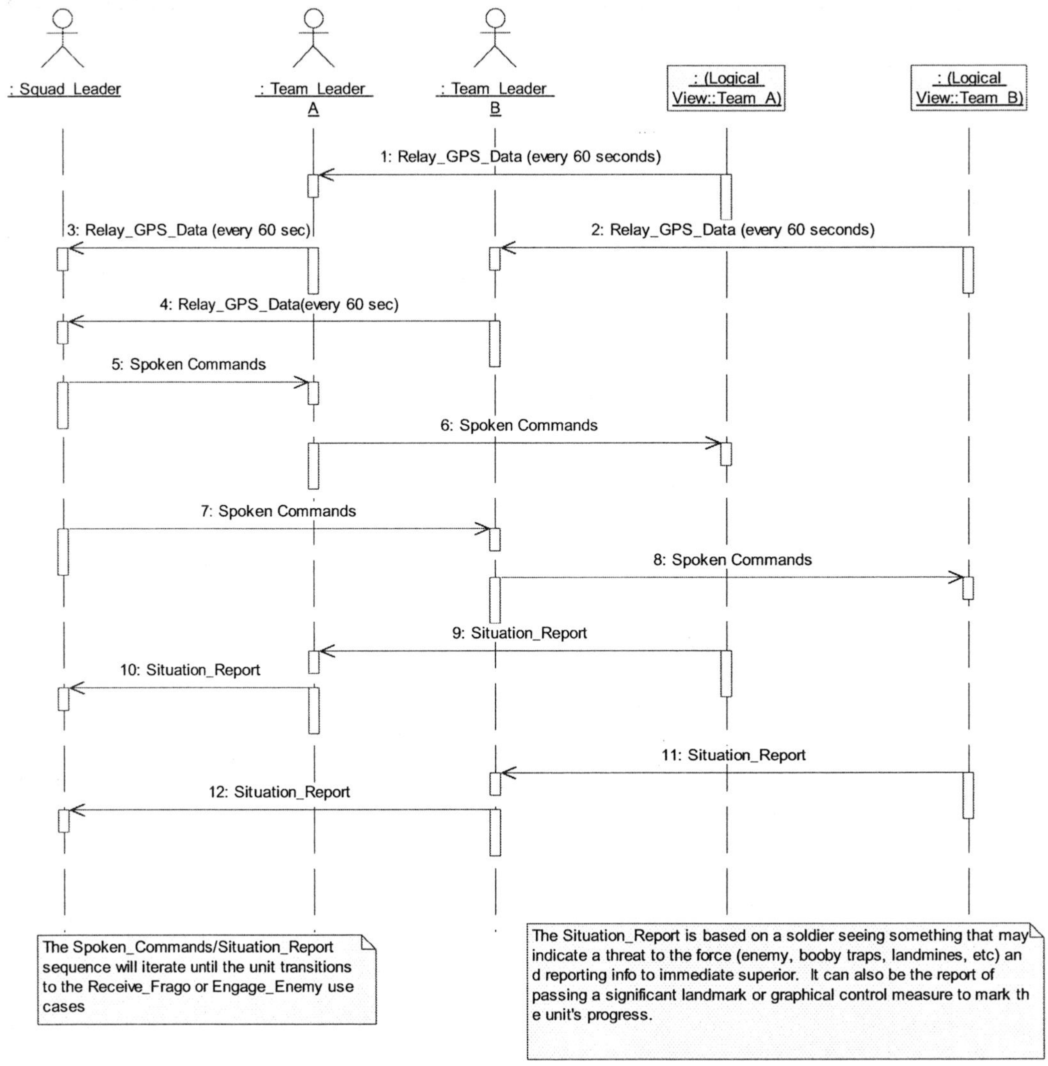

Figure 6. Wireless Application Sequence Diagram Scenario.

We augment the static information about the domain structure with modeling the sequence of interactions resulting in moving these artifacts around the network in accordance with Army procedure. What is important about the artifacts is we have determined, through our analysi, that they represent—in an abstract form—the key transactions of interest in the battlefield domain. It is the transactions executed among the domain actors (identified during the conceptual analysis activity) in the execution of the operational scenarios that represent the salient aspects of the domain for purposes of capturing simulation model information at an appropriate level of detail for simulation of the wireless network. In considering the formulation of an appropriate network model on which to run simulations, some understanding of the network properties is required. To get this understanding, we need to be able to relate domain interactions with network parameters—such as the number of network nodes, the types of traffic (based on the data associated with the various artifact types being transmitted through the network in support of the movement through the battlefield), and the volume of traffic—at any given time during the lifetime of the mission.

3.3. Defining the Wireless Network Model

In general, modeling and simulation give us the opportunity to extend and test, under varying conditions, a networking protocol layer contained in an infrastructure without incurring the cost of building a live system for testing purposes. Without simulation, it has been noted that most communications hardware units can end up discarded due to failure [Bres00]. The simulation models created allow protocol enhancements to be explored, subsequently leading to adaptive wireless protocols to be built in an appropriate hardware delivery platform.

For domain simulation, we select the *ns-2* simulation engine [Fall00] and available wireless networking extensions as part of our platform, as it offers several advantages we will exploit for our model creation and validation via experiments. First, *ns-2* offers a varying amount of abstraction that will allow us to examine the desired aspects of the 802.11 MAC protocol—namely, the behavior of the medium management without considering the details of the protocol layers above or below this layer (other than to specify traffic and mobility properties of the network). In addition, we are able to couple the *ns-2* input and output with a hardware model simulator for purposes of tying a clock cycle-accurate, MAC layer node-level model of the station to that of system-level model of the wireless network as a whole.

In addition, we can develop different simulation scenarios that account for different dynamic properties of the domain model, such as: network node movement following the patterns of troop movement; topology changes that are consistent with battlefield planning scenarios; varying traffic patterns and volume that follow the time-varying nature and mix of data and voice traffic on the battlefield; and, definition of dynamic events that perturb the network infrastructure. These can be explored as a means to test the network model created under situations as close to battlefield conditions as possible.

Finally, it is possible to test proposed adaptation extensions to the 802.11 MAC protocol by changing the definition of the wireless LAN model supported by *ns-2*, in that the simulator source code can be modified to fit our needs, for which scenarios can be re-executed to test the effects of protocol changes [Fall00]. Scenarios modeling a fully functioning WLAN network can provide a baseline for metrics such as quality of service (as measured by delivery time), throughput (frames per second), percentage of frame retries (indicating a level of congestion), and recovery time (mean time to repair), as well as providing data on the impact of human response time on network resilience and level of service. By creating running simulation scenarios that represent changes in the WLAN environment, simulating battlefield conditions and the movement of the fighting force in that environment, we can measure how much traffic is lost and the time required for returning the wireless network to acceptable levels of service and throughput after perturbation events. This time frame is of interest to mission planners.

Some network changes are *location-based*—in that troop movement increases or decreases the distance between nodes and affects the direction of individual node movement with respect to one another. Other network changes are *event-based*, which could result in stations being dropped from the network, either as a casualty of enemy fire, a node powering down, or a node moving out of range. We introduce event-based network changes at different levels of traffic intensity and patterns of movement to see the impact on network resilience.

The *ns-2* simulator is an object-oriented, discrete-event simulator developed to test various networking protocols over wired and wireless networks. *Ns-2* uses C++ for the simulator engine with an Object Oriented Tool Control Language (*Otcl*) script interpreter as the simulator front end to satisfy two disparate objectives. On the one hand, the simulation requires efficient manipulation of bytes and packet headers and implementation of algorithms that run over large data sets. Therefore, detailed protocols requiring fast run times but infrequent changes are programmed into the engine written in C++. On the other hand, the small changes are made to a simulation scenario in the scripting language to determine their effect or explore different scenarios. Scripting the simulation in *Otcl* allows quick, interactive changes to the simulation script, but runs much slower than C++. *Ns-2* links the two languages, allowing simulation scripts to be quickly written [Fall00].

By the very nature of its split-level programming, *ns-2* offers several useful advantages for this effort. First, we were able to quickly write simulation scripts that scaled up the network from fire team to platoon level. Second, we could customize our traffic sources to model the different traffic types generated by the Land Warrior system. Finally, we could introduce network dynamics into the simulation to model lost nodes or links. More importantly, when making changes to the MAC protocol, we have the option of modifying the model of the wireless MAC protocol within the C++ source code. After creating and compiling the new protocol, the baseline simulation scripts can be re-run to validate results of protocol modification.

3.4. Transforming a Conceptual Model into a Network Model

To create an accurate network model for simulation, we start with the domain analysis described earlier—with how unit movement, fighting force roles, and artifact creation and distribution affect network traffic in terms of traffic pattern and traffic timing as the mission unfolds through its stages. Understanding these conceptual attributes of the domain provides guidance for setting these *ns-2* parameters to accurately define the simulation model in software. This understanding also allows us to determine expected network traffic as a result of document artifacts and the effect on network QoS as a result of the patterns of movement.

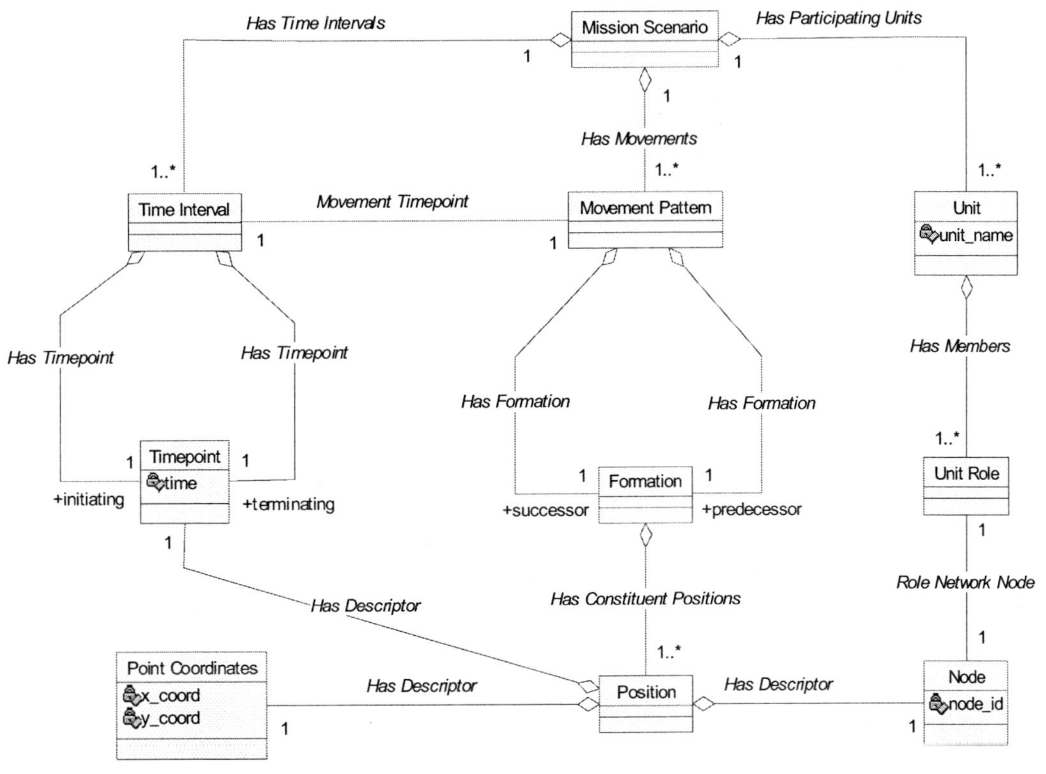

Figure 7. UML Class Diagram depicting mission decomposition.

We have the basic structure of the domain model, but need further drill-down into the domain semantics to be able to define a mapping transition to the network simulation model. This drill-down involves understanding the detailed compositional structure of a mission scenario—as abstracted in the Sequence Diagram shown earlier in Figure 6—into its constituent conceptual elements. It is these elements that will support the instantiation of a particular fighting force in the battlefield through simulation.

Figure 7 depicts the essential meta-model elements that comprise the data elements we must capture in order to transform to a network model consisting of scripts and changes to the simulator engine code. Namely, within the scope of a Mission Scenario, we have a fighting force Unit, the various Movement Patterns, and the Time Interval over which we are concerned at any point in our analysis. A Time Interval is comprised of two discrete Time Points. A Movement Pattern is comprised of two Formation structures and two Movement Technique structures, wherein each Node (say, associated with a soldier in the formation) has a Position expressed as a set of Point Coordinates that are valid for a given Time Point. The Mission Scenario has a set of participating Units, which are comprised of Unit Roles played by soldiers in the unit, where each such role played by a soldier is identified with a network Node and its *node_id* in the formation in both *time* (time point) and *space* (x, y coordinates).

3.4.1. Mapping Network Nodes in Time and Space

From the UML Class diagram in Figure 7, we see the decompositional nature of a scenario comprised of the patterns of movement, each with starting and ending formations and movement techniques. Each formation consists of a number of positions, consisting of a tuple of (*Node*, *Coordinates*, *Timepoint*). A node is associated uniquely with a role that is part of a specific unit in the fighting force. This enables us to derive the entire "script" of a specific simulation scenario from aggregating the constituent positions adopted by each node in the units involved in the scenario, tracking each unit as it moves through the battlefield space, progressing through the movement pattern over the time horizon of a given planned simulation run.

From this understanding, we are able to state the complex orchestration of movement on the battlefield as a sequence of individual movements by the wireless network nodes as they progress through the modeling and simulation scenarios—which is the level of representation required to capture the node movement for purposes of simulation in *ns-2*. We can represent this initial topology of the mobile stations as a nodal diagram, as the example shown in Figure 8. We have the *roles*, *node identifiers* and *positions* occupied by the nodes associated with the unit's roles. In this example, we show the platoon formation, with its 3 squads (forward, center, and trail) aligned in the *Platoon Column* formation and using the *Traveling* movement technique. This node diagram represents a snapshot of the progression of the scenario at one time point. Onto this base representation, we layer the succession of time points comprising the simulation scenario, as well as some designation of the specific network traffic at each of the relevant time points.

3.4.2. Characterizing Network Performance and QoS

Unit formations and movement techniques show the spatial relationship between soldiers and, in the case of Land Warrior, the spatial relationship between wireless communications nodes. Analysis of the *Mission Preparation* and *Execution* processes describes the type and volume of traffic anticipated on the wireless network, as shown in Table 1. At the platoon level and below, executing this process generates most of the network traffic and the greatest number of events capable of disrupting the network.

Table 1. Objective criteria requirements for network traffic types.

Task Name	Process Artifact	Data Type	Typ. size (KB)	QoS Reqs. (ms)	Through-put Reqs. (KB/sec)
(1) Pre-arrange conditions of the fight	Virtual Meeting,	VoIP	500	300	750
	Overlay,	Graphics	750	300	750
	OPORD,	Email	500	300	750
	Position.	GPS	20	300	750
(2) Strike enemy with decisive maneuver	Virtual Meeting,	VoIP	500	300	750
	Overlay,	Graphics	750	300	750
	Intel reports,	Email	500	300	750
	Position.	GPS	20	300	750

On an infantry platoon WLAN, network traffic is derived from four sources: electronic mail (email), graphical map overlays (standardized graphics that can be drawn on a topographical map to convey information such as unit type and locations and control measures to guide unit movement), global positioning system (GPS) data, and voice-over-IP (VOIP). All of the artifacts produced by the mission preparation and execution process fit within one of these four categories. Traffic from these sources can travel up and down the hierarchy and between soldiers within the same unit. From company level, platoons will receive artifacts such as operations orders (OPORDs), fragmentary orders (FRAGOs), situation reports (SITREPs), map overlays, and verbal commands. The soldiers in the teams, squads, and platoon will send up GPS data, SITREPs, enemy reports (SALUTE reports), requests for information (RFIs, used to clarify orders), and contact reports.

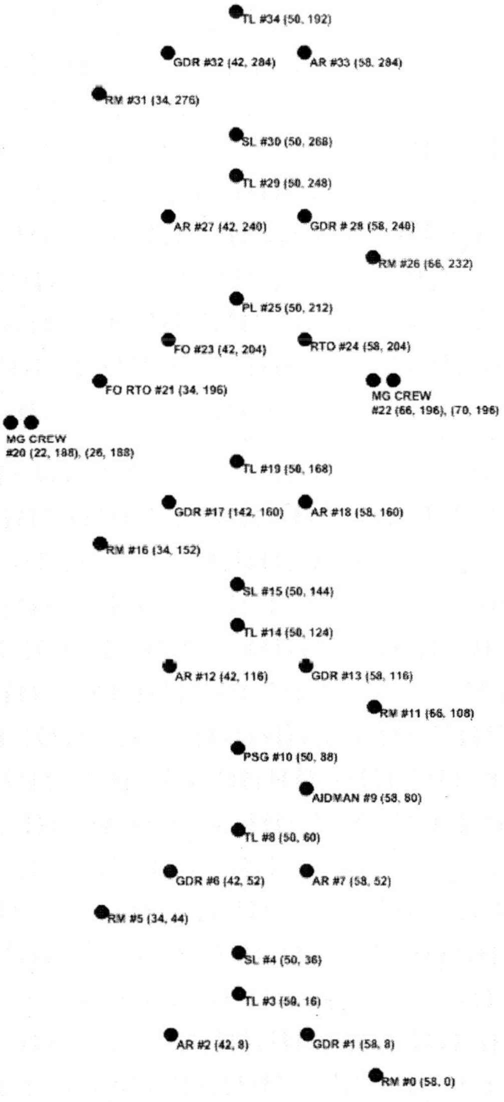

Figure 8. An instantiated formation, with nodes, positions, and roles.

In addition to the bulk of network traffic, network dynamic events occur primarily during the mission execution stage. These events include formation changes, dropped links, and dropped nodes. Nodal spatial relationships can change based on changes in the terrain or changes in the enemy situation. Soldiers moving behind "folds" in the terrain can interrupt network links, or movement behind dense vegetation can degrade it. Nodes can drop completely because of equipment failures or soldier casualties. An additional complication for network management is that time frames when dynamic network events are most likely to occur, which can potentially degrade network QoS, correspond to periods of highest network traffic. Periods of high network traffic with interruptions to some of the nodes results in additional traffic load on the wireless medium as a result of the overhead associated with 802.11 MAC frame retries.

4. Defining the Simulation Framework

4.1. The ns-2 Simulation Model

In describing the software simulation model, there are three components that have been constructed: the base simulation script; the node movement script; and the network traffic script. The base simulation script manages the nodes, manages the trace files, and extracts and displays packet information from the traffic "sinks". The node movement file establishes an initial node position, provides direction and speed for node movement and defines network path information. Finally, the network traffic script establishes agents and applications at each node to simulate traffic on the network. We combined these scripts to simulate the wireless LAN through a number of scenarios to derive a set of base parameters for the simulation model.

For this scenario, we chose to simulate a Land Warrior network-enabled light infantry platoon conducting an attack mission. This mission, while somewhat simplistic in actual mission terms, contains enough detail to adequately capture how the network changes as the platoon maneuvers across the battlefield. For example, it contains each of the three movement techniques, of which the Traveling Overwatch and Bounding Overwatch techniques require the platoon members (and thus the nodes) to change their inter-nodal distances.

The scenario discussed in this paper has five mission steps, each simulated separately by using different node movement files and slightly different base scenarios. In steps 2-5, the initial node position is determined by the final node position in the previous step. The platoon begins Step 1 in a platoon column formation built with the RM in the trail team of the trail squad as the base node located at (100, 0) along an x-y axis. By locating each node 8 meters along the x-axis and 8 meters along the y-axis, we ensured approximately 10 meters between soldiers, which is a standard interval in the US Army designed to guard against area-affect weapons. The platoon is arranged in the *traveling* movement technique, which requires 20-meter spacing between the squads. To complete this step, the platoon moves forward 200 meters at 0.5 m/s. We chose that speed based on one of the author's experience in various Army field exercises. The distance traveled allows the network enough time to stabilize while a longer simulation time would not add value to this particular simulation.

In Step 2, the platoon transitions to Traveling Overwatch by having the lead squad move forward 30 meters. The lead squad also transitions to Traveling Overwatch by increasing the distance between its teams to 50 meters. The remaining squads and the HQ team continue to use the traveling movement technique. The platoon moves forward another 200 meters at 0.5 m/s.

At the beginning of Step 3, an external event triggers the platoon to move into an assault position. The PL directs the 1st (lead) squad to form a squad line to the left of the original direction of travel. One of the MG crews joins the 1st squad and takes a position online and the left of the 1st squad. After the 1st squad is in position and over-watching the enemy, the 2nd (center) squad forms a squad line to the right of 1st squad. The second MG crew positions itself online and to the right of 2nd squad. The PL, RTO, FO, and FO RTO take positions to the rear of 1st squad. The PSG and Aidman place themselves just behind the 2nd squad. After the 2nd squad is in position, the 3rd (trail) squad is directed to an assault position 75m to the right and 50m ahead of 1st and 2nd squads. During this maneuver, we increase the soldiers' speed to 1 m/sec.

In Step 4, the 3rd squad assaults across the objective at 1 m/s by moving from right to left (with respect to the 1st and 2nd squads) to a point about 25 meters to the left and 50 meters ahead of the 1st and 2nd squads. Step 5 finishes the scenario by having 1st and 2nd squads move forward and form a defensive perimeter with the 3rd squad. The HQ team assumes a position in the center of the perimeter.

4.2. Formulating the ns-2 Experimental Framework

The first step in writing *OTcl* scripts corresponding to our scenario is to create Node Movement files. These files contain only the information required to establish an initial node position, provide movement instructions, and next hop information. Actual mobile nodes move in a three dimensional (*x, y, z*) topology; however, the *mobilenode* object defined in the *ns-2* environment [Fall00] is assumed to move only in a flat plane. We wrote each step into a separate movement file, in which we explicitly defined the node's starting position, destination, speed and time to being movement. The simulation engine will allow the node to continue to move until either it reaches its destination or is directed to move to another point.

```
Initial node placement
$node_ set X_ <x1>
$node_ set Y_ <y1>
$node_ set Z_ <z1>

Defining node movement
$ns_ at 10 "$node_ setdest <x2> <y2> <speed>"
```

Figure 8. ns-2 node definition script example.

The code in Figure 8 is an example of defined node movement. After this step, we append the next hop information required by the General Operations Director (GOD) object. The GOD object stores the number of *mobilenodes* and a table of shortest number of hops required to reach from one node to another. It is possible to compute this information during the simulation, but this can be time consuming [Grei98].

```
Setting Next Hop Information
$god_ set-dist <src> <dest> <#hops>

Changing Next Hop Information
$ns_ at <time> "$god_ set-dist <src> <dest> <#hops>"
```

Figure 9. ns-2 GOD script example.

Instead, we use the *calcdest* tool provided with *ns-2* to compute next hop information and append the appropriate code to our node movement files. The appended code provides initial next hop information and any changes to the GOD object at the beginning of the simulation. Figure 9 shows an example.

Next, after node movement, we defined the traffic pattern file, which remains the same for all steps in the scenario. In the Land Warrior system, each platoon is on the same wireless channel, therefore, every soldier in the platoon receives packets from every other soldier [Hall02]. To define the traffic pattern, we create UDP and Loss Monitor agents at each node and attached a CBR traffic generator to each of the UDP agents. However, *ns-2* does not provide a means within the *Otcl* code structure to set the agent to broadcast the CBR traffic to all of the other nodes. Admittedly, wireless packets are broadcast to every other node within range; however, within *ns-2*, the agents are only set up to provide unicast links [Fall00]. Therefore, we compromise to model broadcast communications in this manner.

Our setup for broadcast communications over the 802.11b wireless networking channel is best described as a two dimensional matrix. The source agents would appear along the matrix rows, whereas the sink agents are along the columns. We indicated the pairing of sources to sinks by marking a matrix cell with a '1', representing a connection between one source on one port at one node to one sink on one port on another node. There are no connections where the source index and sink index are equal, as there can be no connection between a source and a sink at the same node.

The *ns-2* simulator models these connections by having each node maintain a list of 34 source agents and 34 sink agents. Each agent is assigned a separate port on the port demultiplexer, through which it communicates with its assigned end points. Since we created a two-dimensional matrix of sources and sinks, it is a simple matter to iterate through the matrix and create the links between the agents and assign the traffic generator to the source agent.

4.3. Defining Experimental Network Traffic Data Sets

We want to create traffic generators that match actual traffic patterns as closely as possible. Therefore, we adopt a data set created for a Land Warrior field test at Fort Polk, Louisiana, subsequently subjected to statistical analysis by researchers at the United States Military Academy at West Point [Hall02]. The research at West Point revealed four major types of packets used in this environment: (1) Voice over IP; (2) Active Soldier, consisting of GPS information; (3) Email; and, (4) Map overlay packets, listed in Table 2. We create CBR traffic generators for each type of packet and use the derived inter-arrival times and packet sizes from the West Point research. Table 3 shows the average inter-arrival times and packets size data for this formulation.

Table 2. Fighting force application configuration parameters.

Application Type	**Packet Size (byte)**	**Inter-arrival Time (sec)**
Email	78	424
GPS	185	60
Overlay	231	10
VoIP	106	0.170

When using all four traffic generators as planned, we observe that the number of generated events for the script to record in the trace files created file sizes that are too large to process. Starting with the traffic generator simulating GPS packets alone indicate the size of the trace files to be a problem for later analysis. As an example, we can show the number of events generated using our connection setup with just the GPS traffic generator sending a packet every 60 seconds.

```
#create UDP sources and LossMonitor sinks at each node
#connect each node to every other node
for {set i 0} {$i<$val(nn)} {incr i} {
  for {set j 0} {$j<$val(nn)} {incr j} {
     if {$i != $j} {    ;#Doesn't allow connection b/n source and sink
at same node
        set agList [$ns_ create-connection-list UDP $node_($i)
LossMonitor $node_($j) 0]
        set src($i.$j) [lindex $agList 0]
        set fid_ $i.$j
        set sink($i.$j) [lindex $agList 1]
     }
  }
}
# Create CBR traffic generators for each node
# Connect the traffic generators to each source port on each node
for {set i 0} {$i<$val(nn)} {incr i} {
  for {set j 0} {$j<$val(nn)} {incr j} {
     if {$i != $j} {
        # Create GPS Application
        set gps($i.$j) [new Application/Traffic/CBR]
        $gps($i.$j) set packetSize_ 185
        $gps($i.$j) set interval_ 60000ms
        $gps($i.$j) attach-agent $src($i.$j)
        $ns_ at [expr $i.$j/1.0] "$gps($i.$j) start"
        # Create email Application
        set email($i.$j) [new Application/Traffic/CBR]
        $email($i.$j) set packetSize_ 79
        $email($i.$j) set interval_ 400000ms
        $email($i.$j) attach-agent $src($i.$j)
        $ns_ at [expr 60 + ($i.$j/1.0)] "$email($i.$j) start"
        # Create overlay application
        set overlay($i.$j) [new Application/Traffic/CBR]
        $overlay($i.$j) set packetSize_ 231
        $overlay($i.$j) set interval_ 10000ms
        $overlay($i.$j) attach-agent $src($i.$j)
        $ns_ at [expr 120 + ($i.$j/1.0)] "$overlay($i.$j) start"
     }
  }
}
```

Figure 10. ns-2 traffic connection script example.

For all 35 nodes, there are 34 traffic generators for a total of 1190 traffic generators in point-to-point configuration to emulate the broadcast capability of the actual 802.11 WLAN.

Additionally, there are 1190 traffic sinks. Every 60 seconds, all the traffic generators will generate a packet causing a send event to be recorded. A receive event will also be recorded when the packet arrives at the sink. Therefore, there are 2380 events generated by the simulated action of each node sending one GPS packet. Over the life span of a 400 second simulation, this translates into 14,280 events (400 sec/60 sec ≈ 6 send events/node where 6 events * 2380 simulation events = 14,280 simulation events). This number is just for events at the agent level and does not include events at the router or MAC layers, nor does it include packet drop events.

It was decided to create a smaller experimental data set based on using the GPS, Email and Overlay applications without adding in the VoIP traffic. In particular, the VoIP application was observed to produce a majority of the network traffic in the field, since this application broadcasts a 106-byte packets every 170 milliseconds, and would thus contribute most heavily to simulation "data overload".

Initial simulation results indicate a large amount of packet drops at the onset of the scenario, deriving from large numbers of DSR protocol and ARP packets broadcast by each node to determine routing as well as for resolving MAC addresses. Since we want to allow the network to achieve steady state before node movement, we start each application at one-minute intervals, and start node movement one minute after the last application starts. This process is very realistic, as military missions will not start without a "communications exercise" to verify the working condition of each communications platform in the unit.

For this scenario, we start the GPS application at time 0, which simulates the system being turned on and trying to find the other nodes in the network. The Email application starts at time 60 seconds, which simulates a function check on the email application in the Land Warrior system. Next, the Overlay application starts at time 120 seconds, representing another function check. Finally, the nodes start movement, simulating a successful "communications exercise" and the start of the mission. Figure 10 shows how these applications are created.

```
Define the node options
set val(chan)          Channel/WirelessChannel  ;# Channel
set val(prop)          Propagation/TwoRayGround ;# Radio wave model
set val(netif)         Phy/WirelessPhy          ;# Physical network
interface
set val(mac)           Mac/802_11               ;# MAC Protocol
set val(ifq)           Queue/DropTail/PriQueue  ;# Type of Queue
set val(ll)            LL                       ;# Link Layer type
set val(ant)           Antenna/OmniAntenna      ;# Propagation
direction
set val(ifqlen)        50                       ;# max packet in ifq
set val(adhocRouting)  DSR                      ;# Ad-hoc Routing
protocol

Set the Parameters
$ns_ node-config -adhocRouting $val(adhocRouting) \
                 -llType $val(ll) \
                 -macType $val(mac) \
                 -ifqType $val(ifq) \
                 -ifqLen $val(ifqlen) \
                 -antType $val(ant) \
                 -propInstance [new  $val(prop)] \
                 -phyType $val(netif) \
                 -channel [new $val(chan)] \
                 -topoInstance $topo \
                 -agentTrace OFF \
                 -routerTrace OFF \
                 -macTrace ON
```

Figure 11. ns-2 base script example.

The final step in writing an *OTcl* script to match our scenario is to create the base script. In this script, we create the *mobilenodes* objects, set their parameters and set up traces. We used the code in Figure 11 to create the nodes and set their parameters. We set channel, propagation, network interface, interface queue and link layer type options to the basic settings *ns-2* requires for wireless nodes. The MAC protocol option is obviously set to 802.11. Land Warrior uses the On-Demand Multicast Routing Protocol (ODMRP) [Hall02]; however, we used the Dynamic Source Routing (DSR) protocol for ad-hoc routing, as ODMRP, a custom-defined military protocol, is not defined within *ns-2*. The DSR routing protocol represents a reasonable approximation for our purposes [Lee00, John96].

After configuring the mobile nodes, scripting code such as shown in Figure 11 is written to create the mobile node objects. When a mobile node object is created, a routing agent is also created as specified along with the network stack, which is connected to the channel [Fall00]. In addition to defining the *mobilenode* characteristics, we open the *ns* and *nam* (network animator) trace files and instantiate the *ns* and *nam* trace objects as shown in Figure 12. The *ns* trace object traces all packet events, which are created when a packet is sent, received, or dropped, that occur in the simulation at the agent, router, and MAC levels and saves the information to a file.

5. Example 1: Adaptive Localization

For our experiments, we're interested in developing a set of heuristics for which we can make MAC layer modifications, according to data patterns it can detect and act upon. One such metric is the number of dropped packets, for which we can draw certain selected inferences—namely about whether all stations in the network domain are within range and able to operate at appropriate power levels. One aspect of mobility on the battlefield is that traveling patterns and environmental obstructions in the terrain can cause some stations to move out of range. One characteristics of resiliency of the wireless network, then, is to be able to recognize when this phenomenon is occurring, and to take adaptive measures to compensate. One compensation scheme is to dynamically create subnets; this attention to subgroups within the broader network is sometimes referred to as *localization* [Megu01]. To assess whether this might be a reasonable strategy to explore, we must create an appropriate simulation model and validate this idea based on our ability to recognize indicative patterns in the data stream. In this case, we'll be looking at data collected from simulation runs as an indicator.

In *ns-2*, drops are classified by reason, so it is possible to examine packet drops that occur at the MAC Layer and make observations based on reasonable root cause. Table 3 lists the codes used in the simulation trace file along with the reason for the drop for those events that are of interest. If packet drops are concentrated for one or two reasons, then we should get an indication as to what is happening in the network and an indication of network quality.

```
# create trace objects for ns and nam
# Assingn file names to variables
set tracefd     [open step1-out.tr w]
set namtrace    [open step1-out.nam w]

# Use new trace format
$ns_ use-newtrace

# Create the ns trace object
$ns_ trace-all $tracefd

# Create the nam trace object
$ns_ namtrace-all-wireless $namtrace $val(x) $val(y)
```

Figure 12. ns-2 configuration script example.

5.1. Experimental Validation Objective

In this example, our objective is to assess, via a priori simulation, whether adjusting the transmission power at the node, or by partitioning the larger network into several smaller subnets by adjusting power output, provides a reasonable means to reduce congestion of the wireless channel. As discussed earlier, one of the motivations of creating "agency" in the wireless nodes is to facilitate coordination behavior among the agents as a strategy to maximize robustness of the network. One such strategy is to segment, or to localize, portions of the network into subnets. Localization has been discussed in the literature as a strategy to reduce bandwidth contention [Meng01].

Table 3. MAC layer error codes of interest reported by ns-2.

Reason Code	Reason
COL	MAC Collision
DUP	MAC Duplicate
ERR	MAC Packet Error
RET	MAC Retry Count Exceeded
STA	MAC Invalid State
BSY	MAC Busy
NRTE	RTR No route available
LOOP	RTR Routing loop
TTL	RTR Time to Live has reached zero
TOUT	RTR Packets has expired
CBK	RTR MAC Callback
DSR	Dropped by DSR Protocol
AREQ	Dropped by ARP

We use *throughput* as a quality of service metric to detect any association between the number of drops at the MAC and RTR level and network quality. In our simulations, *throughput* is a measure of how much real data, excluding routing and address resolution requests, is actually received by a node, per-second of simulation time. For each packet an application sends, we calculate throughput by dividing the number of data bits sent and received by the *elapsed time*. The elapsed time is the time from when the sending application gives the packet to the routing layer to the time when the receiving agent gets the packet from its routing layer. We look at the total throughput for all stations, sampled every ten seconds. This gives us an indication of how the quality of the network changes over the lifetime of the simulation.

For the initial runs, we change the *transmit radius* of the nodes to see how it impacts network quality. In the first run, each node has a radius of 350 meters; the second, 90 meters, and the third, 30 meters. We affect the transmission radius by changing the Receive Power Threshold, which is a threshold value used to determine if a packet has enough power to be correctly received by a node. Table 4 shows the Receive Power Threshold values for each transmission range selected. For wireless simulations in *ns-2*, transmitted packets have an associated power level. The simulator determines the distance between the sending and receiving nodes and uses that distance to calculate how much power the packet has remaining at the receiving node, based on the propagation model used [Fall00]. If the remaining power is below the Receive Power Threshold, the packet is deemed as "not-received". Therefore, as we increase the Receive Power Threshold parameter, we effectively reduce the transmission radius of the nodes.

Table 4. Receive power thresholds at each transmission range.

Transmission Range (meters)	Receive Power Threshold (watts)
350	9.5081×10^{-11}
90	2.1746×10^{-8}
30	1.7615×10^{-6}

By reducing the node transmission radius, our goal is to segregate the network into sub-networks based on unit organization. So, for a transmit radius of 350 meters, we have all soldiers in the platoon competing for access to the same network. For a transmit radius of 90 meters, we reduce the number of nodes competing for access to that of the squad level, therefore creating three sub-networks. In turn, for a radius of 30 meters, we reduce the number of nodes competing for access to that of a fire team level, creating six sub-networks.

Changing the transmission radius is, at best, an imperfect means of segregating the network into distinct sub-networks. However, we are able to reduce the number of neighbor nodes for any particular node, and therefore, reduce the number of nodes competing for medium access. This should have the resulting effect of reducing the number of dropped packets from MAC-Layer collisions on the wireless medium, thereby increasing throughput, as a result of eliminating retries.

5.2. Experimental Apparatus

After simulation runs, we see a majority of packet drops resulted from MAC access collisions. The plot shown in Figure 13 compares the number of dropped packets caused by MAC collisions for the different transmission ranges for one of the five steps in the scenario (we select step 3 in the mission scenario discussed in Section 3 to show by way of example).

Each point in the graphs shows the total number of drops for all nodes in the previous 10 seconds. For all of these plots, we start plotting at time 120 since the first two minutes of simulation time are spent starting the *ns-2* agents. The simulation time from 0 seconds to 120 seconds shows a large amount of packet drops related to network setup, such as route discovery by the routing protocol and MAC address resolution required by the ARP table.

Figure 14 shows the number of drops from packets timing out for the scenario with a transmission range of 30 meters. This scenario doesn't have many drops due to MAC collisions. However, since the transmission range is so small, the network is physically segregated into distinct sub-networks; therefore, packets with a destination outside of a given sub-network timeout and are dropped by the node.

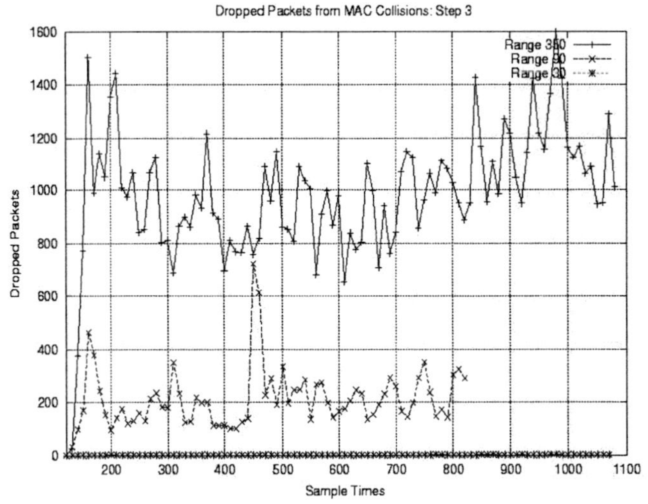

Figure 13. Dropped packets plot for mission step 3.

Figure 15 shows resulting throughput measurements for the simulation for mission step 3. We compare the throughput values for each transmission range at each step. The plot shown for step 3 doesn't show throughput for the 30-meter range, as all of the data packets were dropped during the simulation.

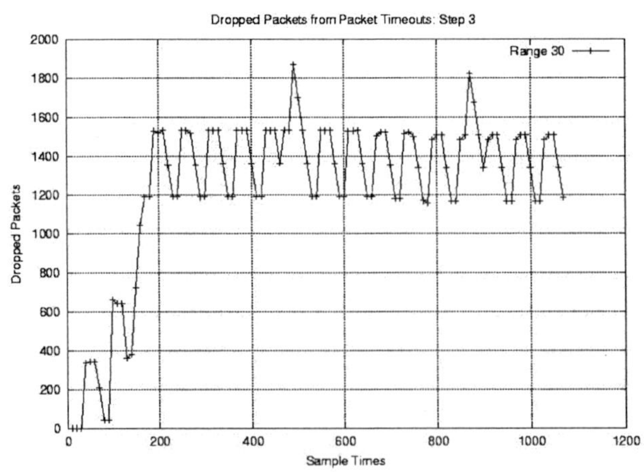

Figure 14. Timed out packets plot for mission step 3.

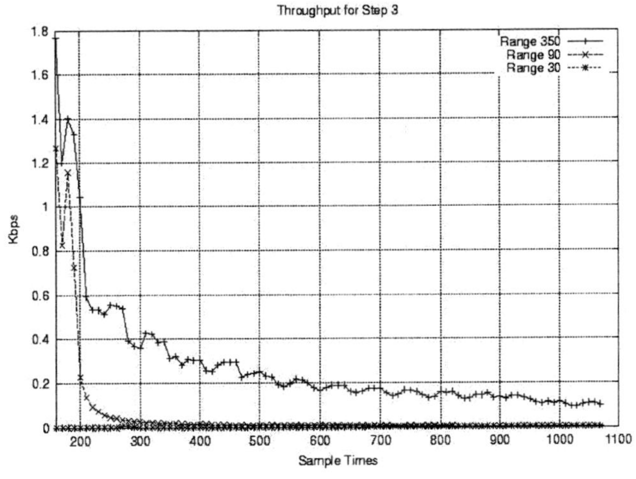

Figure 15. Throughput plot for mission step 3.

From our data, we show a reduction of the number of drops from MAC collisions as we reduce the transmission range. However, the throughput actually decreases when we decrease the transmission range, which is as a result of an increase in the number of hops from source to destination. Additionally, reducing the transmission range too much can result in physically separate networks, as shown when the transmission range is 30 meters.

If our adaptive agents running in the 802.11 MAC nodes were executing this scenario, the neighboring nodes would have made an agreement to drop back in their power and switch to a different 802.11 sub-band channel. However, on evaluating the performance based on reviewing the time lag for routing, or the number of dropped transmissions, the nodes would arrive at some balance between a condition of excessive retries and one involving high routing overhead or dropped transmissions. The simulation analysis indicates that this is a behavior that is promising, and which should be explored through an implementation at the node-level.

6. Example 2: Adaptive Code Selection

Another area we have explored is in the adaptive selection of different error coding schemes for detecting and correcting error bursts in the channel. The problem we are increasingly confronted with in wireless communications is the noise associated with a given channel. In wireless network standards such as IEEE 802.11, we address the need of increasing the robustness of communications over a 2.4 GHz channel by transmitting the data stream along with Cyclic Redundancy Check (CRC) codes. A CRC-32 code is transmitted in the MAC layer frame, which is, in turn, encoded inside a Physical layer frame with a CRC-16 code appended to the PHY frame's bit sequence. This scheme has been borrowed from the IEEE 802.3 standard [IEEE99].

The manner in which 802.11 operates, the Transmitting station is responsible for reliable transmission through use of timers around the frame transmissions and of transmission retries when the time for an ACK response from the remote receiving station has been exceeded.

This presents problems when the medium becomes congested, as increased noise from collisions in the channel, along with the overhead of frame transmission retries, further exacerbates the level of channel congestion—having deleterious effect on overall channel throughput versus capacity.

One solution is to use of alternate coding schemes with the 802.11 MAC layer, namely the Reed-Solomon scheme. The basic structure of the system is shown in Figure 16. We have constructed various hardware simulation models of the MAC and PHY layers to assess the impact of changes to the MAC layer error detection scheme. We use 802.11 as a prototypical protocol test platform, as it is recognized as being representative of other communications protocols.

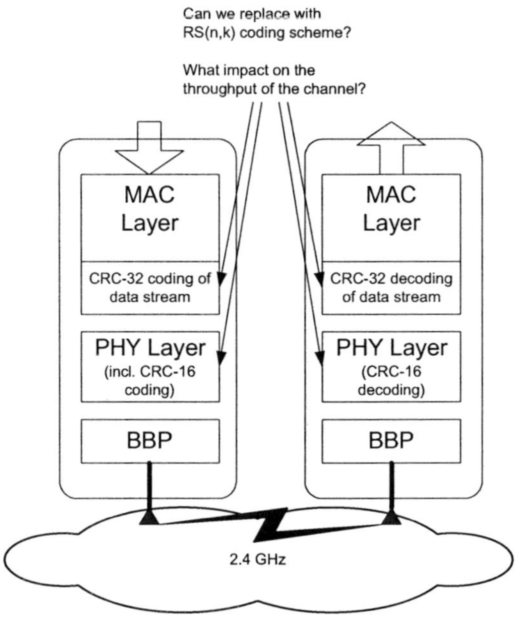

Figure 16. System structure for 802.11 WLAN.

6.1. Experimental Validation Objective

The question being explored is whether using a coding scheme having the power to not only detect, but also to correct, burst errors can improve throughput of the wireless medium. We are interested in using Linear Feedback Shift Register (LFSR) architectures to address coding and decoding using the Reed-Solomon cyclic (i.e., shift-around) block code technique. Selecting coding schemes is a tradeoff between code efficiency and the cost of computing and transmitting the code word [Skla01].

We would like to adopt a coding scheme that provides optimal code efficiency at a computational cost that makes its use feasible within the required throughput envelope for the 802.11 protocol (namely, operating such that the PHY and MAC Layers don't become the bottleneck for achieving close-to-theoretical throughput of 56 Mbits/sec). The WLAN protocol uses a transmission response scheme, whereby it is the Transmitter's responsibility to insure the integrity of the transmission by using a "watchdog" timer around frame transmissions. If the destined Receiver doesn't respond within the allotted time frame, the Transmitter retries the transmission until either the Receiver acknowledges the transmission or the Transmitter's retry count is exceeded, whichever comes first.

The Reed-Solomon codes, denoted by R-S(*n*,*k*), are designed to support error detection and correction of bit bursts that vary in length, depending on the coding scheme chosen. The code scheme is based on use of finite field GF(2^P) defined over 2 raised to a power of a prime number, so that operations of addition and multiplication are possible for the bit values within the field. Thus, it is possible to devise a non-binary code, allowing the encoding of entire words for detection and correction—making it the most powerful coding scheme for digital communications [Skla01]. For an R-S(*n*,*k*) code, we form a "code word" comprised of *n* symbols, each *m*-bits in length, by taking *k*-symbols of the original message stream and calculating a set of *2t* parity symbols to append to the transmission. This structure is depicted in Figure 17. On the receiving end, each *n*-symbol code word is taken and decoded to obtain the original message of *k* *m*-bit symbols.

The set of symbols for the code consists of all the possible binary bit patterns in the finite field for the given symbol's *m* bit-length. For example, if we selected a scheme R-S(*7*,*3*), we would take 3 3-bit "symbols" from the data stream (passed to the MAC layer as a stream of bytes), generate a set of 4 3-bit parity symbols to form a code word of 7 symbols. Given that $m = 3$, we have an extended finite field defined on GF(2^3), giving us 8 elements in the field. This means that the coding of bit patterns is fixed within the range of 0-7, inclusive, and that values generated outside of that range are defined modulo prime number *m*, thereby staying within the bounds of the field. The coding of values for the polynomial describing the field can be done via a table lookup for a small field such as the GF(2^3).

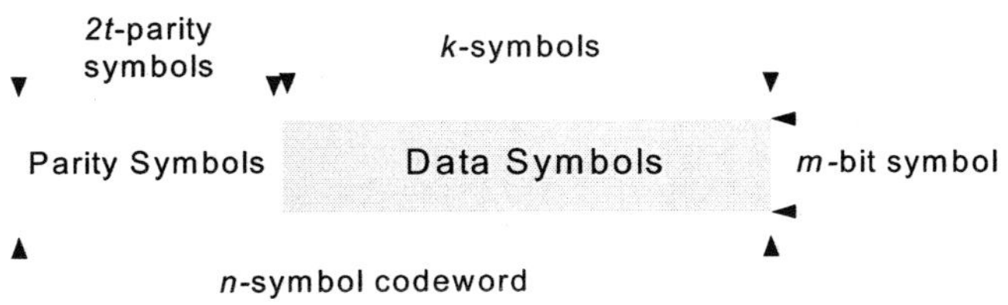

$$0 < k < n < 2^m + 2$$

$$(n,k) = (2^m - 1,\ 2^m - 1 - 2t)$$

Figure 17. Structure of R-S(n,k) code word.

There are a number of possible R-S(*n*,*k*) codes to choose from, which is a matter of trade-off. We can choose a code that has high redundancy—and thus high code efficiency—but at a cost of larger frame size. In a "bursty" noise environment, having greater redundancy in the code word affords greater probability that the decoding circuit can detect and correct multi-word errors spanning many tens of bits. The issue, however, is whether this enhanced capability of the protocol Link layer to correct errors degrades the throughput of the channel, and which code provides the best tradeoff of robustness and throughput. There is no one answer, and it depends on the conditions of the channel [Zhan05].

The source of burst noise in the channel may be related to natural atmospheric disturbance, channel crosstalk, or even malicious attack. Our interest is to be able to monitor the conditions of the channel, compiling statistics as to frame retry and failure rates related to channel noise, and to negotiate among the nodes to move to a different channel-coding scheme that fits the current conditions of the channel. If we are operating with a coding scheme that is not strong enough, the inability to correct error bursts will result in increased retries. This increases the normal load on the channel, causing congestion, leading ultimately to failure of the wireless network.

On the other hand, if we are operating with a coding scheme that has a high code-strength (in terms of number of burst symbols that can be detected and corrected), in conditions where the channel is not subjected to noise bursts, then we are wasting channel bandwidth with unnecessary parity words in the transmission. This leads us to the conclusion that the selection of an error code should be dynamic and adaptive—based on sensing of channel conditions by compiling frame data on all traffic sensed by each node, and by agent-based negotiation and quorum voting to select a new coding scheme as conditions warrant. Our initial study of this problem assumes that the thresholds for changing coding schemes is static; however, over time, the agents executing in the nodes can effectively learn what constitutes appropriate thresholds by monitoring feedback of channel throughput and performance of the network over time.

6.2. Experimental Apparatus

In Figure 18, we plot the code efficiency [Skla01], represented as the *code rate*, a ratio k/n, indicating the percentage of the code word that is comprised of the original message symbols. A code efficiency approaching 1 indicates that a small number of parity words are used, limiting the corrective capability of the given R-S(n,k) code scheme. The code schemes to the right side of the graph have a code efficiency approaching 1. By contrast, a code rate approaching 0 indicates that there are a large number of parity words appended to the message symbols to form the code word. Given that there is a high redundancy in the code word for these codes (shown towards the left side of the graph), a larger frame size must be transmitted in order to move the data across the communications channel.

For purposes of our experiments, we set the 802.11 data frame to be of fixed size, with the byte count for the various fields within the frame indicated below:

802.11g MAC frame for Data (ad-hoc mode): 2k payload + 2-byte FCH header + 3 6-byte Addr + 2-byte DID time value + 2-byte Sequence/Frag. No. + 4-byte CRC-32 = 2076 bytes.

Of these bytes in the standard data frame, 4 bytes are dedicated to CRC-32 coding for error detection. These 4 bytes are replaced with some number of R-S(n,k) parity words. The number of bytes dedicated to parity encoding depends on the scheme selected.

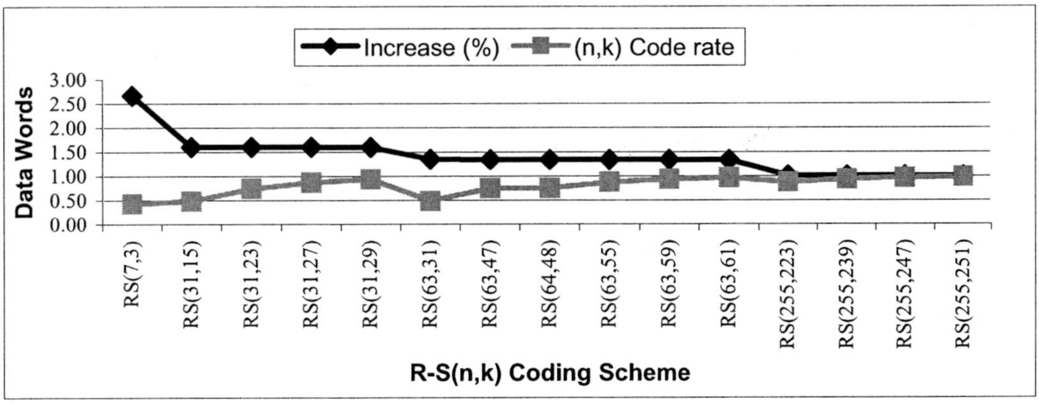

Figure 18. Code rate versus Data overhead for different R-S(n,k) codes.

We "design" a coding scheme by: (1) selecting a value for symbol size m; (2) selecting a corresponding code word size n (given the relationship between n and m as shown in Figure 17, namely that $n < 2^m + 1$); (3) selecting the symbol correcting capability of the code, namely the value of t, where $t = (n - k)/2$; and, thereby arriving at the number of m-bit symbols that must be encoded to form the code word. Using this method, we develop models for checking efficacy of the coding technique. The data for the different methods is not being presented here; rather, we want to focus on our method used for creating the set of candidate coding system models.

Creating a co-simulation environment that allows network-level models to interact with node-level models is a novel approach. The network-level models capture information about the following network properties:

- The size of the network (in terms of the number of nodes);
- The level of data load, and the types of application data, generated by each node in the network;
- The specific time-based traffic patterns generated by each node (over each node's duty cycle) which, when taken as a composite of all nodes, gives a picture of anticipated load;
- The specific patterns of mobility of each node relative to one another, which allows consideration of nodes moving into and out of transmission range in the network; and,
- The transmission strength of the nodes, relative to their power delivered to their antenna, which indicates the transmission ratio of each node in the network.

From these parameters, we can explore the impact of different wireless network link layer topologies and traffic patterns, and at what points the network might become congested. Running a simulation results in being able to evaluate the resilience and performance of the network, in terms of frame retries and dropped frames (i.e., transmissions that were not able to complete due to congestion or because specific nodes were out of range), for example.

Creating a cycle-accurate model of a reconfigurable digital embedded system that carries out the processing of the MAC and PHY layers allows system architects to explore alternate architectures and algorithms for implementing the network protocol at the node-level. The node-level models capture information on expected processing times to perform each required computing task, from which accurate throughput numbers can be obtained. The node-level models allow very detailed behaviors to be modeled in the protocols, so that a wide range of operating scenarios at the node-level can be examined, such as:

- The processing time to each step in the transmission and reception of PHY and MAC frames of all types, from which distributions can be drawn based on the relative occurrences of different types of frames in different modes of ad hoc operation;
- The processing times associated with various forms of error and exception handling and recovery, which is generally not considered in the overall modeling of the network, but which is essential to understand in terms of the impact exception processing has on node-level throughput; and,
- The impact of different algorithms and architectures employed for various forms of sub-processing (such as error detection/correction coding, compression coding, encryption coding, gaussian noise generation, FFT processing, etc.)

In the following figure, we present the basic structure of the co-simulation environment, and label the components and their primary interfaces. Using the *ns-2* network simulator, a model of a network is created according to the methodology described earlier, and is scripted using *Otcl* scripting language. This model is read by the simulator, which executes the simulation run according to the set of execution assumptions embodied in the *ns-2* C++ code containing the compiled simulation engine. As the simulation executes in *ns-2*, a Trace file is created, and entries are written into this file. The simulation of network traffic and mobility patterns is executed in a batch mode. Ordinarily, the completion of this simulation run would be the end of the modeling effort; however, we augment this wireless system-level simulation by extracting parameters from the trace and passing these to a hardware model simulator.

A simulation of a hardware model is based on the representation of concurrent threads modeled in either the VHDL or Verilog hardware description language. The HDL source code (both languages are standardized by the IEEE) is "analyzed" by an HDL model simulator, which creates a model consisting of operations that are scheduled for execution at different points during simulation. We use the ModelSim® simulator for this purpose, a commercial product. Given the analyzed hardware model, the user provides some stimuli to the model, in the form of values on input signals and buses, which causes the model to execute its complement of functionality. Before being able to execute such a simulation, we first need to have created an appropriate complement of coding models that can be selected, based on statistical data passed from the *ns-2* simulation. As shown in the previous example, trace data contains information on packet retries and dropped packets. This information, along with information kept in the hardware model about the specific frame transmission set sent during the previous simulation window, is used to make the selection of the code scheme.

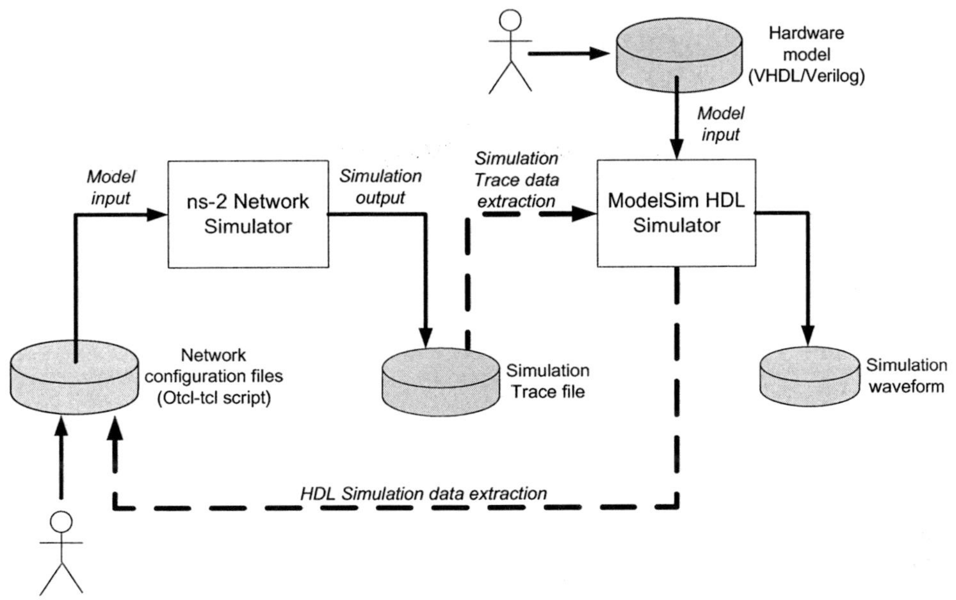

Figure 19. Co-simulation Architecture and Components.

The hardware model simulator is *event-driven*, meaning that it responds according to small incremental time steps, wherein events cause execution of functional portions of the HDL model, resulting in additional events being generated. The model executes until there are no more events to process, whereupon the simulator enters a quiescent state. The values of each signal are tracked and output in the form of a signal waveform, where the binary value of each signal is tracked as the simulation time clock is advanced. The passage of time conforms to the notions of clocking and delay in digital systems, allowing timing models of varying accuracy to be explored.

These two simulators deal with the notion of the networking "system" at very different level of abstraction. From the standpoint of *ns-2*, nodes are simply processes that execute according to the underlying execution model of the *ns-2* engine, being initialized to a particular network model—number of nodes, types of traffic, and pattern of mobility of the nodes over simulation time—according to the specification of the input script files. The HDL model simulator operates on events presented to the engine from either outside the simulator or from test "driver" components of the hardware model. The common denominator is the use of events to frame the input-response of both simulators. This common notion of event has different semantics between the two simulation models; it is overcome by a mapping between the meta-models of the two simulators—where this meta-modeling was carried out, *a priori*, .as part of our conceptual analysis as described earlier.

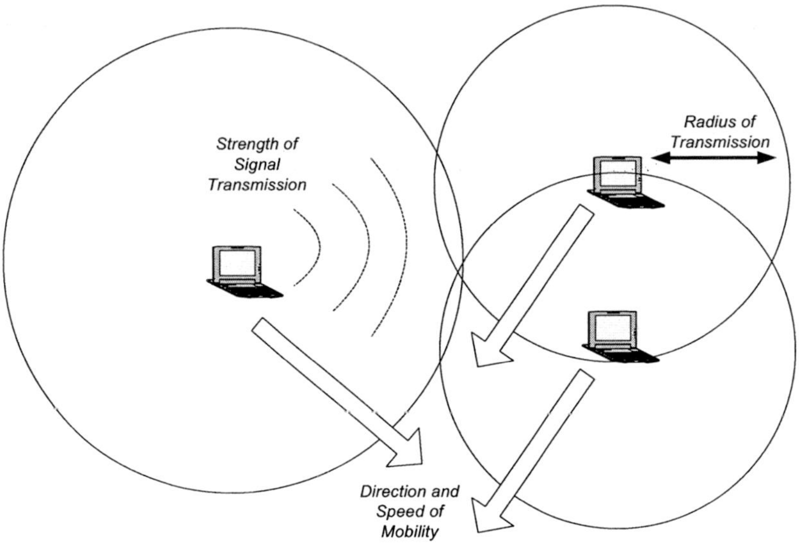

Figure 20. Network Simulation Semantics.

To envision how the two simulators work together, we start with an abstraction of the mobile ad hoc network, as shown in the figure above. We have 3 nodes in motion with respect to one another. Two of the nodes (on the right side of the figure) are initially within transmission distance of one another, and are exchanging link layer frames as part of some higher-level application activities. An example of this mobility scenario would be soldiers moving on the battlefield, as discussed earlier, in constant communication with one another via a high-speed wireless LAN network. In this scenario, we have another node that is moving in a vector that would bring it within transmission range of the other two nodes at some point into the simulation. Given that both existing nodes in the ad hoc network may not see this new node, it may set up a situation of interference (assuming all nodes are transmitting on the same 2.4 GHz sub-band). The intruding node could be operating at a power level that is greater than that of the other nodes. As this simulation scenario unfolds in *ns-2*, the Trace file would have data written to it as to the nature of the transmission medium—in the form of frame retries, dropped frames (as a result of exceeding the 802.11 MAC transmission retry count).

From the standpoint of a hardware model for the 802.11 PHY and MAC Layers, the HDL model would have a component whose function is to provide a serial data stream representing the data frame bits acquired over the 2.4 GHz channel. Given the architecture and behavior of the hardware model, the frame data stream is processed (on reception), the data payload is passed up the protocol stack (modeled as another abstract test point), and an appropriate response frame is built and transmitted, once the medium is acquired by the node. This basic model behavior is captured in the UML sequence diagram depicted in the figure below.

Here, we see the three nodes in a Sequence diagram, and the scenario corresponding to the discussion thus far. Inside the model for a given node in this configuration, we have all the elements of the protocol implemented in the HDL model, which is subjected to a series of input events corresponding to the scenario. Therefore, the HDL node-level model would be subjected to a series of good frames from the associated node in the network, followed by reception of a series of frames that are corrupted, in which the station must carry out exception handling. This scenario can be modeled from either side (or both sides) of the two-party transaction. If the node being modeled were the source of frames that are corrupted, then the node's processing would consist of transmission ACK timeout events followed by a sequence of frame transmission retries. If the scenario were such that the offending node continued to transmit interfering frames, the source node would exceed its retry count and report to higher layer components (represented in the HDL model by test components) that the transmission failed.

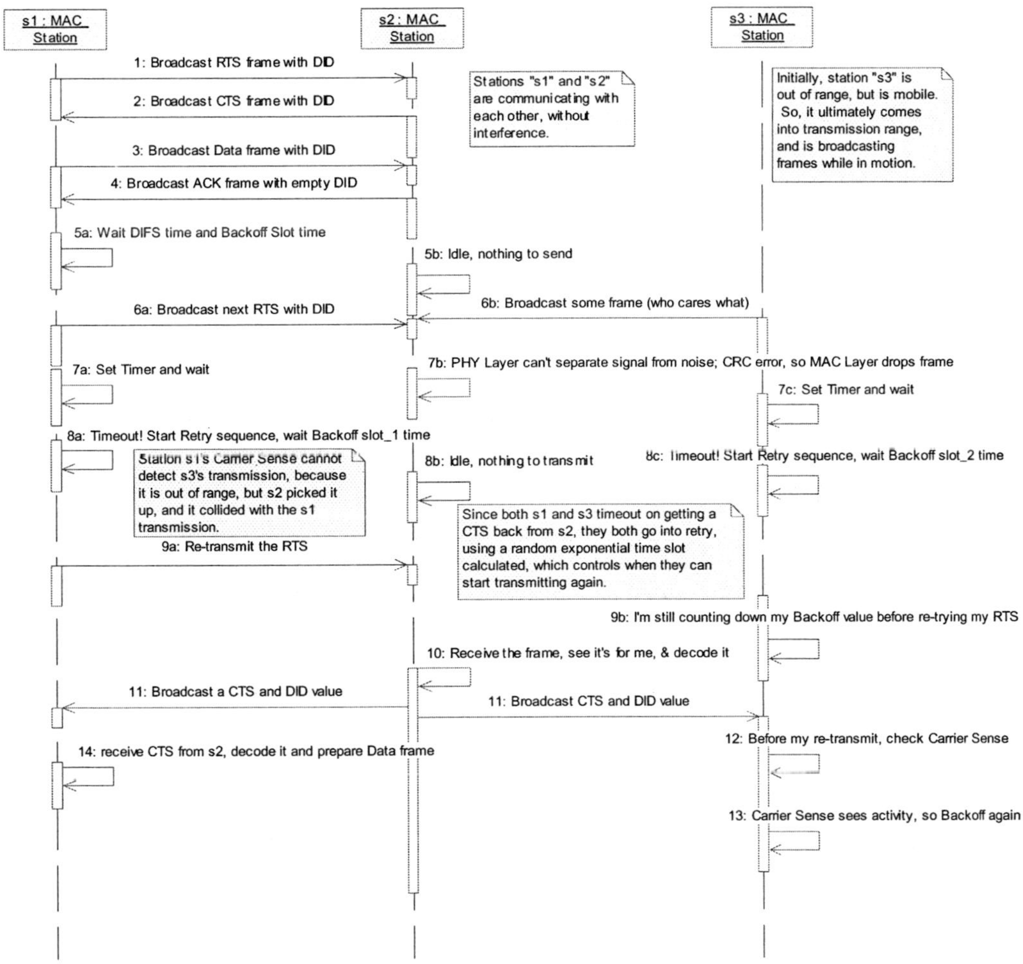

Figure 21. Node-level Simulation Semantics.

Given this type of scenario, we can discuss how the two simulators interact. The *ns-2* simulator generates traffic according to specifications, and all mobility vectors are defined relative the "reference" node being executed in the HDL simulator. Each specified event—whether mobility related, traffic related, or that associated with other model parameters—results in it being recorded in the Trace file. The key to bridging the simulators is to parse the Trace file and extract events that are relevant to the state of the reference node executing in the HDL simulator. The HDL simulator has a PLI (programming language interface), an Application Programming Interface (API) that allows external programs to provide stimulus to a HDL model. Special C/C++ code is written to parse the *ns-2* Trace file, extract the required data, and pass it across the simulator boundaries by making an appropriate sequence of PLI API calls into the HDL simulator. Inside the simulator model, there are HDL components that wait on externally supplied events, which causes events to be passed to the rest of the HDL model, according to a specially defined protocol. An example of the PLI code written to drive MAC frames across the co-simulation boundary is shown in Figure 22.

```
//File: Test bench of frame control decoder
`include "FCD.v"
module FCS_test;
  reg [15:0]FCD_IN ;
  reg [3:0] FCD_EN ;
  reg FCS_ERROR;
  reg CLK;
  reg RESET;
  wire  RTS;
  wire  CTS;
  wire  ACK;
  wire  DATA;
  wire  RETRY;
  wire  MFRAG;
  wire  [1:0]VERSION;

  parameter sim_time = 0;
FCD fcd_ports(FCD_IN, FCD_EN, FCS_ERROR, CLK, RESET,
           RTS, CTS, ACK, DATA, RETRY, MFRAG, VERSION);
 initial
  begin
// reset the FCD from PLI
   $FCD_reset();
   CLK = 0;
  end

 always
  begin
  #5 CLK = ~CLK;
  end

initial
 begin
 #3 RESET = 0;
 #1  $FCD_sendframe(FCD_IN,FCD_EN,FCS_ERROR);    // RTS
 #10 $FCD_sendframe(FCD_IN,FCD_EN,FCS_ERROR);    // CTS
 #10 $FCD_sendframe(FCD_IN,FCD_EN,FCS_ERROR);    // ACK
 #10 $FCD_sendframe(FCD_IN,FCD_EN,FCS_ERROR);    // data frame
 #10 $FCD_sendframe(FCD_IN,FCD_EN,FCS_ERROR);    // multiple fragments
 #10 $FCD_sendframe(FCD_IN,FCD_EN,FCS_ERROR);    // retry
 end

always
  begin
  #4 $FCD_displaysignals_OUT(RTS,CTS,ACK,DATA,MFRAG,RETRY);
  $display($time," RTS=%b CTS=%b ACK=%b DATA=%b MFRAG=%b RETRY=b",RTS,CTS,ACK,DATA,MFRAG,RETRY);
  end

initial
#65 $finish;
endmodule
```

Figure 22. Hardware simulator PLI code for generating frame traffic.

The HDL simulator executes its model code, updating the state of its internal data structures. Considerable number of hardware clock cycles may elapse before the node-level model has some response to the stimuli it received from the network-level simulation model.

Once some response has been generated, the appropriate signals are read and data are converted and passed back out through PLI interface codes specifically for this purpose. C/C++ code residing external to both simulators reads these "instructions" and constructs a set of *Otcl* script codes and the command sequence to re-execute the script model in *ns-2* with the updated changes. The flow of these transformations between the two simulators is shown in Figure 23.

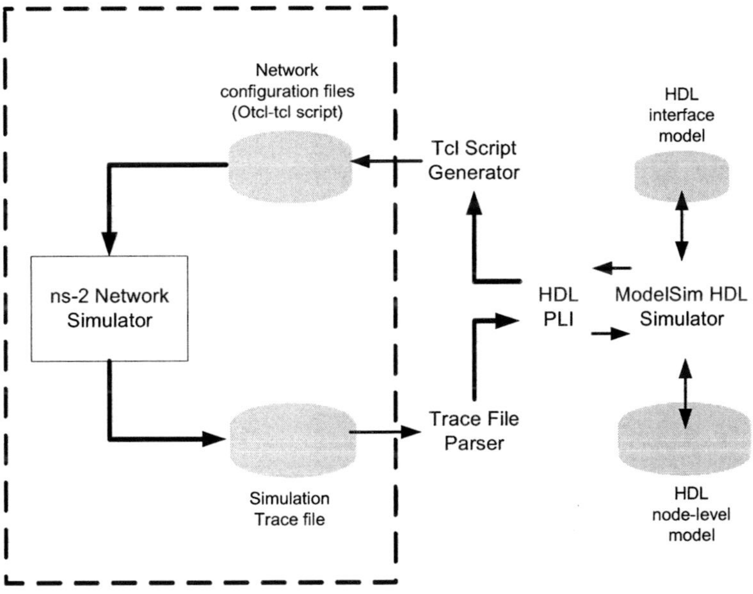

Figure 23. Interfacing and Data Transformation Between Simulator Runs.

The *ns-2* simulator executes in batch mode, in that its parameterized model is run to completion, and the results are dumped to the Trace file. The HDL simulator, on the other hand, may continue to execute, where its state is continually modified by new stimuli provided through the external interface. Two pieces of external code are created, one that parses the Trace file for extracting semantically relevant information for the HDL simulation, and another to generate appropriate *Otcl* script code to modify the network simulation model parameters for re-execution in *ns-2*.

Here we have looked at the potential of incorporating adaptation on the basis of error correction code selection. Assessing the impact of such changes to the MAC protocol are not possible using system-level simulation of the wireless network alone; we must also have the ability to simulate candidate models of the MAC node as an instance of one of many such nodes. The models of the coding application and the MAC Layer itself have been implemented in a hardware description language (HDL) that is cycle accurate within the node. This method allows node-level models to be developed and run against different sets of system-level assumptions to explore the impact of possible adaptive behaviors.

7. Summary

In this chapter, we have presented a simulation modeling methodology and experimental framework, along with the conceptual model design methodology used to arrive at the simulation model, for exploring adaptive, self-regulating wireless network behaviors. We started with a discussion of an application domain, the Army's hierarchical fighting force infantry unit structure, onto which a wireless LAN-based communications infrastructure is being developed.

To create an accurate network model for simulation, one must understand how concepts in the problem domain map onto constructs for using simulators. We started with concepts of the battlefield domain—unit movement, fighting force roles, and artifact creation and distribution—and mapped these onto a simulation-scripting model for the *ns-2* network simulator. We presented how the properties of the conceptual domain are used to explore effects on network behavior in terms of traffic pattern and traffic timing as the simulation unfolds. To that end, we adopted a conceptual modeling method and its associated notations, namely, UML, as a means to capture and codify knowledge that, heretofore, had only been available in document form. This codification using domain models was critical to understanding what elements of the fighting force domain would be required for constructing network simulation models of appropriate detail to understand the impact of parameter settings on network performance, robustness and QoS.

A key aspect of the analysis component of this research has concerned how to map from a domain model representing the fighting force to a model appropriate for use in constructing network simulation scripts in *ns-2*. In addition, it involved developing an appropriate mapping between two different sets of simulator semantics—one involving system-level constructs and another using node-level constructs. The primary issues here are as follows: (1) moving from a perspective of modeling the enterprise-oriented aspects to one based on considering the domain artifacts as a collection of network nodes with certain traffic and movement properties; (2) translating the enterprise semantics of entities, relationships and static constraints to those appropriate for creating topological models useful for simulating network dynamics; and, (3) mapping between system-level and node-level simulation constructs for "round trip" simulation.

To make this transition, we considered a number of ways to represent the model but, ultimately, the process consisted of the following: (1) understanding the semantics of the *ns-2* networking and the station hardware simulation models; (2) creating meta-models of these semantic concepts and relations as UML Class diagrams; and, (3) using this understanding of the respective domain semantics to dictate creating a set of mapping policies between the problem domain and the simulation model domains. Since one of the considerations of this research problem has been to create a simulation model that could be extended by creating specific battlefield scenarios, it was important to be able to make the transition between the models as straightforward as possible. Furthermore, our analysis indicates that this process can be automated with appropriately-constructed model capture and compilation tools, created as "plug-in" extensions to UML-based design tools through the use of COM component add-in technology [IBM01].

Our research objective has been to use our simulation baseline model development effort to provide a platform on which to explore answers to issues posed in regards to evolving the

802.11 WLAN protocol and architecture at the MAC layer. Namely, we have been interested in codifying a methodology and creating a co-simulation framework allowing us to explore inclusion of adaptive self-organizing extensions to wireless network MAC layer protocols and validate these extensions through analysis and simulation. We are making such modifications to add adaptive capabilities to Land Warrior-like networks as a whole, such that we can add self-regulating and self-management capabilities to the wireless network on the battlefield in support of the Army's fighting force "workflow".

Assessing the types of changes we might posit making to the MAC layer is based on collecting large amounts of simulation data, across a number of different model runs, then attempting to analyze the data to identify patterns in the behavior of the network. Battlefield scenarios can be diverse, as can be the conditions under which the network can be expected to operate. To be able to explore such changes easily, we need a model platform that is well codified and easily modifiable, to support easy creation of model scenarios that can be subjected to simulation experiments. Our approach has, thus, been to focus on creating a robust simulation model infrastructure for such exploratory work, where the necessary domain knowledge has been articulated, first, as a robust collection of domain models representing both the application-specific problem domain and the simulation abstractions of this domain. We found that this objective could be validated, and then transformed into a set of extensible *Otcl* scripts, which were used as the basis for subsequent exploratory simulation runs to define baseline behaviors against which we could run validation exercises. Early indications tell us that this methodology and experimental framework can be adapted to other domains in a straightforward manner—which is part of our ongoing work.

References

[Appl99] Appleby, S., and S. Steward, "Mobile Software Agents for Control in Telecommunications Networks", Hayzelden, A., and J. Bigham (eds.), Software Agents for Future Communication Systems, Springer Verlag: Berlin, 1999, pp. 270-286.

[Army01a] U.S. Army, "United States Army White Paper: Concepts for the Objective Force", Headquarters, Department of the Army, Washington, DC, U.S. Government Printing Office, November, 2001.

[Army01b] U.S. Army, "Operational Requirements Document for Land Warrior, ACAT I," United States Army, October 2001, www.natick.army.mil/soldier/wsit/LW_ORD.pdf.

[Aror03] Arora, A., P. Dutta, S. Bapat, V. Kulathumani, and H. Zhang, "GS3: Scalable Self-configuration and Self-healing in Wireless Sensor Networks", *Computer Networks*, **43**(4), Elsevier Science Publishers, 2003, pp. 459-480.

[Axel97] Axelrod, R., *The Complexity of Cooperation: Agent-Based Models of Competition and Cooperation,* Princeton University Press, 1997.

[Bak96] Buell, D. A., J. M. Arnold, and W. J. Kleinfelder, eds., Splash 2: FPGAs in a Custom Computing Machine, *IEEE Computer Society Press*, 1996.

[Barb96] Barbacci, M.R., M.H. Klein, and C.B. Weinstock, Principles for Evaluating the Quality Attributes of a Software Architecture, *Technical Report CMU/SEI-96-TR-036*, Software Engineering Institute, 1996.

[Baue02] Bauer, B., "UML Class Diagrams Revisited in the Context of Agent-Based Systems", in Wooldridge, M. J., G. Weiss, and P. Cianciarini (eds.), *Agent-Oriented Software Engineering - AOSE 2001*, LNCS 2222, Springer-Verlag, Berlin, 2002, pp. 101-118.

[Baum99] Bäumer, C., and T. Magedanz, "Grasshopper: A Mobile Agent Platform for Active Telecommunication Networks", *Proceedings of the 3rd International Workshop on Intelligent Agents for Telecommunication Applications (IATA-99)*, LNAI, Vol. 1699, Springer Verlag: Berlin, 1999, pp. 19-32.

[Bec02] Becker, A. S., "Wearable Computing is Silently Infiltrating the Main stream", *Portable Design Magazine*, January 2002, pp. 14-16.

[Bell02] Bellows, P., J. Flidr, T. Lehman, B. Schott, K. Underwood. GRIP: A Reconfigurable Architecture for Host-Based Gigabit-Rate Packet Processing, *Tenth Annual IEEE Symposium on Field Programmable Custom Computing Machines*, Napa Valley, California, April 2002.

[Bens98] Ben-Shaul, I., and G. Kaiser, "Coordinating Distributed Components Over the Internet", *IEEE Internet Computing*, Vol. 2, No. 2, 1998, pp. 83-86.

[Bern98] Bernardo, L., and P. Pinto, "Scalable Service Deployment on Highly Populated Networks", *Proc. 2nd International Workshop on Intelligent Agents for Telecommunication Applications (IATA-98),* LNAI, Vol. 1437, Springer Verlag: Berlin, 1998, pp. 29-44.

[Bon00] Bonabeau, E., M.Dorigo, and G.Theraulaz, "Inspiration for Optimization from Social Insect Behaviour", *Nature*, **406**, 2000, pp. 39–42.

[Bor01] Nikita Borisov, Ian Goldberg and David Wagner, "Intercepting Mobile Communications: The Insecurity of 802.11", *Proceedings 7th Annual International Conference on Mobile Computing and Networking*, July 16-21, 2001, Association of Computing Machinery

[Bres00] Breslau, L., D. Estrin, K. Fall, S. Floyd, J. Heidemann, A. Helmy, P. Huang, S. McCanne, K. Varadhan, Y, Xu, and H. Yu, "Advances in Network Simulation", *IEEE Computer*, Vol. 33, No. 5, May 2000, pp. 59-67.

[Breu98] Breugst, M., and T. Magedanz, "Mobile Agents -- Enabling Technology for Active Intelligent Network Implementation", IEEE Network, 12(3), Aug. 1998, pp. 53-60.

[Busu99] Busuioc, M.N., "Distributed Intelligent Agents – A Solution for the Management of Complex Telecommunications Services", Hayzelden, A., and J. Bigham (eds.), *Software Agents for Future Communication Systems*, Springer Verlag: Berlin, 1999, pp. 112-129.

[Bre00] Breslau, L., D. Estrin, K. Fall, S. Floyd, J. Heidemann, A. Helmy, P. Huang, S. McCanne, K. Varadhan, Y. Xu, and H. Yu, "Advances in Network Simulation", *IEEE Computer*, Vol. 33, No. 5, May 2000, pp. 59-67.

[Bre02] Brebner, G., "Single-chip Gigabit Mixed-Version IP Router on Vertex-II Pro", in K. Pocek and J. Arnold, eds., *Proceedings IEEE Symposium on Field-Programmable Custom Computing Machines FCCM-02*, IEEE Press, 2002.

[Bru91] Brustolini, J. C., "Autonomous Agents: Characterization and Requirements", *Technical Report CMU-CS-91-204,* Carnegie Mellon University, 1991.

[Cam91] Camposano, R., and W. Wolf (eds.), High-level VLSI Synthesis, Kluwer Academic Publishers, 1991.

[Coop99] Cooperative Distributed Problem Solving for Communications Network Management", Hayzelden, A., and J. Bigham (eds.), Software Agents for Future Communication Systems, Springer Verlag: Berlin, 1999, pp. 247-269.

[Davi05] Davis, J., and R. Bonnell, "A Taxonomy of Propositional Logic Constraint Patterns for the Unified Modeling Language", to appear in *IEEE Transactions of Knowledge and Data Engineering.*

[Den95] Dennett, D. C., Darwin's Dangerous Idea: *Evolution and the Meanings of Life.* Simon and Schuster, New York, 1995.

[Dhb98] Davis, J. P., ", M. N. Huhns, and R. D. Bonnell, "A Method and Architecture for Building Compliance Agents", *Proceedings AAAI-98 Workshop on Intelligent Agents*, AAAI Press, 1998.

[Die00] Diehl, S., "Finding the Right ASIC Formula", *Portable Design Magazine*, June 2000, pp. 22-32

[Dign02] Dignum, V., H. Weigand, and L. Xu, "Agent Societies: Towards Frameworks-Based Design", Wooldridge, M., G. Weiss, and P. Ciancarini (eds.), Agent-Oriented Software Engineering II, AOSE 2001, Springer Verlag: Berlin, 2002, pp. 33-49.

[Dil00] Diehl, S., "IP Reuse Drives SoC Design", *Portable Design Magazine*, May 2000, pp. 26-35.

[Dil02] Dilpert, B., "Global Codec Warming", *EDN Magazine*, May 2, 2002, p. 25.

[Dori96] Dorigo, M., V. Maniezzo, V. and A. Colorni, "The Ant System: Optimization by a Colony of Cooperating Agents", *IEEE Transactions on Systems, Man, and Cybernetics–Part B,* **26**, 1996, pp. 29–41.

[Dula02] Dulay, N., N. Damianou, E. Lupu, and M. Sloman, "A Policy Language for the Management of Distributed Agents", Wooldridge, M., G. Weiss, and P. Ciancarini (eds.), Agent-Oriented Software Engineering II, AOSE 2001, Springer Verlag: Berlin, 2002, pp. 84-100.

[Dzc02] Davis, W. R., N. Zhang, K. Camera, F. Chen, D. Markovic, N. Chan, B. Nikolic and R. W. Brodersen, "A Design Environment for High Throughput, Low Power Dedicated Signal Processing Systems", *IEEE Journal of Solid State Circuits*, Vol. 37, No. 3, March 2002, pp. 420-431

[Econ02] The Economist, "Computers that Run Themselves", The Economist, Sept. 19, 2002.

[Elda99] El-Darieby, M., and A. Bieszczad, "Intelligent Mobile Agents: Towards Network Fault Management Automation", *Proc. 6th IFIP/IEEE International Symposium on Integrated Network Management*, May 1999, pp. 611-622.

[Ete01] Etemad, K., "Enhanced Random Access and Reservation Scheme in CDMA2000", *IEEE Personal Communications*, April 2001, pp. 30-36.

[Etz94] Etzioni and Weld, "A Softbot-Based Interface to the Internet", Communications of the ACM, Vol. 37, No. 7, July 1994, pp. 72-79.

[Fall00] Fall, K. and K. Varadhan, editors, The ns Manual, The VINT Project, www.isi.edu/nsnam/ns/ns-documentation.html, 24 August 2000.

[Feng81] Feng, T-Y., "A Survey of Interconnection Networks", *IEEE Computer*, December 1981, pp. 12-27.

[FIPA97] Foundation for Intelligent Physical Agents, *FIPA 97 Specification Part 7: Network Management and Provisioning*, 1997.

[FIPA98] Foundation for Intelligent Physical Agents, *FIPA 98 Specification Part 10: Agent Security Management October*, 1998.

[Frei98] Frei, C., and B. Faltings A Dynamic Hierarchy of Intelligent Agents for Network Management *Proc. 2nd International Workshop on Intelligent Agents for Telecommunication Applications (IATA-98)*, LNAI, Vol. 1437, Springer Verlag: Berlin, 1998, pp. 1-16.

[Forr93] Forrest, S., and M. Mitchell, "What Makes a Problem Hard for a Genetic Algorithm? Some Anomalous Results and Their Explanation", *Machine Learning* **13**, 1993, pp. 285-319.

[Gaj94] Gajski, D., and L. Ramachandran, "Introduction to High-level Synthesis", *IEEE Design & Test of Computers*, Vol. 11, No. 4, Winter 1994, pp. 44-54

[Gava99] Gavalas, D., D. Greenwood, M. Ghanbari, and M. O΄Mahony, "Using Mobile Agents for Distributed Network Performance Management", *Proc. 3rd International Workshop on Intelligent Agents for Telecommunication Applications (IATA 99),* LNAI, Vol. 1699, Springer Verlag: Berlin, 1999, pp. 96-112.

[Gen94] Genesereth, M. and Ketchpel, "Software Agents", *Communications of the ACM*, Vol. 37, No. 7, July 1994, pp. 48-53.

[Geor02] George, S., D. Evans, and L. Davidson, "A Biologically Inspired Programming Model for Self-Healing Systems", *Proceedings WOSS '02,* Nov 18-19, 2002, Charleston, SC, USA.

[Gerb99] Gerber, C., B. Bauer, and D. Steiner, "Resource Adaptation for a Scalable Agent Society in the MoTiV-PTA Domain", Hayzelden, A., and J. Bigham (eds.)*, Software Agents for Future Communication Systems*, Springer Verlag: Berlin, 1999, pp. 183-206.

[Goo00] Goodman, D. J., "The Wireless Internet: Promises and Challenges", *IEEE Computer Magazine*, Vol. 33, No. 7., July 2000, pp. 36-41

[Gord99] Gordon, D., *Ants at Work: How an Insect Society is Organized*, The Free Press, 1999.

[Gra00] Bob Gray, "Soldiers, Agents and Wireless Networks: A Report on a Military Application", *Proceedings PAAM 2000*, Manchester, England, http://actcomm.dartmouth.edu.

[Griv98] Grivas, M., and S. Turner, "Agent Technology in Load Balancing for Network Applications", *Proc. Workshop on Intelligent Agents on the Internet and Web, 4th World Congress on Expert Systems*, 1998.

[Grei98] Greis, Marc. "Tutorial for the Network Simulator ns", available at URL http://www.isi.edu/nsnam/ns/tutorial/index.html.

[Gvs01] Golmie, N., R. E. Van Dyck and A. Soltanian, "Interference of Bluetooth and IEEE 802.11: Simulation Modeling and Performance Evaluation", in *Proceedings ACM* MSWiM 2001, Rome, Italy, pp. 11-18

[Hall02] Hall, A., J. Surdu, F. Maymi, A. Deb, and K. Freberg, "Modeling the Communications Capabilities of the Infantry Soldier," *Communication Networks and Distributed Systems Modeling and Simulation Conference (CNDS 2002),* Society of Computer Simulation International, January 2002.

[Hayz98] Hayzelden, A., and J. Bigham, "Heterogeneous Multi-Agent Architecture for ATM Virtual Path Network Resource Configuration", *Proceedings 2nd International Workshop on Intelligent Agents for Telecommunication Applications (IATA-98),* LNAI, Vol. 1437, Springer Verlag: Berlin, 1998, pp. 45-59.

[Hayz99] Hayzelden, A., J. Bigham, and Z. Luo, "Multi-Agent Interactions for a Network Management System (Tele-MACS Approach)", Hayzelden, A., and J. Bigham (eds.), *Software Agents for Future Communication Systems*, Springer Verlag: Berlin, 1999, pp. 160-182.

[Hei01] John Heidelmann, Nirupama Bulusu, Jeremy Elson, Chalermek Intanagonwiwat, Kun-chan Lan, Ya Xu, Wei Ye, Deborah Estrin and Ramesh Govindan, "Effect of Detail in Wireless Network Simulation", *In Proceedings of the SCS Multiconference on Distributed Simulation*, pp. 3-11. Phoenix, Arizona, USA. January, 2001.

[Heu98] Heusse, M., S. Guérin, D. Snyers, and P. Kuntz, "Adaptive Agent-driven Routing and Load Balancing in Communication Networks", *Adv. Compl. Syst.* **1**, 1998, pp. 237-254.

[Hol92] Holland, J., Adaptation in Natural and Artificial Systems, 2nd Ed., MIT Press, 1992.

[Hos01] Hoskin, J., "Robust Implementations of Java Technology for Information Appliances and Embedded Devices", *Handheld & Wireless Solutions Journal*, Volume 1, Fall 2001, Intel Corporation, pp. 22-28

[Huhn01] Huhns, M., "Interaction-Oriented Programming", Wooldridge, and P. Ciancarini (eds.), *Agent-Oriented Software Engineering*, AOSE 2000, Springer Verlag: Berlin, 2001, pp. 29-44.

[IBM01] IBM Corporation, Rational Rose 2001 Extensibility Interface Guide, http://www.rational.com, 2001.

[IEEE99] LAN MAN Standards Committee of the IEEE Computer Society, ANSI/IEEE Standard 802.11, Part 11: *Wireless LAN Medium Access Control (MAC) and Physical Layer (PHY) Specifications,* 1999.

[Int00] Intanagonwiwat, C., Ramesh Govindan and Deborah Estrin, "Directed Diffusion: A Scalable and Robust Communication Paradigm for Sensor Networks", In *Proceedings of the Sixth Annual International Conference on Mobile Computing and Networks (MOBICOM),* Boston, Massachusetts, ACM, August 2000.

[Jft01] Jobin, J., M. Faloutsos and S. K. Tripathi, "Performance Evaluation of Mobile Wireless Networks: A New Perspective, in *Proceedings ACM MSWiM 2001*, Rome, Italy, pp. 65-72.

[Jic00] Jin, K. T. and D. H Cho, "A MAC Algorithm for Energy-Limited Ad-Hoc Networks", *Proceedings IEEE VTC 2000*, pp. 219-222.

[Joh01] Johnson, S., Emergence: The Connected Lives of Ants, Brains, *Cities and Software*, Scribner Publishing, Inc., 2001.

[John89] Johnson, N. E., "Mediating Representation in Knowledge Elicitation", in Diaper, D. (ed.), *Knowledge Elicitation: Principles, Techniques, and Applications*, Ellis Horwood, Chichester, UK, 1989, pp. 179-194.

[John96] Johnson, David B., and D. Maltz, "Dynamic Source Routing in Ad Hoc Wireless Networks," *Mobile Computing*, Kluwer Academic Publishers, Dordrecht, The Netherlands, 1996.

[Kur03] Kurose, J. F., and K. W. Ross, Computer Networking: *A Top-Down Approach Featuring the Internet,* Addison Wesley Publishers, Inc., 2003

[Lee00] Lee, S, W. Su, and M. Gerla, "On-Demand Multicast Routing Protocol in Multihop Wireless Mobile Networks," *Mobile Networks and Applications, special issue on Multipoint Communication in Wireless Mobile Networks*, ACM/Baltzer, 2000.

[Mae94] Maes, P., “Agents that Reduce Work and Information Overload”, Communications of the ACM, Vol. 37, No. 7, July 1994, pp. 31-40.

[Mae95] Maes, P., “Artificial Life Meets Entertainment: Lifelike Autonomous Agents”, Communications of the ACM, Vol. 38, No. 11, November 1995, pp. 108-114

[May02] Maymi, F., J. Surdu, A. Hall, R. Beltramini, “Modeling the Wireless Network Architecture of Land Warrior,” in *Proceedings of the 2002 Winter Simulations Conference*, Society of Computer Simulation International, 2002.

[Megu01] Meguerdichian, S., S. Slijepcevic, V. Karayan, and M. Potkonjak, “Localized Algorithms In Wireless Ad-Hoc Networks: Location Discovery and Sensor Exposure”, *Proc. 2nd ACM International Symposium on Mobile Ad hoc Networking & Computing*, Long Beach, CA, USA, 2001, pp. 106 – 116.

[Mina99] Minar, N., K. Kramer, and P. Maes, “Cooperating Mobile Agents for Dynamic Mobile Routing”, Hayzelden, A., and J. Bigham (eds.), Software Agents for Future Communication Systems, Springer Verlag: Berlin, 1999, pp. 287-304.

[Mit01] Mittag, L., “Mobile IP”, Embedded Systems Programming, July 2001, pp. 47-56.

[Mitc96] Mitchell, M., J. Crutchfield, and R. Das, “Evolving Cellular Automata with Genetic Algorithms: A Review of Recent Work”, *Proceedings of 1st International Conference on Evolutionary Computation and Its Applications*, EvCA-96, Moscow, Russia, 1996.

[Mms02] Memik, G., S. O. Memik and W. H. Mangione-Smith, “Design and Analysis of a Layer Seven Network Processor Accelerator Using Reconfigurable Logic”, in K. Pocek and J. Arnold, eds., Proceedings *IEEE Symposium on Field-Programmable Custom Computing Machines FCCM-02,* IEEE Press, 2002

[Nas02] Nass, R., “What’s Next for Portable Power?”, *Portable Design Magazine*, January 2002, pp. 24-30.

[Noon02] Noonan, R. and P. Cuviello, “C4 and ISR for the Objective Force,” United States Army, www.us.army.mil.

[OMG03] Object Management Group, OMG Unified Modeling Language Specification Version 1.5, formal/03-03-01, March 2003, http://www.omg.org.

[Pfl97] Pfleeger, C. P., Security in Computing, 2nd Ed., Prentice Hall Publishers, 1997.

[Puli99] Puliafito, A., and O. Tomarchio, “Advanced Network Management Functionalities through the Use of Mobile Software Agents”, *Proceedings 3rd International Workshop on Intelligent Agents for Telecommunication Applications (IATA-99),* LNAI, Vol. 1699, Springer Verlag: Berlin, 1999, pp. 33-45.

[Qhu01] G. Quan and X. (Sharon) Hu, "Enhanced fixed-priority scheduling with (m,k)-firm guarantee", IEEE Real-Time Systems Symposium, pp 79-88, 2000.

[Rrh00] Ramanathan, R. and R. Rosales-Hain, “Topology Control of Multihop Wireless Networks using Transmit Power Adjustment”, in *Proceedings IEEE Infocom* 2000, pp. 404-413

[Rumb99] Rumbaugh, J., I. Jacobson, and G. Booch, The Unified Modeling Language Reference Manual, Addison-Wesley Longman, Inc., 1999.

[Rum86] Rumelhart, D. E. and J. L. McClelland, *Parallel Distributed Processing – Explorations in the Microstructure of Cognition: Fundamentals*, MIT Press, 1986.

[Sgv02] Stitt, G., B. Grattan, J. Villarreal and F. Vahid, “Using On-Chip Configurable Logic to Reduce Embedded System Software Energy”, in K. Pocek and J. Arnold, eds.,

Proceedings IEEE Symposium on Field-Programmable Custom Computing Machines FCCM-02, IEEE Press, 2002.

[Sch02] B. Schott, S. Crago, R. Parker, L. Carter, C. Chen, J. Czarnaski, M. French, I. Hom, T. Tho, and T. Valenti, "Reconfigurable Architectures for System-Level Applications of Adaptive Computing," submitted to IEEE Transactions on VLSI Design, 2002.

[Scho99] Schoonderwoerd, R., and O. Holland, “Minimal Agents for Communications Network Routing: The Social Insect Paradigm”, Hayzelden, A., and J. Bigham (eds.), Software Agents for Future Communication Systems, Springer Verlag: Berlin, 1999, pp. 305-325.

[Shin99] Shinseki, E. “The Army Vision: Soldiers on Point for the Nation, Persuasive in Peace, Invincible in War,” *Office of the Chief of Staff*, Washington, DC, October 1999 https://www.army.mil/vision/Documents/The%20Army%20Vision.PDF.

[Sing99] Singh, M. P., “Conceptual Modeling for Multiagent Systems: Applying Interaction-Oriented Programming”, in Chen, P. P. (ed.), *Conceptual Modeling, LCNS 1565*, Springer-Verlag, Berlin, 1999, pp. 195-210.

[Sing03] Singh, M., and V. Prasanna, “A Hierarchical Model for Distributed Collaborative Computation in Wireless Sensor Networks”, *Proc. IEEE International Parallel and Distributed Processing Symposium*, April 22-26, 2003, pp. 10-16.

[Skla01] Sklar, B., *Digital Communications: Fundamentals and Applications, 2nd ed.*, Prentice-Hall Publishers, Inc., 2001.

[Slt01] Staehle, D., K. Leibnitz and K. Tsipotis, “QoS of Internet Access with GPRS”, in *Proceedings ACM MSWiM 2001*, Rome, Italy, pp. 57-64

[Sms02] Shaikh, F. A., S. McClellan, M. Singh and S. K. Chakravarthy, “End-to-End Testing of IP QoS Mechanisms”, *IEEE Computer*, May 2002, pp. 80-87

[Sxt01] Xu, S. and T. Saadawi, “Does the IEEE 802.11 MAC Protocol Work Well in Multi-hop Wireless Ad Hoc Networks?”, *IEEE Communications Magazine*, June 2001, pp. 130-137.

[Tar02] Tarasewich, P. and P. R. Mc Mullen, “Swarm Intelligence: Power in Numbers”, *Communications of the ACM, August 2002*, Vol. 45, No. 8, pp. 62-67.

[Toh01] Toh, C. K., “Maximum Battery Life Routing to Support Ubiquitous Mobile Computing in Wireless Ad Hoc Networks”, *IEEE Communications Magazine*, June 2001, pp. 138-147.

[Tint99] Tintin, R., and D. Lee, “Intelligent and Mobile Agents over Legacy, Present and Future Telecommunication Networks”, *Proc. 1st International Workshop on Mobile Agents for Telecommunication Applications (MATA'99),* World Scientific Publishing Ltd., October 1999, pp. 109-126.

[Vand02] Van Dyke Perunak, H., and J. Odell, “Representing Social Structures in UML”, Wooldridge, M., G. Weiss, and P. Ciancarini (eds.), Agent-Oriented Software Engineering II, AOSE 2001, Springer Verlag: Berlin, 2002, pp. 1-16.

[Vasc02] Vasconcelos, W., D. Robertson, and J. Agusti, “A Lifecycle for Models of Large Multi-Agent Systems” Wooldridge, M., G. Weiss, and P. Ciancarini (eds.), Agent-Oriented Software Engineering II, AOSE 2001, Springer Verlag: Berlin, 2002, pp. 297-317.

[Whit98] White, T., A. Bieszczad, B. Pagurek, G. Sugar, and X. Tran, “Intelligent Network Modeling using Mobile Agents”, *Proceedings IEEE Global Telecommunications Conf., GLOBECOM '98*, Sydney, Australia, November 8-12, 1998, pp. 1082-1087.

[Witt85] Witt, B.I., "Communicating Modules: A Software Design Model for Concurrent Distributed Systems", *IEEE Computer, January* 1985, pp. 67-77.

[Wool00] Wooldridge, M., N. Jennings, and D. Kinny, "The Gaia Methodology for Agent-oriented Analysis and Design", *Journal of Autonomous Agents and Multi-agent Systems*, Vol. 3(3), 2000, pp. 285-312.

[Wub01] Wu, C. and D. P. Bertsekas, "Distributed Power Control Algorithms for Wireless Networks", *IEEE Transactions on Vehicular Technology*, Vol. 50, No. 2, March 2001, pp. 504-514.

[Zhan02] Zhang, H., and A. Arora, "GS3: Scalable Self-Configuration and Self-healing in Wireless Networks", *Proc. 21st ACM Symposium on Principles of Distributed Computing, PODC 2002*, pp. 58-67.

[Zhan05] Zhang, H., A. Arora, Y. Choi, and M. Gouda , "Reliable Bursty Convergecast in Wireless Sensor Networks, *Proc. 6th ACM International Symposium on Mobile Ad Hoc Networking and Computing, MobiHoc 2005,* Association for Computing Machinery, 2005.

In: Adaptation Techniques in Wireless Multimedia Networks ISBN 1-59454-883-8
Editors: Wei Li and Yang Xiao, pp. 191-213

Chapter 9

OPTIMAL DOWNLINK RESOURCE ALLOCATION FOR CELLULAR CDMA NETWORKS*

Keivan Navaie[1,2]†, ***Delfin Y. Montuno***[1,3]‡,
Halim Yanikomeroglu[2§] ***and Yiqiang Q. Zhao***[1¶]
[1]School of Mathematics and Statistics, Carleton University,
Ottawa, Ontario, Canada, K1S 5B6
[2]Broadband Communications and Wireless Systems (BCWS)
Center, Department of System and Computer Engineering,
Carleton University, Ottawa, Ontario, Canada, K1S 5B6
[3]Nortel Networks, Nepean, Ontario, Canada, K2H 8E9

Abstract

In this Chapter we study the optimal downlink resource allocation problem for data traffic in a CDMA/TDMA cellular network. We use the dynamic pricing approach based on utility-functions to formulate the optimal downlink resource allocation as an optimization problem. A utility-function is used as an indicator of the network benefit earned by serving a packet of a user from a given base-station. This utility-function is a function of the corresponding base-station load, the packet experienced delay before its transmission, and wireless channel quality. We then formulate the downlink resource allocation problem in each time-slot as an optimization problem with the objective of maximizing the total network utility that is defined as the aggregation of individual users' utilities in each time-slot. This optimization problem should be solved for each time-slot *to adaptively exploit* wireless link variations, users' quality of service flexibility, and base-station load conditions. We then show that this optimization problem is a *0–1 multi-dimensional multiple-choice knapsack problem* that simultaneously gives base-station assignment and packet scheduling as the output; therefore, it can be

*This work was supported partly by Nortel Networks,Ottawa, Canada, National Capital Institute of Telecommunications (NCIT), Ottawa, Canada, Mathematics of Information Technology and Complex Systems (MITACS), Canada, and Natural Sciences and Engineering Research Council (NSERC), OSR grant for participation in Wireless World Initiative New Radio (WINNER) project.

†E-mail address: keivan@sce.carleton.ca
‡E-mail address: delfin@nortel.com
§E-mail address: halim@sce.carleton.ca
¶E-mail address: zhao@math.carleton.ca

adapted to consider not only wireless link dynamics but also network load variations. We also propose a heuristic algorithm with polynomial-time complexity to solve this problem. Simulation results indicate a significant performance improvement. Specifically, a higher performance gain is achieved in the network coverage area with greater non-uniform user spatial distribution.

Keywords: adaptive resource allocation, base-station assignment, cellular networks, packet scheduling

AMS: 90-02, 91A30, 91B16, 93-02, 94-02, 94A40, 94A99, 90B35, 90B36, 68M20, 90B22

1. Introduction

The increasing demand for data services in cellular networks along with the inherent differences between the Quality–of–Service (QoS) requirements of data services and those of the conventional voice services necessitates the design of more efficient resource management mechanisms for the next generation cellular networks. To adaptively allocate resource over a wireless link, data scheduling methods are used. There is a very rich literature that addresses the crucial topics such as fairness and delay bounds (see for example [1], [2], [3] and references therein).

A novel approach to providing efficient resource allocation in data services is to opportunistically utilize the QoS flexibility of data services (e.g., delay tolerance), alongside the resource fluctuations in the physical layer (see e.g., [4], [5], [6]). In wireless cellular communications, delay tolerance of elastic traffic [7] can be exploited in transmission scheduling to improve system throughput (see, e.g., [5], [6]). In such techniques the transmission is carried out when time varying channel capacity happens to be at (or near) its peak. The resulting throughput improvement is referred to as *multi-user diversity* gain. This approach has been employed in highspeed downlink standards for cellular communications such as HSDPA [8] and 1xEV-DV [9]. In these methods, it is assumed that there exists a function that assigns a base-station to each user as a server[1].

In order to facilitate the base-station assignment, each base-station broadcasts a pilot channel. A user may receive the pilot signals of various base-stations with the corresponding bit-energy to the interference-plus-noise-spectral-density ratios (E_c/I_0) greater than a performance threshold. The base-station with the highest E_c/I_0 is then assigned to the user as the server [10]. In cellular networks, specially in the dense areas with small-size cells, a large number of users are able to receive pilot channels with acceptable quality from more than one base-station (e.g., nearly 40% of UMTS users are reported to be in soft-hand-off mode of transmission [10]). For each user, its *active-set* is the set of base stations having acceptable quality level to the user.

In a cellular network, the required resource to be allocated to a user for providing a required QoS level is related not only to the corresponding wireless link quality, but also to the serving base-station load condition. This fact would be more important for cellular networks with non-uniform spatial distribution of the users in the network coverage area. For example, consider a user that is located somewhere in the network coverage area such

[1]When a base-station is appointed to transmit/receieve data to/from a user, we say that the base-station is *assigned* to that user.

that there are two base-stations in its active-set. The wireless links with acceptable quality level with two neighboring base-stations could have different base-station load conditions: a very good wireless link quality to a heavily loaded base-station and a moderate wireless link quality to a lightly loaded base-station. Allocating the available resource of the base-station with better channel quality to a user would results in a greater benefit because a larger amount of data could be sent in the same given amount of time, but at the same time, it could likely be that a user with critical delay condition is associated the highly loaded base-station such that not being able to serve that user may result in a big loss. This condition would be even worse if the heavily loaded base-station does not have enough free resource to allocate. This load difference between the base-stations as well as the independent time variations of the channels between a user and the base-stations in its active-set introduces a type of diversity which we refer to in this chapter as *multi-access-point diversity*. This kind of diversity is very important for the case of non-uniform spatial distribution of users in the network coverage area. Note that the conventional approaches to the optimal resource allocation problem in cellular networks assume the base-station assignments are already given (see e.g., [11], [12], [13], [14], [15], [16]).

In cellular networks, the time scale of packet transmission is smaller than that of base-station assignment. Therefore it is quite likely that at some point in time, a channel between a user and a base-station in its active-set will be better than the one between that user and its currently assigned base-station. This better channel (if and when it exists) can be used for more efficient transmission. For this reason, a fast cell selection mechanism is introduced in 1xEV-DV, in which a base-station assignment method based on the pilot signal strength is carried out with a frequency higher than that of the conventional methods [9]. However this mechanism does not consider base-station load and it may result in the assignment of a user to an overloaded base-station [17]. Such an inefficient assignment may happen despite the fact that some other base-stations in the active-set of a user may have available resources that could be allocated to the user. Therefore exploiting multi-access-point diversity based on a network-wide knowledge can improve system throughput.

In this Chapter we focus on the downlink since in cellular data networks, due to asymmetric data services, the downlink performance is more critical [5]. We assume a cellular CDMA/TDMA data network. This system uses the *time domain scheduling*, in which, over each time interval, the total power is allocated to a single user with the rest of the users kept inactive. We formulate the optimal downlink resource allocation using the utility-based framework for dynamic pricing presented in [18]. Dynamic pricing method has been used in the recent works such as [4] ,[12], [13], [19].

For each user being served by a base-station, we consider a general utility-function as a combination of a *network utility-function*[2] and a *service utility-function*. Network utility-function depends on the base-station load and service utility-function depends on the corresponding channel quality and the experienced delay. The total network utility (the sum of the utilities of all users) is then used as a total network performance indicator. The objective is to maximize the total network utility. The existing approaches, such as that presented in [13], only focus on maximizing the individual total utility of a base-station and thus are not able to exploit multi-access-point diversity. Unlike previous approaches, in this

[2]Note that network utility-function is different from "total network utility", the latter is defined in Subsection 2.2.

Chapter, we maximize the total network utility.

We show that the maximizing total network utility results in joint packet scheduling and base-station assignment optimization; this approach exploits simultaneously multi-user and multi-access-point diversities, enabling it to adaptively allocate resource based on system dynamics. We further demonstrate that this problem can be mapped into a 0–1 Multi-dimensional Multiple-choice Knapsack Problem (0–1 MMKP) known to be NP-hard. We then propose a heuristic algorithm based on Lagrange Multipliers method with polynomial-time computational complexity. The proposed method can be implemented in a radio network controller (RNC) which controls a set of base-stations.

Joint base-station assignment and resource control for downlink is considered in [13] and [20] as well. However, in both these methods, separate base-station assignment and resource allocation algorithms are proposed. These algorithms are then combined in a sequential manner for more efficient resource utilization. The distinguishing feature of our method compared to the previous works is that our proposed method gives in one shot both the packets scheduled for transmission and the corresponding base-stations to transmit those packets based on network-wide knowledge. Moreover, since base-station assignment is performed in the same time-scale of the packet scheduling, the proposed method will be able to exploit multi-user and multi-access-point diversities. Furthermore, by adopting a heuristic polynomial-time algorithm, our proposed method can be implemented with an acceptable level of complexity.

To study the performance of the proposed method for optimal downlink resource allocation, we model a two tier cellular network with non-uniform users' spatial distribution. We then compare the performance of the proposed method with those of an opportunistic scheduler and a first-come-first-serve scheduling strategy. Simulation results show that the proposed scheme manages more efficiently the fundamental coverage-capacity trade off in terms of achieved utility per unit transmit power. Also it is more robust against the spatial distribution of users in the network coverage area. Simulation results also show that using the proposed method improves network throughput and reduces packet dropping ratio.

The outline of this Chapter is as follows: in Section 2. we define the system model. In Section 3. we formulate the downlink resource allocation as an optimization problem, and show that this optimization problem for each time slot is a 0–1 MMKP. We then propose a polynomial time heuristic algorithm for this problem in Section 4.. Simulation results are presented in section 5.. At the end the concluding remarks of this Chapter are presented in Section 6..

2. System Model

We consider the downlink of a cellular CDMA/TDMA data network which uses time domain scheduling[3]. The time t, is slotted with each slot containing M frames[4]. The length of each time slot is T_w seconds and the length of each frame is T_f seconds, therefore $T_w = MT_f$. In time domain scheduling within each frame, the total available power of base-station j, denoted by P_j, is allocated to a single user while the rest of the users are

[3]For an overview of a CDMA cellular system see e.g., [10], [21].

[4]These frames can be considered as an air interface frame.

kept inactive. T_w is chosen such that the channel fluctuation within this time is negligible.

Data traffics destined to the users [5] are generated and packetized into fixed L bits packets. The average number of generated packets per second is the same for all users.

Each packet is transmitted over the air-interface in an integer multiple of T_f second. Each packet of the user i has a maximum delay tolerance of D_i. We assume that $D_i \gg T_w$ and that there are $N(t)$ users[6] waiting in the system to be served at time t.

For user i, $d_i(t)$ denotes the total delay incurred by the head-of-line packet since its arrival to the system up to time t. The head-of-line packet of user i with $d_i(t) > D_i$ is dropped and removed from the system.

Let $\mathcal{B} = \{BS_1, \ldots, BS_B\}$ be the set of the B base-stations in the network. Each base-station transmits its own pilot signal. For user i, its active-set $\mathcal{A}_i(t)$ is defined as the set of base-stations in $\mathcal{B}$ which can be assigned to that user as a server [10]:

$$\mathcal{A}_i(t) = \{j | j \in \mathcal{B}, \gamma_{ij}(t) \geq \gamma_{min}\}, \tag{1}$$

where $\gamma_{ij}(t)$ is the bit-energy to the interference-plus-noise-spectral-density ratio of the pilot channel of base-station j received by user i, and γ_{min} is the minimum required E_c/I_0. The maximum number of base-stations in $\mathcal{A}_i(t)$ denoted by $N_{\mathcal{A}_i(t)}$, is usually limited by the air interface protocol; we denote this maximum by $N_{\mathcal{A}}$.

Note that if the number of base-stations in $\mathcal{A}_i(t)$, $N_{\mathcal{A}_i(t)} > 1$, then the user is in soft-hand-off condition and there is more than one base-station that can serve that user with acceptable link quality.

For each user i at time t, corresponding to each base-station j in its active-set, $\mathcal{A}_i(t)$, $\rho_{ij}(t)$ denotes the average bit energy to the interference spectral density (E_b/I_0). $\rho_{ij}(t)$ is the wireless link performance metric that indicates the quality of the wireless link between user i and base-station j;

$$\rho_{ij}(t) = \frac{WP_j(t)g_{ij}(t)T_{ij}(t)}{LI_{ij}(t)}, \tag{2}$$

where W is the chip rate, $P_j(t)$ is the transmit power of base-station j, $g_{ij}(t)$ is the stable channel gain between base-station j and user i, $I_{ij}(t)$ is the received interference by user i when served by base-station j, and $T_{ij}(t) \leq T_w$ is the time extent in which user i is served by base-station j using time domain scheduling.

For a given $\rho_{ij}(t)$, $T_{ij}(t)$ can be determined from (2). We denote the number of air interface frames that should be allocated to user i in base-station j as

$$m_{ij}(t) = \left\lceil \frac{T_{ij}(t)}{T_f} \right\rceil, \tag{3}$$

where $\lceil . \rceil$ is the rounding up to the nearest integer.

We define the *assignment matrix* $\mathbf{b}(t) = \{b_{ij}(t)\}_{N(t)\times(B+1)}, i = 1, \ldots, N(t), j = 0, \ldots, B$, where $b_{ij}(t)$ is the assignment indicator at time t; that is $b_{ij}(t) = 1$, if base-station j is assigned to user i, and $b_{ij}(t) = 0$ otherwise. Note that for a user i, $b_{ij}(t) = 0$

[5]In this Chapter, by "user" we refer to an active user in the network coverage area.

[6]Here for simplicity of the presentation we assume that for each user there is only one packet for transmission. The proposed method can be easily modified for the case that has more than one packet waiting to be transmitted to each users.

for the base-stations which are not in $\mathcal{A}_i(t)$. We also define $j = 0$ corresponding to a virtual NULL base-station. For a user i, assignment to the NULL base-station, $b_{i0}(t) = 1$, means that no transmission is scheduled for user i at time t. At each time, t, one and only one base-station is assigned to each user, i.e.,

$$\sum_{j=1}^{B} b_{ij}(t) = 1, \text{ for } i = 1, \ldots, N(t). \tag{4}$$

For a base-station in the network coverage area, we define $0 < \eta_j(t) \leq 1$ as an indicator of the base-station load at time t. In a cellular CDMA network, for a base-station j, $\eta_j(t)$ is a function of the allocated power to the users in the downlink as well as maximum transmit power of that base-station (see, e.g., [10], [22]). We assume that the base-station will be *overloaded* at time t, if $\eta_j(t) > 1$, and not be able to serve new users.

2.1. Users' Utility Function

For a user i being served by base-station j at time t, we associate a utility-function, $u_{ij}(t)$, that shows the benefit that network earns from serving that user. This benefit earned by serving a user i, $u_{ij}(t)$, is a function of allocated network resource to that user as well as its experienced quality of service.

In a cellular network the required resource to be allocated to a user for providing a required level of quality of service is related not only to the corresponding wireless link quality, but also to the serving base-station load. For example, consider a user that is located somewhere in the network coverage area and has wireless links with two neighboring base-stations; a very good wireless link quality to a heavily loaded base-station and a moderate wireless link quality to a lightly loaded base-station. Allocating the available resource of the base-station with better channel quality to a user would enable a larger amount of data to be sent, but at the same time if the base-station is heavily loaded and the user is in critical delay condition, not being able to serve that user may result in a big loss.

As it is seen in the above example, different factors should be taken into account during resource allocation. In this Chapter we consider a general utility-function, u_{ij} for a user i served by base-station j as a function of the wireless link quality between user i and base station j, modelled by $\rho_{ij}(t)$, corresponding base-station load condition, modelled by $\eta_j(t)$, and users' experienced quality of service modelled by $d_i(t)$[7]. Here, the objective of defining the utility-function is to capture for later optimization the different effects of these parameters on the benefit that network earn from serving a user i by base-station j. We define $u_{ij}(t)$ as follows:

$$u_{ij}(t) \triangleq \begin{cases} \Upsilon(\eta_j(t))\Theta(d_i(t), \rho_{ij}(t)) & \text{if } j \in \mathcal{A}_i(t), \\ 0 & \text{otherwise,} \end{cases} \tag{5}$$

where $\Upsilon(.)$ indicate the *network utility-function* and $\Theta(.,.)$ is the *service utility-function*. The network and service utility-functions are defined based on the network resource allocation strategy.

[7]Here we assume that the average rate of packet generation destined to the users in the network coverage area are equal.

The network utility-function is defined such that for $\eta_j(t) > 1$, $\Upsilon(\eta_j(t)) = 0$; this indicates that an overloaded base-station is not able to serve that user. On the other hand, for $\eta_j(t) \leq 1$, we consider $\Upsilon(\eta_j(t))$ as a decreasing function of $\eta_j(t)$[8]. The decreasing property of the network utility-function, $\Upsilon(\eta_j(t))$, indicates that, under the assumption of the equality of all other contributing factors, the benefit earned by the network through serving a user by a lightly-loaded base-station is more than that of serving that user by a highly-loaded base-station. Such a network utility-function can be used for designing load balancing algorithms that are able to exploit soft-capacity of CDMA networks [9].

The benefit earned by the network serving a user is an increasing function of its wireless link quality. Also for two users with the same channel quality at time t, serving the one with larger experienced delay would be more beneficial for the network. Due to these facts, we define service utility-function, $\Theta(d_i(t), \rho_{ij}(t))$, as an increasing function of $\rho_{ij}(t)$[10] and $d_i(t)$.

2.2. Total Network Utility

In a general case, the total network utility at time t, can be defined as a function of the individual utility-functions of the users, that is

$$Q(\mathbf{u}; t) : \mathbf{u}(t) \rightarrow \mathbb{R}. \tag{6}$$

$Q(\mathbf{u}; t)$ is a causal function that is an indicator on the network performance. Note that $Q(\mathbf{u}; t)$ is directly related to the resource management strategy adopted by the service provider. In this Chapter, we define $Q(\mathbf{u}; t)$ as the total benefit that network earns by serving the users:

$$Q(\mathbf{u}; t) \triangleq \sum_{i=1}^{N(t)} \sum_{j=0}^{B} u_{ij}(t) b_{ij}(t). \tag{7}$$

The higher $Q(\mathbf{u}; t)$ results in the higher network benefit. As it is seen in (7), $Q(\mathbf{u}; t)$, is a function of the assignment matrix. The assignment matrix indicates the users scheduled to be transmitted in time t and their corresponding assigned base-station [11].

3. Downlink Optimal Resource Allocation

Here the objective is to find an assignment matrix, $\mathbf{b}(t)$ such that total network utility, $Q(\mathbf{u}; t)$ is maximized.

Considering above system model, maximizing $Q(\mathbf{u}; t)$ results in the following optimization problem:

---

[8] $\Upsilon(x)$ is a decreasing function of x if and only if $x \leq y$, results in $\Upsilon(x) \geq \Upsilon(y)$.

[9] Since CDMA networks are interference limited, when the average traffic load in the network coverage area is in a moderate level, there is typically extra capacity available in its neighboring cells that a cell can borrow. This characteristics of cellular CDMA networks is referred to as *soft capacity* [10, 21]

[10] Here we implicitly assume that the achieved bit-rate during time slot t is a continuous and concave function of $\rho_{ij}(t)$ (see e.g., [21]).

[11] Alternatively, $Q(\mathbf{u}; t)$ may be defined as e.g., $\max\{u_{ij}\}$, $\min\{u_{ij}\}$, $\max\min\{u_{ij}\}$.

Problem $\mathcal{O}$:

$$\max_{b_{ij}(t)} \quad \sum_{i=1}^{N(t)} \sum_{j=0}^{B} u_{ij}(t) b_{ij}(t), \tag{8}$$

$$\text{s.t.} \quad \sum_{i=1}^{N(t)} m_{ij}(t) b_{ij}(t) \leq M, \quad j = 1, \ldots, B, \tag{9}$$

$$\sum_{j=0}^{B} b_{ij}(t) = 1, \quad i = 1, \ldots, N(t), \tag{10}$$

$$b_{ij}(t) \in \{0, 1\}. \tag{11}$$

where (9) indicates the network downlink resource constraint and (10) shows that each user should be served by only one base-station.

The output of the optimization problem $\mathcal{O}$, is the assignment matrix $\mathbf{b}(t)$. Therefore, at time t a non-zero $b_{ij}(t)$, $j \neq 0$, indicates that user i is scheduled to be transmitted during the corresponding time slot t, by base-station j. The total transmission time for this user at base-station j is $m_{ij}T_f$ seconds. If $b_{i0} = 1$, then user i cannot be served at this time slot, for the head-of-line packet of this $d_i(t+1) = d_i(t) + T_w$. Therefore the outputs of problem $\mathcal{O}$ are base-station assignments and packet schedulings.

For the special case of a given base-station assignment, e.g., existence of an independent base-station to users assignment [12], the optimization problem $\mathcal{O}$ results in maximizing the utility for each base-station $\acute{j}$, for $\acute{j} = 1, \ldots, B$:

Problem $\acute{\mathcal{O}}$:

$$\max_{b_{i|\acute{j}}(t)} \quad \sum_{i=1}^{N_{\acute{j}}(t)} u_{i|\acute{j}}(t) b_{i|\acute{j}}(t), \tag{12}$$

$$\text{s.t.} \quad \sum_{i=1}^{N_{\acute{j}}(t)} m_{i|\acute{j}}(t) b_{i|\acute{j}}(t) \leq M, \tag{13}$$

$$b_{i|\acute{j}}(t) \in \{0, 1\}, \tag{14}$$

where $b_{i|\acute{j}}(t)$, $m_{i|\acute{j}}(t)$ and $\rho_{i|\acute{j}}(t)$ are the corresponding values of $b_i(t)$, $m_i(t)$ and $\rho_i(t)$ for a given base-station assignment $\acute{j}$, respectively. $N_{\acute{j}}$ is the number of users which base-station $\acute{j}$ is assigned to them. The output of $\acute{\mathcal{O}}$ is a scheduling discipline in which, for time t, packets of the user with the highest value of utility-functions are transmitted. In this case the resulting scheduling only exploits the multi-user diversity in the cell coverage area and another mechanisms would have to be considered for load balancing and soft capacity exploitations.

In the following, we show that the optimization problem $\mathcal{O}$ is a classical combinatorial problem namely, the 0–1 Multi-Dimensional Multi-Choice Knapsack Problem (0–1

[12]Conventional base-station assignments usually use received pilot power quality to assign the base-station with the highest pilot strength to the user and this assignment is made independently of the network load condition.

MMKP). In the followings, we first define the 0–1 MMKP and subsequently we show that the optimal downlink resource allocation $\mathcal{O}$ is a 0–1 MMKP. Since the optimization problem $\mathcal{O}$ should be solved for each time slot t, for brevity, in the following we drop the time index t.

3.1. Knapsack Problem

Knapsack Problems have applications in many area where tasks must be scheduled or budgeted, such as management, business, and defense. The objective of the original Knapsack Problem is to optimally distribute a fixed amount of resource among several tasks to obtain a maximum payoff [23].

For a knapsack of fixed size and a set of elements each characterized by a weight and a value, a subset of elements must be chosen such that the sum of the weights does not exceed the knapsack size while the sum of the values is optimal. 0–1 Multidimensional Multiple-choice Knapsack Problem (0–1 MMKP) is an extension of the original Knapsack Problem. Here we restate the definition of the 0–1 MMKP.

0–1 Multidimensional Multiple-choice Knapsack Problem [23]: Consider a knapsack with B distinct resources represented by $(M_1, M_2, \ldots, M_B)$. Assume that there are N groups each with K_i items in group i. Each item j of a group i has a particular value, u_{ij}, and requires B distinct resources represented by the vector $(m_{ij1}, m_{ij2} \ldots, m_{ijB})$. The objective of the 0–1 MMKP is to select one item from each group to maximize the total value of the collected items, subject to B resource constraints:

Problem $\mathcal{MMKP}$:

$$\max_{b_{ij}} \quad \sum_{i=1}^{N}\sum_{j=1}^{K_i} u_{ij} b_{ij} \tag{15}$$

$$\text{s.t.} \quad \sum_{i=1}^{N} m_{ijk} b_{ij} \leq M_k, \quad k = 1, \ldots, B, \tag{16}$$

$$\sum_{j=1}^{K_i} b_{ij} = 1, \quad i = 1, \ldots, N, \tag{17}$$

$$b_{ij} \in \{0, 1\}, \tag{18}$$

where b_{ij} is the selection variable; $b_{ij} = 1$ if item j of group i is selected and $b_{ij} = 0$, otherwise, (16) indicates knapsack resource constraints and (17) shows that only one item from each group is selected.

In the following proposition we show that the optimal downlink resource allocation problem $\mathcal{O}$ is a 0–1 MMKP problem.

Proposition 1: Optimal downlink resource allocation problem, $\mathcal{O}$, is a 0–1 multi-dimensional multiple-choice knapsack problem.

Proof: We consider the *extended active-set* of each user i, $\acute{\mathcal{A}}_i$, as a group, that is associated with $K_i = N_{\mathcal{A}_i} + 1$ base-stations as items including the base-stations in $\mathcal{A}_i$ plus the

NULL base-station. For each item in $\acute{\mathcal{A}}$ its value is the corresponding utility-function u_{ij}. For each item the required resource is

$$m_{ijk} \triangleq \begin{cases} m_{ij} & \text{if } j = k, \\ 0 & \text{otherwise,} \end{cases} \tag{19}$$

where m_{ij} is as in (3).

The objective of the optimal downlink resource allocation problem, $\mathcal{O}$, is to select exactly one base-station i.e., item, from each group, i.e., users' extended active-set, to maximize total network utility, subject to $B+1$ resource constraints.

Therefore (8), (9), (10) and (11) in Problem $\mathcal{O}$ are corresponding to (15), (16), (17) and (18) in $\mathcal{MMKP}$. Therefore, the downlink resource allocation in (8) is a 0–1 MMKP. ■

Using above method it is simple to show that problem $\acute{\mathcal{O}}$ is a KP.

3.2. Solutions of 0–1 Multidimensional Multiple-choice Knapsack Problem

Proposition 1 shows that the downlink resource allocation in (8) is a 0–1 MMKP problem, which is NP-Complete [23]. Therefore, the optimal downlink resource allocation problem, $\mathcal{O}$, is NP-Hard and requires a solution time which is exponential in the problem size. That is certainly not practical for resource control mechanisms.

To solve a 0–1 MMKP, there exist two distinct approaches: *exact* and *heuristic*. Exact solutions based on the branch-and-bound algorithm and dynamic programming algorithms can be found in the literature (see i.e. [24], [23]). However, these solutions are NP-complete and the their computational complexity is exponential in the order of the problem size (i.e., $O(2^{NN_AB})$), known as "the curse of dimensionality" [25]. Therefore, the corresponding algorithm will be too complex to be useful for actual applications.

The alternative is a heuristic approach. There are some heuristic algorithms as published in [26], [23], [25], [27]. Usually these algorithms are based on an approximation of the exact solution with an acceptable level of computational complexity (e.g., polynomial-time computational complexity).

In the next Section we present a heuristic algorithm for the optimal downlink resource allocation problem. In Section 4.2., we show that the computational time complexity is polynomial.

4. A Heuristic Algorithm for Optimal Downlink Resource Allocation

The heuristic algorithm presented in this Chapter is based on the following basic Theorem 1 that is a modified version of the classical Lagrange Multipliers Theorem in [28].

Theorem 1 : Let $\lambda_1, ..., \lambda_B$, be B non-negative Lagrange multipliers, and $b^*_{ik} \in \{0,1\}$ be the solution of

$$\max\Big\{\sum_{i=1}^{N}\sum_{k=1}^{K_i} b_{ik}u_{ik} - \sum_{j=1}^{B}\lambda_j\sum_{i=1}^{N}\sum_{k=1}^{K_i} b_{ik}m_{ijk}\Big\}. \tag{20}$$

Then, the binary variables b^*_{ik} are also the solution to the following optimization problem:

$$\max_{b_{ik}} \quad \sum_{i=1}^{N}\sum_{k=1}^{K_i} u_{ik}b_{ik} \tag{21}$$

$$\text{s.t.} \quad \sum_{i=1}^{N}\sum_{k=1}^{K_i} b_{ik}m_{ijk} \le \sum_{i=1}^{N}\sum_{k=1}^{K_i} b^*_{ik}m_{ijk}. \tag{22}$$

Theorem 1 shows that the solution of the unconstrained optimization problem (20) is also the solution of the constraint optimization problem (21). The corresponding unconstrained optimization problem of $\mathcal{O}$ is

$$\max\Big\{\sum_{i=1}^{N}\sum_{k=1}^{N_A}(u_{ik} - \sum_{j=1}^{B}\lambda_j m_{ijk})b_{ik}\Big\}, \tag{23}$$

where the solutions are

$$b^*_{ik} = \begin{cases} 1 & \text{if } u_{ik} - \sum_{j=1}^{B}\lambda_j m_{ijk} > 0, \\ 0 & \text{otherwise.} \end{cases} \tag{24}$$

Considering (10), among the solutions in (24) we should find the one for which (10) is satisfied that is still an optimal solution. Therefore regardless the constraint in (10), Theorem 1 is valid for the optimal downlink resource allocation problem, therefore the constraint in (10), only restrict the possible choice for the b_{ik}. Consequently, if the multipliers λ_j are known, the optimization problem $\mathcal{O}$ can be easily solved.

Therefore, we should compute the Lagrange multipliers λ_j, $j = 1, \ldots, B$. If these multipliers are computed such that the terms $M - \sum_{i=1}^{N}\sum_{k=1}^{K_i} b^*_{ik}m_{ijk}$ are non-negative, the solution satisfies (9) and thus is feasible. Moreover the solution is optimal, if

$$\sum_{j=1}^{B}\lambda_j(M - \sum_{i=1}^{N}\sum_{k=1}^{K_i} b^*_{ik}m_{ijk}) = 0. \tag{25}$$

Based on above discussion a heuristic algorithm is presented in the following subsection.

4.1. Heuristic Algorithm

The proposed 3-step heuristic algorithm for optimal downlink resource allocation is presented in Fig. 1. The heuristic algorithm uses Lagrange Multiplier approach as the basis for its approximation of the optimization solution and is a modified version of that presented in [26], where modifications are made to match the requirements of our specific problem.

The algorithm starts in Step 1, *Initializing* by setting the Lagrange multipliers to zero and assigning each user i to the corresponding base-station in its extended active-set with the maximum achieved utility, $\hat{K}_i$, where, $\hat{K}_i = \arg\max u_{ij}$, such that the constraint in (10) and (25) are satisfied.

If the base-station resource constraints is violated, the initial base-station assignments will be adapted in Step 2, *Relaxing the Resource Control Violations* , until resource constraints hold. For adaptation, we repeatedly improve the most offending constraint violation, $\hat{J}$. For improving $\hat{J}$ in Step 2, we consider the users whose assigned base-stations is $\hat{J}$. For each base-station k in the extended active-set of these users, we compute the increase Δ_{ik} of Lagrange multiplier $\lambda_{\hat{J}}$, that results from exchanging the assigned base-station in the extended active-set of user i. We then choose the base-station K^* in the extended active-set of user I^*, which causes the least increase of multiplier $\lambda_{\hat{J}}$ for exchange. This new assignment minimizes the widening of the gap between the optimal solution characterized by (25) and the previous solution returned by this algorithm. The procedure in Step 2 is repeated until for each user a base-station has been assigned and the resource constraints, (9), are also satisfied. Since each user has always a NULL base-station in its extended active-set with corresponding achieved utility equals to zero, the solution is always feasible. However, assigning a user to the NULL base-station results in not scheduling the user in the solution of the optimization problem for the current time slot. Note that the choice tried in each iteration is eliminated from user's extended active set.

After the completion of Step 2, there might be some available unused resources in the base-stations in the network coverage area. In Step 3, *Improving the Solution*, we utilize this remaining resource to improve the solution by replacing some of the assigned base-stations with other base-stations such that their assignments lead to a larger achieved utility. For this reason in Step 3 we check each base-station k in the extended active-set of every user i against the assigned base-station of that user in the last step, $\hat{K}_i$. We then test that whether assignment of base-station k will result in greater achieved utility than the assigned base-station, and if base-station k can replace the assigned base-station without violating the resource constraints. Among all exchangeable base-stations, the base-station K' in the active-set of user I' causing the largest increase of the total achieved utility is exchanged with the assigned base-station to that user ($\hat{K}_{I'}$). We repeat this procedure until no more exchanges are possible. The resulting solution of the assigned base-stations is feasible, and even optimal, if (25) is satisfied.

At each time t, the inputs of the algorithm are $u_{ij}(t), m_{ij}$ for $i = 1, \ldots, N(t), j \in \{k|BS_k \in \mathcal{A}_i + \text{NULL}\}$, and the output is the assignment matrix, $\mathrm{b}(t)$. Since time is slotted and this assignment matrix is valid for time slot t, for the sequel time slot we solve the optimization problem again, but in this case we start the algorithm in Fig. 1 with the last assignment matrix $\mathbf{b}(t)$.

4.2. System Complexity

In this section we study the complexity of the proposed optimal downlink resource allocation scheme. A simple block diagram of the system has been presented in Fig. 2

At time t, each user i measures the pilot channel E_c/I_0, and generates its active-set $\mathcal{A}_i(t)$ and the corresponding channel qualities $\rho_{ij}(t)$ and send them to the base-station. The base-station also sends this information to the optimal resource allocator where the corresponding $d_i(t)$ for the users is also available. The optimal assignment matrix is then ob-

Heuristic Downlink Resource Allocation Algorithm

1. **Initialization**

   a. Resetting the Lagrange multipliers:

   $\lambda_j = 0 \; j = 1, \ldots, B$

   b. Normalized required resources:

   $m_{ijk} = m_{ijk}/M, \; i = 1, \ldots, N, \; k = 1, \ldots, K_i$

   c. Assign base-station with the maximum achieved utility:

   $\hat{K}_i = argmax_k(u_{ik}) \; and \; b_{i\hat{K}_i} = 1 \; i = 1, \ldots, N$

   d. Compute resource constraint violation:

   $\pi_j = \sum_{i=1}^{N} m_{ij\hat{K}_i} \; j = 1, \ldots, B$

2. **Relaxation of the Resource Constraint Violation**

   Repeat until $\pi_j \leq 1, j = 1, \ldots, B$, *or no more assignments can be exchanged*:

   while ($\pi_j > 1$ for any j) **do**

   a. Determine the most offending resource constraint violation:

   $\hat{J} = \arg\max_j\{\pi_j\}$

   b. Find the assigned base-station to exchange:

   for $\{i|\hat{K}_i = \hat{J}\}$

   for $k = 1 : K_i$

   $\Delta_{ik} = \frac{1}{(m_{i\hat{J}\hat{J}} - m_{ikk})}(u_{i\hat{J}} - u_{ik} - \lambda_{\hat{J}}(m_{i\hat{J}\hat{J}} - m_{ikk}))$

   end

   end

   c. Reevaluate the assigned base-stations:

   $(K^*, I^*) = \arg\min_{ik}\{\Delta_{ik}\}, \; \forall i, k,$

   $\lambda_{\hat{J}} = \lambda_{\hat{J}} + \Delta_{I^*K^*},$

   $b_{I^*\hat{K}_{I^*}} = 0,$

   $\hat{K}_{I^*} = K^*, b_{I^*K^*} = 1,$

   $\pi_{\hat{J}} = \pi_{\hat{J}} - m_{I^*\hat{J}\hat{J}},$

   $\pi_{K^*} = \pi_{K^*} + m_{I^*K^*K^*}$

   end

3. **Improving the Solution**

   Repeat until no more assignments can be exchanged:

   while more assignments can be exchanged

   a. Compute total utility increases:

   for $i = 1 : N$

   for $k = 1 : B$

   $$\mu_{ik} = \begin{cases} u_{ik} - u_{i\hat{K}_i} & \text{if } u_{ik} - u_{i\hat{K}_i} > 0 \\ & \text{and } \pi_k - m_{\hat{K}_{I'}I'\hat{K}_{I'}} + m_{ikk} \leq 1 \\ 0 & \text{otherwise} \end{cases}$$

   end

   end

   b. Find the best exchangeable assignment:

   $I'K' = \arg\max_{ik}\{\mu_{ik}\} \; i, k$

   c. Exchange the selected assignment:

   $\pi_{\hat{K}_{I'}} = \pi_{\hat{K}_{I'}} - m_{\hat{K}_{I'}I'\hat{K}_{I'}}$

   $\pi_{K'} = \pi_{K'} + m_{I'K'K'}$

   $b_{I'\hat{K}_{I'}} = 0$

   $\hat{K}_{I'} = K', b_{I'K'} = 1$

   end

Figure 1. Heuristic algorithm for optimal downlink resource allocation.

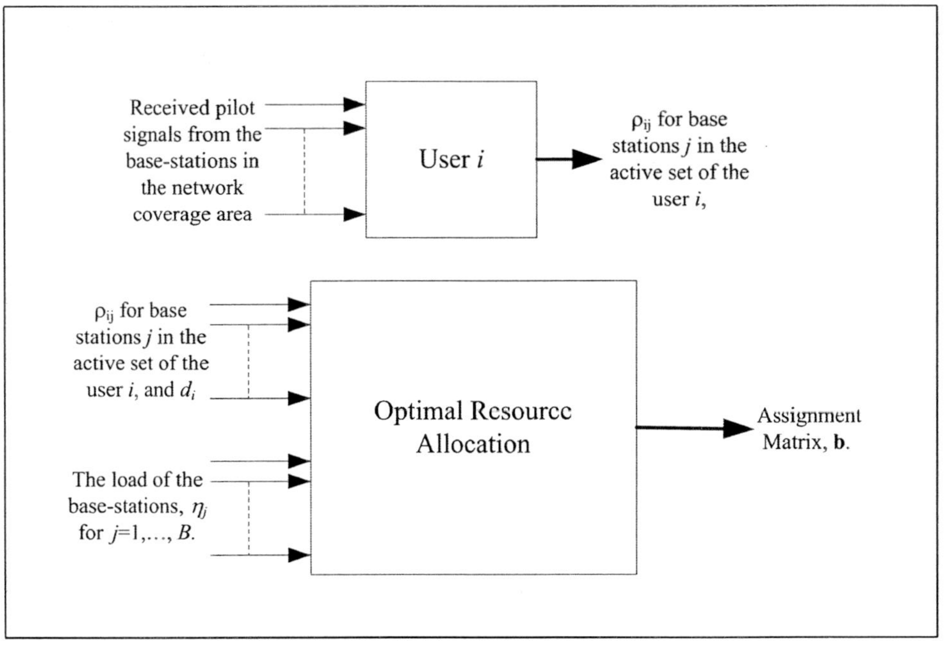

Figure 2. **System block diagram.**

tained using the heuristic algorithm presented in Section 3.. This optimization is performed in a radio resource controller that controls a set of base-stations.

Proposition 2 gives the complexity order of the heuristic algorithm that has been presented in this Chapter for optimal downlink resource allocation.

Proposition 2: The proposed heuristic algorithm in subsection 4.1. for optimal downlink resource allocation has a computational complexity of $O(BN^2N_A^2)$ where N is the number of users, B is the number of base-stations and N_A is the maximum number of base-stations in the active-sets of all the users.

Proof: If the corresponding base-stations with the greatest utility-functions satisfy the resource constraint for each user, then Steps 2 and 3 are not executed. In this case 1.c has at most a computational complexity of the order of NBN_A.

If it is not the case, then the entire algorithm should be executed. 1.a to 1.d have the complexity order of $O(B)$, $O(NN_A)$, $O(BNN_A)$ and $O(BN)$, respectively.

2.a and 2.c have the complexity order of $O(N)$. In 2.b for each of the N users, at most N_A non selected base-stations are in the corresponding extended active-set that should be tested. Each test has a maximum complexity of B. Therefore, there is one iteration for each base-station in the active-set of each user, resulting in $O(BNN_A)$ for 2.b. In every iteration in 2.c, one assigned base-station is removed from one user, therefore the outer *while* loop of Step 2 is executed NN_A times in the worst case. This results in an overall complexity of $O(BN^2N_A^2)$ for Step 2.

In Step 3, 3.b is in order of $O(N_A)$, while 3.c has a constant computational complexity, $O(c)$. In 3.a for each of the N users, at most N_A non selected base-stations are in the corresponding extended active-set that total utility increase should be computed for them.

Each computation has complexity of B. Therefore there is one iteration for each base-station in the active-set of each user, resulting in $O(BNN_A)$ for 3.a. The outer loop of Step 3 is executed at most NN_A times. Since for each user there can be at most N_A base-stations in its active-set that could have higher value than the assigned base-station. This gives an overall complexity of $O(BN^2N_A^2)$ for Step 3. Thus the computational complexity of the heuristic algorithm for the optimal downlink resource allocation presented in Fig. 1 is $O(BN^2N_A^2)$. ■

Proposition 2 shows that the proposed heuristic algorithm presented in this Chapter has a computational complexity of polynomial-time.

5. Simulations Study

In this section we first present the simulation setting and then simulation results.

5.1. Simulation Setting

The simulated cellular network is a two-tier hexagonal cell configuration consisting of 19 cells ($B = 19$); the wrap-around technique is also employed. The number of active users in the network is N. The simulation parameters are presented in Table 1. The pilot channel powers are adjusted such that 40% of users receive the pilot channel of more than one base-station with acceptable quality. Non-uniform user distribution is expressed by a non-uniformity factor, μ_D, such that, $(1 - \mu_D)N$ users are distributed uniformly and the rest of the users are distributed within randomly located hot-spots.

For each user, fixed length packets are generated by a Poisson arrival process with an average rate of Λ packets per second. For a packet i transmitted by base-station j in time slot t, we define the following utility-function, $u_{ij}(t), j \neq 0$:

$$u_{ij}(t) \triangleq \frac{1 - \eta_j^2(t)}{m_{ij}(t)} \exp\big(\tau_i(t) - \overline{\tau}(t)\big), \tag{26}$$

where $\overline{\tau}(t) = 1/N(t)\sum_i \tau_i(t)$ is the average delay. In the definition of the users' utility-function, (26), we consider $\Upsilon_j(t) = 1 - \eta_j^2(t), 0 < \eta_j(t) \leq 1$ as the network utility-function and $\Theta(t) = \frac{1}{m_{ij}(t)} \exp(\tau_i(t) - \overline{\tau}(t))$ as the service utility-function. The service utility-function is similar to that given in [29] which attempts to provide fairness in delay. Note that in (26), a user is given a large utility when either the corresponding user experiences a good channel condition or it experiences a bad relative delay status. The utility-function also weights more for taking service from the lightly loaded base-stations. Note that different utility-functions can be designed to satisfy various design objectives.

For comparison, we consider the three different systems listed in Table 2. In *System I*, the base-station with the strongest received pilot signal is assigned as the server for each user. Packets are then transmitted in first-come-first-serve fashion using a time domain scheduling scheme. In *System II*, the base-station assignment is similar to that of System I; however packets are scheduled such that the total utility of each cell is maximized. Note that in System II only multi-user diversity is exploited. Finally, *System III* uses our proposed optimal downlink resource allocation scheme.

Table 1. Simulation Parameters

Parameter	**Value**
Physical layer	Based on UMTS
Number of base-stations (B)	19
Cell radius	1000 m
Base-stations transmit power	10 W
$N_{\mathcal{A}}$	4
Standard deviation of log-normal fading	8 dB
Thermal noise density	-174.0 dBm/Hz
Propagation loss exponent	4
Frame length (T_f)	10 ms
Number of frames in each time-slot (M)	10
Required E_c/I_0 (γ_{min})	2 dB

Table 2. Compared Systems.

System	Sys. I	Sys. II	Sys. III
Scheduling method	First-come-first-serve	Based on $\acute{\mathcal{O}}$	Based on $\mathcal{O}$
Base-station assignment method	Pilot power strength	Pilot power strength	Based on $\mathcal{O}$

5.2. Simulation Results

We compare the total achieved network utility in Systems I, II and III for different number of users in the coverage area. Λ is assumed to be 5 packets per second. Fig. 3 illustrates the total network utility normalized by the achieved total utility of the network using System I scheme (conventional method), versus N/B. Two different cases are considered: heavy ($\mu_D = 0.5$) and light ($\mu_D = 0.2$) non-uniform spatial distribution of users. The presented result is the average of the results of 100 independent runs of the simulation program. As shown, System III outperforms System II in both cases. Also shown is that the performance gain for System III increases with μ_D. Therefore, the results show that the proposed optimal downlink resource allocation algorithm can exploit soft-capacity.

We also obtain the average achieved utility per power for the network with $N/B = 20$. The results are presented in Fig. 4. It shows that the average achieved utility per Watt is larger in System III. The average values are normalized to the corresponding values achieved from simulation of System I. The performance is better for larger values of μ_D.

We have also obtained the standard deviation (σ) of the total achieved utility in each run of simulations using 100 independent snapshots. The results are shown in Fig. 5. It is seen that System III outperforms System II and is more robust for non-uniform spatial distribution of the users in the network coverage area. It shows that System III is more robust to varying demand of radio resources.

We also study the effect of proposed optimal downlink resource allocation algorithm

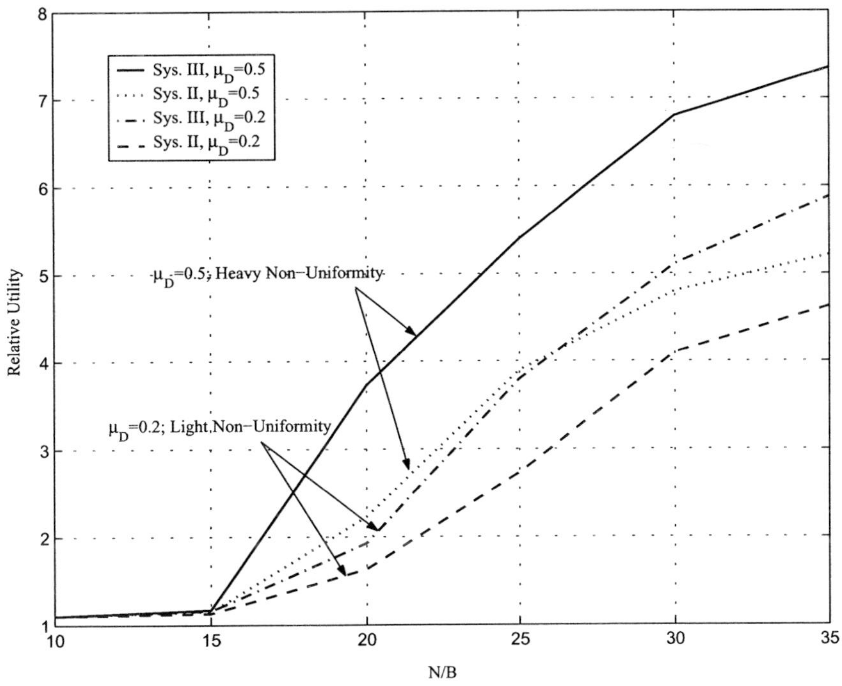

Figure 3. The average achieved total network utility for System II and System III normalized by the average achieved total network utility of System I versus average number of users per base-station (N/B)for two non-uniformity cases: $\mu_D = 0.2$ and $\mu_D = 0.5$. System is simulated for N/B of 10, 15, 20, 25, 30, 35.

on the packet drop ratio (PDR). Packet drop ratio is defined as the probability that a packet is not served within its maximum tolerable delay D_i. The PDR shows the service level performance. We obtain PDR for the above network setting, with $N = 380$ $(N/B = 20)$ in two cases with $\mu_D = 0.2$ and $\mu_D = 0.5$. The results are illustrated in Fig. 6. The figure shows that System III outperforms the other two base-station assignment algorithms.

Fig. 7 illustrates the normalized average achieved throughput versus average number of users in each cell, N_u/B, where the average achieved throughput of Systems II and III is normalized by the average achieved throughput of System I. Two different user spatial distributions are considered: uniform ($\mu_D = 0.2$) and non-uniform ($\mu_D = 0.5$). Note that the throughput gain of System II is due to the multi-user diversity alone. The difference between the throughput gains of System II and III indicates the gain due to multi-access-point diversity. Furthermore, as it is seen, multi-access-point diversity gain is increased by the non-uniformity of the users' spatial distribution.

6. Conclusion

In this Chapter we studied the optimal downlink resource allocation problem for data traffic in a CDMA/TDMA cellular network. We used the dynamic pricing approach in [18] based

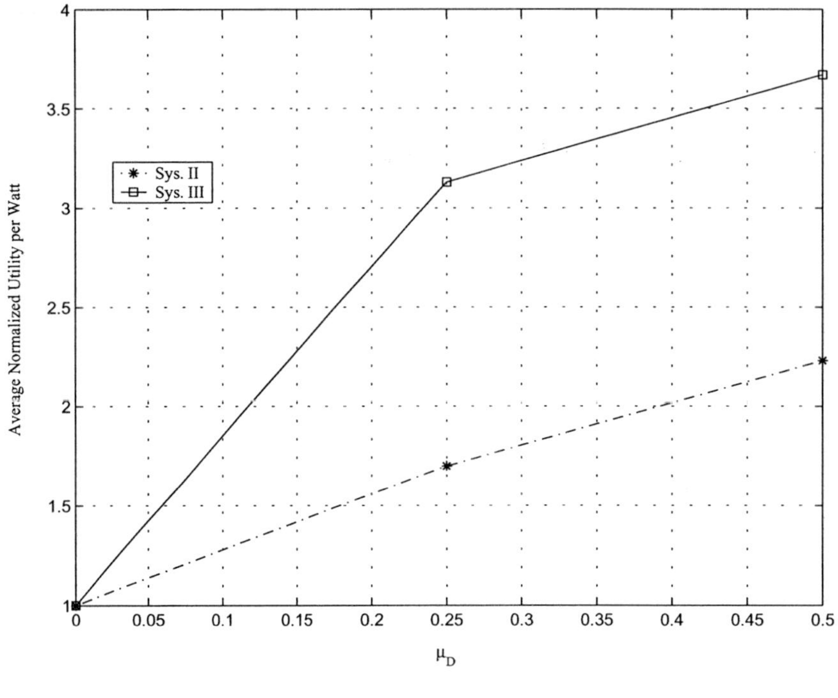

Figure 4. Average normalized utility per Watt versus μ_D.

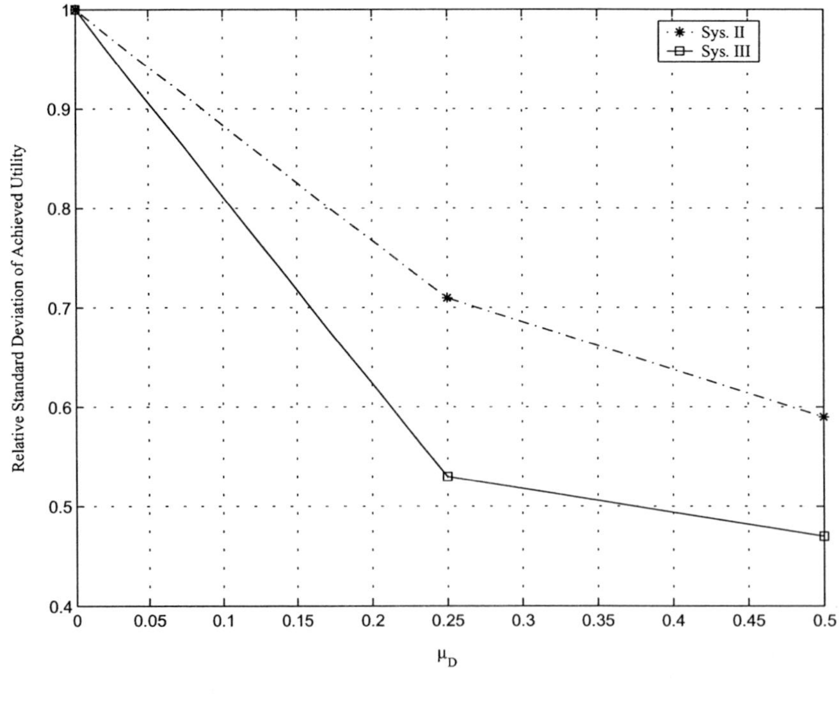

Figure 5. Standard deviation of total achieved utility versus μ_D.

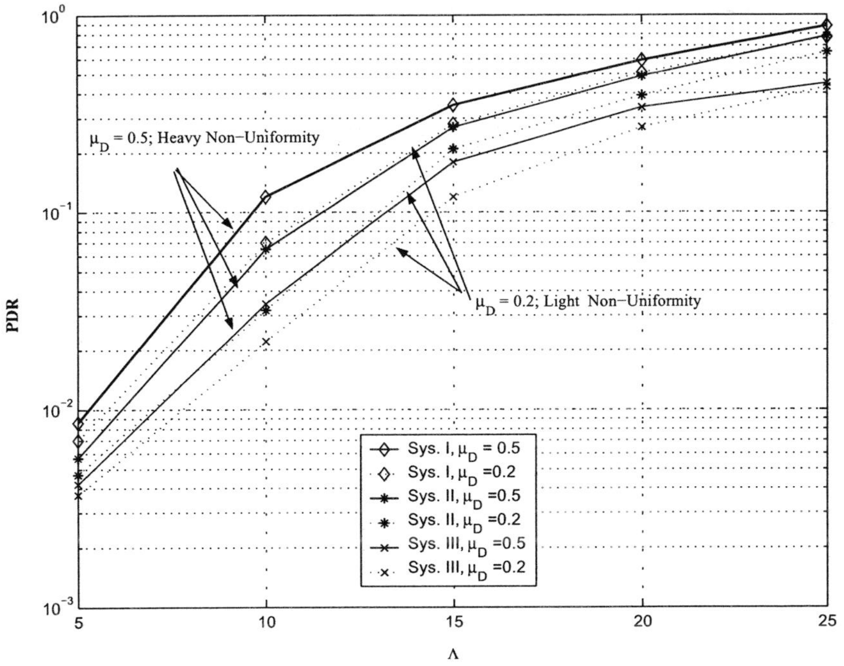

Figure 6. The packet drop ratio versus packet arrival rate.

on utility-functions and formulate the optimal downlink resource allocation as an optimization problem. We defined a utility-function that indicates the network benefit earned from serving a packet of a user from a given base-station. This utility-function is a function of the base-station load, the delay experienced by a packet before its transmission and the wireless channel quality.

We then formulated the downlink resource allocation problem for each time-slot as an optimization problem with the objective of maximizing the total network utility defined as the aggregation of individual user's utility in each time slot. This optimization problem should be solved for each time slot to adaptively exploit wireless link variations, users' quality of service flexibility, and base-station load conditions. We then showed that this optimization problem is a 0–1 multi-dimensional multiple-choice knapsack problem. A heuristic algorithm is then proposed with polynomial-time complexity to solve this problem.

In this Chapter, we also introduced the concept of *multi-access-point diversity* where a user can be served by several base-stations. The multi-access point diversity concept is a form of diversity inherent in cellular networks, provided by independent time-varying channel across a user with different neighboring base-stations, each with different load conditions. We showed that our proposed downlink optimal resource allocation results in a base-station assignment and packet scheduling method where the multi-access point and multi-user diversity benefits are exploited by tracking users' channel fluctuations and their delay constraint along with the load condition of the base-stations. Exploiting multi-access point diversity exploits soft-capacity that is more beneficial for non-uniform users spatial

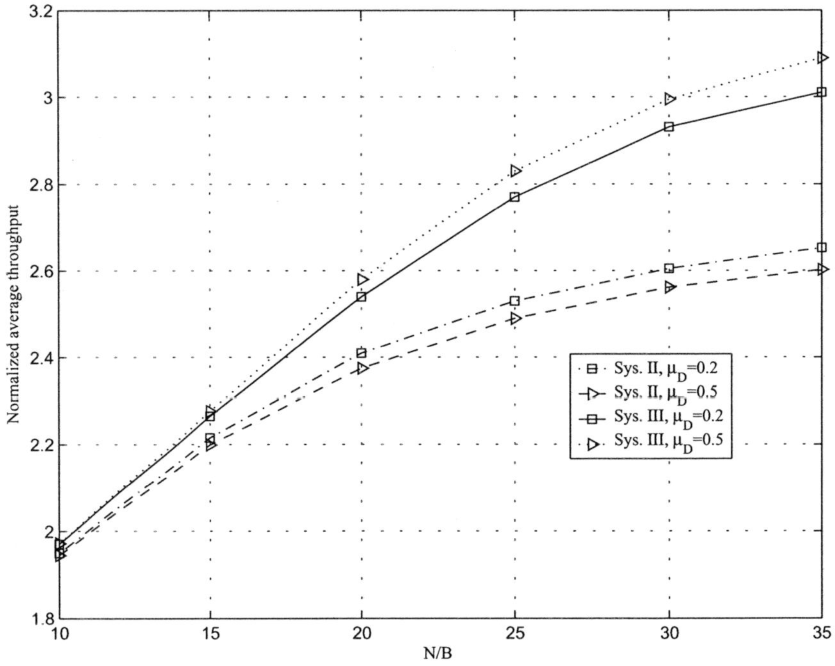

Figure 7. Normalized average achieved network throughput of Systems II and III versus average number of users in each cell for $\mu_D = .2$ and $\mu_D = 0.5$.

distribution.

Our proposed method which uses the channel status and the base-station load condition as the physical layer performance indicators, and the delay as the QoS indicator of the application layer performance is a cross-layer design [30].

Simulation results show that the proposed scheme results in more efficient management of fundamental coverage-capacity trade-off in terms of achieved utility per unit transmit power. Also it is more robust against spatial distribution of users in the network coverage area. It also shows that using the proposed method improves network throughput and reduces packet dropping ratio.

Acknowledgment

The authors would like to express their gratitude to the reviewers and the editors for a wealth of comments and suggestions that improved the presentation of this chapter.

References

[1] S. Lu, V. Bharaghavan, and R. Srikant, "Fair scheduling in wireless packet networks," *IEEE/ACM Transactions on Networking*, vol. 7, no. 4, pp. 473–489, August 1999.

[2] Y. Cao and V. Li, "Scheduling algorithms in broadband wireless networks," *Proceedings of the IEEE*, vol. 89, no. 1, pp. 76–87, January 2001.

[3] K. Navaie, D. Montuno, and Y. Q. Zhao, *Resource Allocation in Next Generation Wireless Networks*. Nova Science Publishers, 2005, ch. Fairness of resource allocation in cellular networks: A Survey.

[4] X. Liu, E. K. P. Chong, and N. B. Shroff, "Opportunistic transmission scheduling with resource-sharing constraints in wireless networks," *IEEE Journal on Selected Areas in Communications*, vol. 19, no. 10, pp. 2053–2064, October 2001.

[5] P. Bender, P. Blake, M. Grob, R. Padovani, N. Sindhushyana, and A. J. Viterbi, "CDMA/HDR: a bandwidth-efficient high-speed wireless data service for nomadic users," *IEEE Communications Magazine*, vol. 38, no. 7, pp. 70–77, July 2000.

[6] P. Viswanath, D. Tse, and R. Larioa, "Opportunistic beamforming using dumb antennas," *IEEE Transactions on Information Theory*, vol. 48, no. 6, pp. 1277–1294, June 2002.

[7] S. Shenker, "Fundamental design issues for the future Internet," *IEEE Journal on Selected Areas of Communications*, vol. 13, no. 7, pp. 1176–1188, September 1995.

[8] 3GPP, "Technical specification group radio access network. High Speed Downlink Packet Access: overall UTRAN description," *3GPP TR 25.855 version 5.0.0*.

[9] "Introduction to 1xEV-DV air interface," *3GPP*, 2002.

[10] H. Holma and A. Toskala, *WCDMA for UMTS: Radio Access for Third Generation Mobile Communications*. John Willy and Sons, 2000.

[11] R. D. Yates, "A framework for uplink power control in cellular radio systems," *IEEE Journal on Selected Areas of Communications*, vol. 13, no. 7, pp. 1341–1247, September 1995.

[12] C. U. Saraydar, N. B. Mandayam, and D. J. Goodman, "Pricing and power control in a multicell wireless data network," *IEEE Journal on Selected Areas of Communications*, vol. 19, no. 10, pp. 1883–1892, October 2001.

[13] J. W. Lee, R. R. Mazumdar, and N. B. Shroff, "Joint resource allocation and base-station assignment for the downlink in CDMA networks," to appear in *IEEE/ACM Trans on Networking*, February 2006.

[14] F. Berggren, S. L. Kim, R. Jäntti, and J. Zander, "Joint power control and intracell scheduling of DS-CDMA non-real time data," *IEEE Journal on Selected Areas of Communications*, vol. 19, no. 10, pp. 1860–1870, October 2001.

[15] S. V. Hanly, "An algorithm for combined cell-site selection and power control to maximize cellular spread spectrum capacity," *IEEE Journal on Selected Areas of Communications*, vol. 13, no. 7, pp. 1332–1340, September 1995.

[16] R. D. Yates and C. Y. Huang, "Integrated power control and base station assignment," *IEEE Transactions on Vehicular Technology*, vol. 44, no. 3, pp. 638–644, August 1995.

[17] F. Santucci, M. Ventulini, and F. Graziosi, "Capacity bound of a multimedia CDMA wireless system," in *in Proceedings of IEEE WCNC'99*, vol. 3, New Orleans, LA, USA, September 1999, pp. 1110–1113.

[18] F. P. Kelly, "Charging and rate control for elastic traffic," *European Transactions on Telecommunications*, vol. 8, pp. 33–37, 1997.

[19] X. Liu, E. K. P. Chong, and N. B. Shroff, "Transmission scheduling for efficient wireless network utilization," in *in Proceedings of IEEE INFOCOM'01*, vol. 2, Anchorage, AK, USA, April 2001, pp. 776–785.

[20] C. Makaya and S. Aissa, "Joint scheduling and base station assignment for CDMA packet data networks," in *in Proceedings of IEEE VTC 2003-Fall*, vol. 3, Montreal, Canada, October 2003, pp. 1693–1697.

[21] A. J. Viterbi, *CDMA: Principles of Spread Spectrum Communication*. Addison-Wesley, 1995.

[22] K. Navaie and A. R. Sharafat, "A framework for UMTS air interface analysis," *Canadian Journal of Electrical and Computer Engineering*, vol. 28, no. 2, pp. 113–129, July/October 2003.

[23] H. Keller, U. Pferschy, and D. Pisinger, *Knapsack Problems*. Springer-Verlag, 2004.

[24] W. Shiih, "A branch and bound method for multiconstraint knapsack problem," *Journal of the Operational Research Society*, vol. 30, pp. 369–378, 1979.

[25] D. Bertsimas and R. Demir, "An approximate dynamic programming approach to multidimensional knapsack problems," *Management Science*, vol. 48, no. 4, pp. 550–565, April 2002.

[26] M. Moser, D. P. Jokanovic, and N. Shiratori, "An algorithm for multidimensional multi-choice knapsack problem," *IEICE Transactions Fundamentals*, vol. E80-A, no. 3, pp. 582–589, March 1997.

[27] M. M. Akbar, E. G. Manning, G. C. Shoja, and S. Khan, "Heuristic solutions for the multiple-choice multi-dimension knapsack problem," in *ICCS '01: Proceedings of the International Conference on Computational Science-Part II*. London, UK: Springer-Verlag, 2001, pp. 659–668.

[28] H. Everett, "Generalized Lagrange multiplier method for solving problems of optimum allocation of resources," *Operation Research*, vol. 11, no. 3, pp. 399–417, May-June 1963.

[29] S. Shakkottai and A. Stolyar, "Scheduling for multiple flows sharing a time-varying channel: The exponential rule," in *American Mathematical Society Translations, A volume in memory of F. Karpelevich*, ser. 2, Y. M. Suhov, Ed., 2002, vol. 207, pp. 185–202.

[30] S. Shakkottai, T. S. Rappaport, and P. C. Karlsson, "Cross-layer design for wireless networks," *IEEE Communications Magazine*, vol. 41, no. 10, pp. 78–80, October 2003.

In: Adaptation Techniques in Wireless Multimedia Networks ISBN 1-59454-883-8
Editors: Wei Li and Yang Xiao, pp. 215-240

Chapter 10

ADAPTATION OF WIRELESS MOBILE MULTI-CARRIER SYSTEMS*

***Hüseyin Arslan*† and Tevfik Yücek**
Department of Electrical Engineering, University of South Florida, 4202 E. Fowler Avenue, Tampa, FL, 33620

Abstract

Adaptation of wireless communication systems has gained strong momentum recently due to the scarcity of resources. New generation wireless standards are designed to be more aware of the variations (and characteristics) of the channel and other variations in the user and service requirements. One of the new generation transmission techniques is the multi-carrier modulation, where the transmission bandwidth is divided into several narrow sub-channels and data is transmitted in parallel over these sub-channels. This chapter focuses on the common adaptation techniques in multi-carrier systems that improve system performance and capacity. First, adaptation methods specific to multi-carrier systems are given. Then, some parameter measurements that make the adaptation techniques possible are discussed in detail. Finally, adaptive OFDM system design for mobile systems is studied and adaptive features in the new IEEE WMAN standard are presented as a case study.

Keywords: multi-carrier modulation, OFDM, adaptation

1. Introduction

Multi-carrier modulation is a transmission scheme which is robust to time dispersion (frequency selectivity) of the channel for high data rate applications. In multi-carrier systems, the transmission bandwidth is divided into several narrow sub-channels and data is transmitted in parallel over these sub-channels. Data in each sub-channel is modulated at a relatively low rate so that the delay spread of the channel does not cause any degradation as

*This work was supported by LOGUS Broadband Wireless Solutions Inc., Ontario, Canada.
†E-mail address: arslan@eng.usf.edu

each of the sub-channels will experience a flat response in frequency. Although, the principles were known since early sixties [43], multi-carrier modulation techniques gained more attention in the last ten years due to the increased power of digital signal processors.

A special case of multi-carrier modulation is orthogonal frequency division multiplexing (OFDM) which can overcome many problems that arise with high bit rate communication, the biggest of which is the time dispersion. In OFDM, the carrier frequencies are chosen in such a way that there is no influence of other carriers in the detection of the information in a particular carrier when the orthogonality of the carriers is maintained. The data bearing symbol stream is split into several lower rate streams and these streams are transmitted on different carriers. Since this increases the symbol period by the number of non-overlapping carriers (sub-carriers), multipath echoes will affect only a small portion of the neighboring symbols. Remaining inter-symbol interference (ISI) can be removed by cyclically extending the OFDM symbol. The length of the cyclic extension should be at least the maximum excess delay of the channel. By this way, OFDM reduces the effect of multipath channels encountered with high data rates, and avoids the usage of complex equalizers.

OFDM is used as the modulation method for digital audio broadcasting (DAB) [6] and terrestrial digital video broadcasting (DVB-T) [5] in Europe, and in asymmetric digital subscriber line (ADSL) [2]. The wireless local area network (WLAN) [3, 4, 7], and wireless metropolitan area network (WMAN) standards [8] use OFDM as their physical layer transmission technique as well. OFDM is also a strong candidate for IEEE wireless personal area network (WPAN) standard [1] and for forth generation (4G) cellular systems (see *e.g.* [50]).

In designing mobile wireless systems, the variation of radio channel characteristics needs to be taken into account. Traditional designs build head-room so that the mobile subscribers at the worst-case channel (link quality) conditions can still maintain the communication. This would translate to lower data rates and inefficient usage of the available resources (like power and spectrum). While this ensures reliable communications from the services and consumer points of view, scarce resources are not exploited efficiently, yielding to a sub-optimal communications system. Therefore, recently, adaptation of wireless communication systems has gained a strong momentum and the new generation wireless standards are designed to be more aware of the variation (and characteristics) of the channel and other variations in the user and service requirements.

In general, adaptive system design improves the overall system performance. For example, for a given average transmit power, adaptation allows the users to experience better signal qualities. Adaptation reduces the average interference observed from other users, as they do not transmit extra power unnecessarily. As a result, the received signal quality will be improved over a large portion of the coverage area. These higher quality signal levels can be exploited to provide increased data rates through rate adaptation. For a desired received signal quality, this might also translate into less transmit power, leading to improved power efficiency for longer battery life. On the other hand, for a desired minimum signal quality, this might lead to an increased coverage area or better frequency re-use. In addition, adaptive receiver designs allow the receiver to work with reduced signal quality values, *i.e.* a desired bit-error-rate (BER) or frame-error-rate (FER) performance can be achieved with a lower signal quality. Adaptive receivers can also enable reduced average computational complexities for the same quality of service, which again implies less power consumption. As can be seen, adaptation algorithms lead to improved performance, increased capacity, lower power consumption, increased radio coverage area, and eventually better overall wireless communication system design.

In terms of adapting the transmission parameters, OFDM offers many possibilities. Adapting the transmit power, cyclic prefix size, modulation and coding, and the number of sub-carriers are some of these transmission parameters. In addition to adaptation over each packet (as in the case of single carrier systems), OFDM also offers adaptation of the parameters for each carrier or over a small group of carriers. In other words, adaptation can be done independently over narrower bands rather than employing adaptation over the whole transmission band. Similarly, reception of an OFDM signal can bring about several new designs and approaches for adaptive receiver.

In this chapter, an overview of commonly used adaptation techniques in multi-carrier systems and their applications for wireless mobile radio systems are discussed. Many of these adaptation techniques require measure (or estimate) of the related channel parameters. Some of the commonly used channel parameters and their estimation using baseband signal processing techniques are explained. Note that there has been a significant amount of research on adaptation of multi-carrier systems. This chapter is not intended to cover all these developments, rather, it is intended to provide the readers an overview and conceptual understanding of adaptation techniques in multi-carrier systems. A more general study of adaptation techniques in wireless communication systems, including single carrier systems, can be found in [12].

2. Characterization of Mobile Transmission Channel

In wireless communication systems, information is transmitted to a receiver through a radio channel (see Figs. 1 and 2). Transmitted signals are typically reflected, diffracted, and scattered, arriving at the receiver along multiple paths with different delays, amplitudes, and phases. Multipath propagation affects the signal significantly, corrupting the signal and often placing limitations on the performance of the system.

Depending on the transmission bandwidth (or symbol duration) and the type of the environment that the communication takes place, the multipath can cause various problems. When the relative delays are small compared to the transmitted symbol period, different "images" of the *same* symbol arrive at the same time, adding either constructively or destructively. The overall effect is a random fading channel response. When the relative path delays are on the order of a symbol period or more, then images of *different symbols* arrive at the same time. For example, when a particular symbol arrives at the receiver along one path, the previous symbol is arriving along another, delayed path. This is analogous to an acoustic echo and results in a more complicated channel response, which is often referred as time dispersive channel (or frequency selective channel). For example, in office buildings, the average delay spread is around $50ns$ and maximum delay spread is around $300ns$ in 5GHz band. Therefore, if the transmitted symbol duration is less than the maximum excess delay of the channel, the dispersion will cause intersymbol interference (ISI). Traditionally, in single carrier systems, ISI is handled by equalizers. For example, in the popular GSM cellular receivers, 3-5 tap equalizers are used to handle the ISI due to channel and filter effects. As the demand for higher and higher data rates increase, the transmission bandwidth

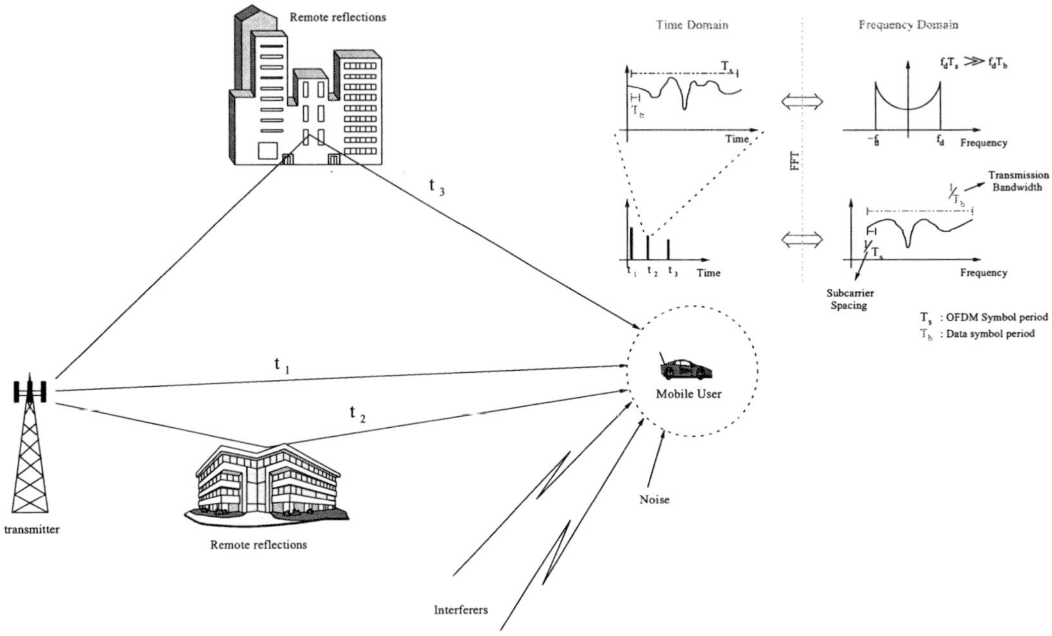

Figure 1. Illustration of some of the effects of radio channel: Local scatterers cause fading; remote reflectors cause multipath and time dispersion, leading to ISI; mobility of user or scatterers cause time varying channel, leading to frequency dispersion (Doppler spread); reuse of frequencies, adjacent carriers etc. cause interference.

to accommodate this high data rates also increases. Hence, the transmitted symbol duration (which is about the inverse of the transmission bandwidth) decreases as well. As a result, for a given environment, the amount of dispersion and equalizer complexity increases dramatically. Multi-carrier transmission and OFDM solves this problem, by dividing the wide frequency band into narrower bands.

In addition to the above, in wireless mobile radio systems, due to the mobility of the transmitter, the receiver, or the scattering objects, the channel response can change rapidly with time, causing spectral broadening which is also referred as Doppler spread. Doppler spread affects the signal differently depending on the transmission bandwidth. As the transmission bandwidth increases, the relative broadening of the channel with respect to the transmission bandwidth will be insignificant. In other words, the time variation within the transmission of a symbol is negligible. This brings about a common trade-off between high mobility and high data rate.

Finally, the interference conditions in wireless systems change rapidly. Most important of all, the radio channels of the desired and interfering users are highly random and the statistical characteristics of the channel are environment dependent. Adaptive techniques can be used to address all these changing conditions as will be described throughout the rest of the chapter.

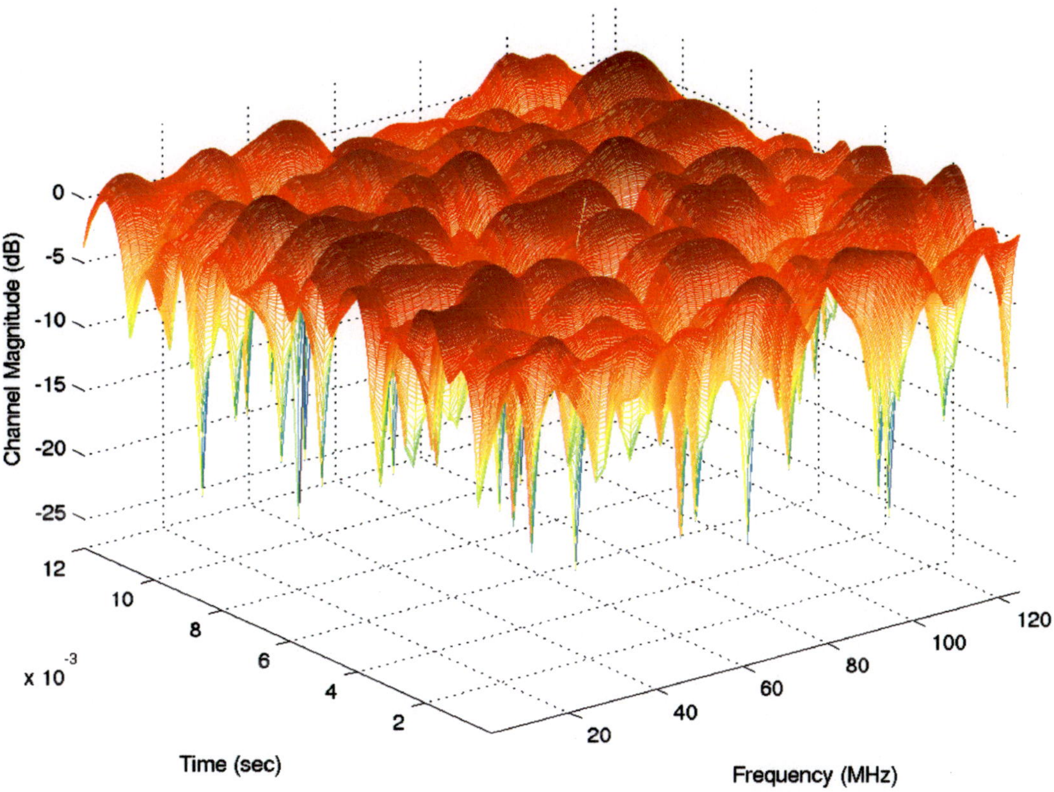

Figure 2. An example 2-dimensional channel response: An exponential PDP with RMS delay spread of $16\mu s$ is used, mobile speed is assumed to be 100km/h, and the center frequency was 5.2GHz. Channel taps are obtained using modified Jakes' model [22].

3. System Model

A block diagram of a basic OFDM system is given in Fig. 3. In a multipath fading channel, due to the frequency selectivity, each subcarrier will have different attenuation. The power on some subcarriers may be significantly less than the average power because of deep fades (see Fig. 2). Therefore, the overall BER may be largely dominated by a few subcarriers with the smallest amplitudes. To avoid this problem, channel coding can be used prior to the modulation of the bits. By using coding, errors can be corrected up to a certain level depending on the code rate and other factors. Interleaving is also applied to randomize the occurrence of bit errors.

Coded and interleaved data is then mapped to the constellation points to obtain data symbols. These steps are represented by the first block of Fig. 3. The serial data symbols are then converted to parallel and inverse fast Fourier transform (IFFT) is applied to these parallel blocks to obtain the time domain OFDM symbols. Time domain samples can be represented as

$$x(n) = IDFT\{X(k)\}$$

$$= \sum_{k=0}^{N-1} X(k) e^{j2\pi nk/N} \qquad 0 \leq n \leq N-1 \,, \tag{1}$$

where $X(k)$ is the transmitted symbol on the kth sub-carrier and N is the number of sub-carriers. Time domain signal is cyclically extended to avoid residual ISI from the previous OFDM symbols, converted to analog signal and passed through a pulse shaping filter. Then, the signal is up-converted to the RF frequencies using mixers, amplified by using a power amplifier (PA), and transmitted through antennas.

In the receiver side, the received signal is passed through a band-pass noise rejection filter and downconverted to baseband. The analog signal is, then, converted to digital and re-sampled. After frequency and time synchronization, cyclic prefix is removed and the signal is transformed to the frequency domain using fast Fourier transform (FFT) operation. The baseband model of the received symbols in the frequency domain can be written as

$$Y(k) = H(k)X(k) + W(k) \,, \tag{2}$$

where $Y(k)$ is the received symbol on the kth sub-carrier, $H(k)$ is the frequency response of the channel on the same sub-carrier, and $W(k)$ is the additive noise plus interference sample which is usually assumed to be a Gaussian variable with zero mean and variance of σ_k^2. Note that OFDM converts the convolution in time domain into multiplication in frequency domain, and hence simple one-tap frequency domain equalizers can be used to recover the transmitted symbols. After FFT, the symbols are demodulated, deinterleaved and decoded to obtain the transmitted information bits.

In the above, a single user system model is represented, where the available channel is used by a single user. Note that OFDM by itself is not a multi-access technique. However, it can be combined with existing multiple accessing methods to allow multiple users to access to the available channel. The most common examples are OFDM time division multiple accessing (OFDM-TDMA), OFDM carrier sense multiple accessing (OFDM-CSMA) [3], OFDM frequency division multiple accessing (OFDM-FDMA) (or OFDMA) [62], and code division multiple access (CDMA) based schemes (multi-carrier code division multiple access (MC-CDMA), direct spread code division multiple access (DS-CDMA)) [26]. Note also that in the above system model, the interference from other users and other technologies (like co-channel interference, adjacent channel interference, narrowband interference *etc.*) are all folded into the noise term for the sake of simplicity. However, in reality, when the received signal is impaired by a dominant interferer, a more accurate model needs to be used, where the noise term will be colored effecting each carrier differently.

4. Adaptation Techniques in Multi-Carrier Systems

In this chapter, some of the popular adaptation techniques that are suitable for multi-carrier systems will be discussed. Adaptation of the link between the transmitter and receiver depending on the variation of the channel, adaptation of the system and transmission parameters depending on the user and channel requirements, and adaptation of the multi-carrier signal receiver will be discussed briefly.

Link adaptation plays an important role in wireless communication systems because of the dynamic nature of the channel. Link adaptation ensures that the signal at the receiver

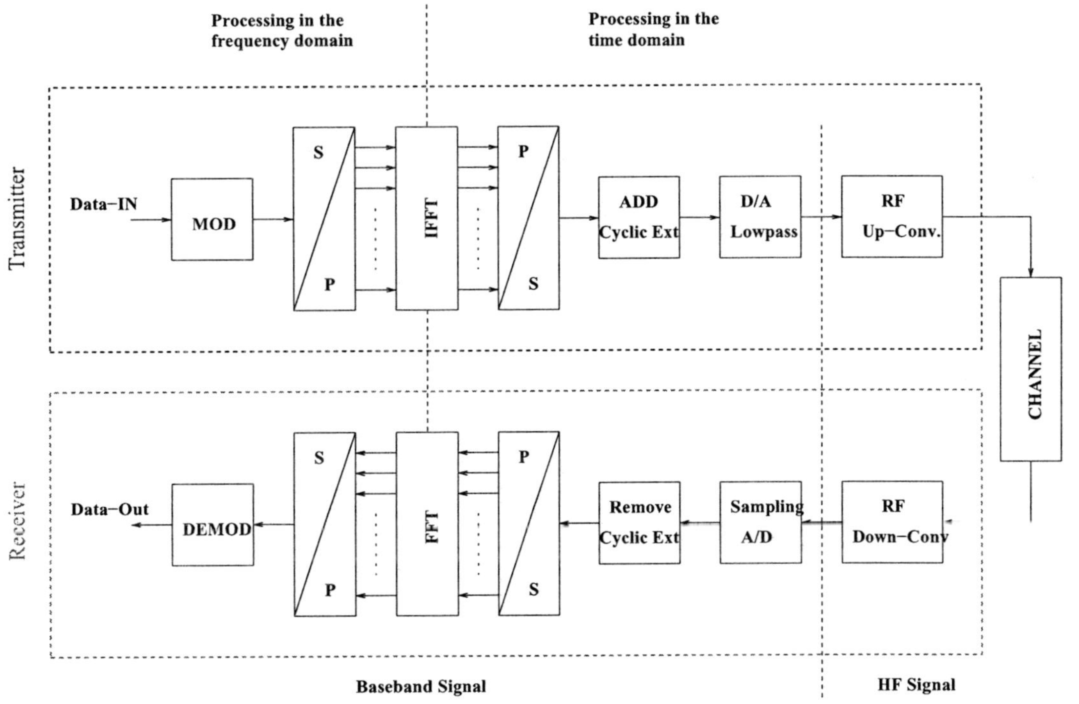

Figure 3. Block diagram of an OFDM transceiver.

side is in a form that can be processed and detected properly. It leads to efficient utilization of the resources including the bandwidth and power, as well as making sure that the communication is maintained to satisfy a desired service quality. The cost function can be multi-dimensional. Also, the number of factors (or parameters) that affect the link quality can be more than one. Among these factors power control, variable coding and modulation are very popular.

Adaptive modulation and coding provide a framework to adjust modulation level and forward error correction (FEC) coding rate depending on the link quality. Higher order modulations allow more bits to be transmitted for a given symbol rate. On the other hand, they are less power efficient, requiring higher energy per bit for a given BER. Therefore, higher order modulations should be used only when the link quality is high, as they are less robust to channel impairments. Similarly, strong FEC and interleaving provide robustness against channel impairments at the expense of lower data rate and spectral efficiency. As a result, by properly adapting the coding and modulation depending on the link quality, the average throughput can be maximized, which makes it very attractive for wireless packet data communication. Note that for mobile users, the instantaneous data rates will vary depending on the link quality.

Without power control, adaptive coding and modulation increase the variation of throughput among users such that users with good link quality always have high throughput, but users with bad link quality always have low throughput [10]. On the other hand, *power control* tries to equalize the link quality of the users so that all the receivers will have a constant signal-to-interference ratio (SIR). This allows to reduce the interference from other sources while making sure that each user gets just enough power to be able to detect what is being transmitted by the desired transmitter. If the transmitter sends more power for a specific user, the user benefits from it by having a better link quality, but the level of interference for the other users increases accordingly. On the other hand, if the user does not receive enough power, a reliable link cannot be established. In order to establish a reliable link while minimizing interference to other users, the transmitter should continuously control the transmitted power level. From a user perspective, power control ensures a constant service quality across the network, and such adaptation is suitable for real time audio and video communication applications. Power control can be combined with adaptive coding and modulation scheme to achieve both benefits for better resource utilization [9].

In OFDM systems, it is also possible to adapt the modulation and power for each subcarrier individually. Note that since the bandwidth of OFDM systems are much larger than the coherence bandwidth of the channel, different carriers will experience different signal-to-interference-plus-noise ratio (SINR). Therefore, the modulation level or the power on different carriers can be changed depending on the link quality observed at that carrier [32].

Transmitting different modulation on each carrier requires a large overhead for signaling. Therefore, approaches that group the neighboring carriers into subsets and use the same modulation in each group of carriers are preferred. This is called subband adaptive OFDM and illustrated in Fig. 4. The signaling can also be avoided in time division duplexing (TDD) systems under certain assumptions. Unlike frequency division duplexing (FDD) systems, where the channel on the downlink and uplink are different, in TDD systems, using the assumption of the reciprocal and slow varying channel, the transmitter and receiver can be assumed to experience the same channel response. Therefore, this might eliminate the need for signaling of the channel state information to the transmitter if the channel estimates are used as the link quality measures. However, the receiver still needs to know which modulation is used at the transmitter for each group. Blind modulation detection techniques can be used for this purpose [38, 64] as there is enough *a priori* information to estimate the modulation type. Note that although the channels could be the same, the interference observed in transmitter and receiver are not necessarily the same because the interferer may be physically closer to transmitter or receiver.

Even though adaptive power control has a long and rich history for wireless communication systems, the adaptive coding and modulation recently become popular due to the increased interest for high data rate communications, and they are becoming an integral part of most of the new generation standards. Adaptive modulation and coding are employed for WLAN and WMAN systems, which use OFDM as the transmission technique, successfully. IEEE 802.11a physical layer, for example, allow four different modulation options (BPSK, QPSK, 16-QAM, and 64-QAM) with different coding rates. The coding rates are obtained with different puncturing patterns to a mother convolutional code, resulting in 8 different modulation/coding options [3]. Therefore, a data rate ranging from 6 to 54 Mbit/s can be obtained by using various modes. BPSK, QPSK, and 16-QAM are used as mandatory modulation formats, whereas 64-QAM is applied as an optional mode.

In addition to the link adaptation, other **transmission parameters** can also be adjusted in a multi-carrier systems. Some of these parameters include the cyclic prefix size, interleaving type and length, subcarrier spacing, pilot patterns, antenna configurations for multiple transmitter antennas, and transmitted carrier frequency. In the rest of this section,

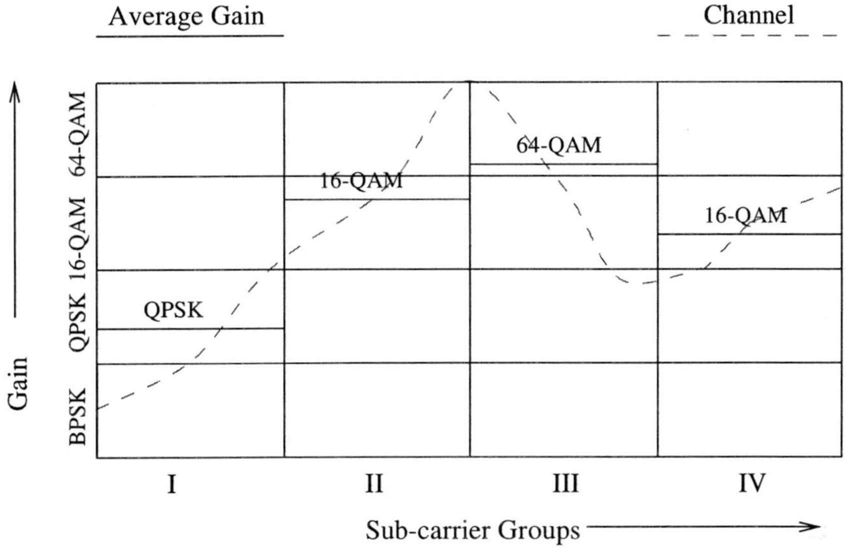

Figure 4. Modulation mode selection based on channel quality in subband adaptive OFDM systems.

we will discuss these adaptation techniques briefly.

4.0.1. Cyclic Prefix

The cyclic prefix (CP) provides multipath immunity and tolerance to symbol time synchronization errors. In order to remove the distortion due to multipath, the length of the CP should be larger than the maximum excess delay of the channel. Conventional OFDM system uses fixed-length cyclic prefix to combat the ISI caused by multipath fading. This may cause considerable performance degradation when the CP length is less than the channel delay spread [21, 28]. On the other hand it may decrease the system power and spectral efficiency when it is much larger.

The CP length may be changed adaptively depending on the channel conditions, specifically maximum excess delay of the channel. Adaptive CP is studied in [27, 67]. In the proposed algorithms the CP length is varied according to the current delay spread of the channel and the length information is conveyed to the receiver by signaling. This adaptation requires the knowledge of the maximum excess delay of the channel, estimation of which is studied in Section 5.2..

4.0.2. Adaptive Interleaving

Interleaving is a method that is used in communication systems to randomize the bursty channel errors. This can greatly improve the performance of channel coding which works better when the errors do not occur in bursty fashion. In OFDM, interleaving can be achieved in both time and frequency. Time interleaving is similar to the single carrier systems. Frequency interleaving is achieved by changing the order of symbols in frequency domain. The interleaving pattern is another parameter that may be changed adaptively to

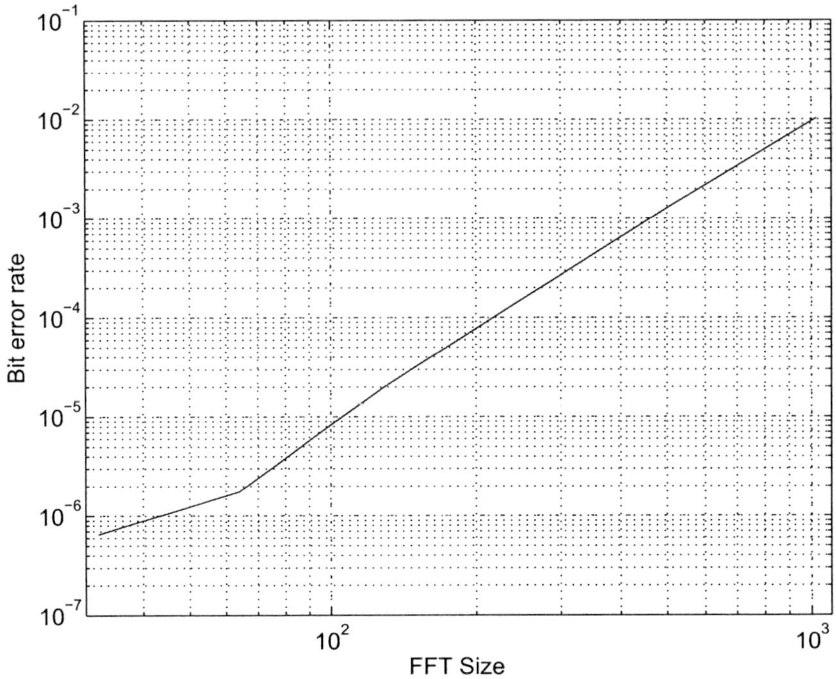

Figure 5. Bit error rate of an OFDM system in time-varying channel as a function of FFT size. The transmission bandwidth is 3.5MHz, the center frequency is 5.8GHz and the mobile speed is 100km/h.

obtain better shuffling of the symbols in time and frequency domain. The symbols may be rearranged according to instantaneous channel state information (CSI) of the OFDM subcarriers so as to minimize the BER after decoding. This approach is proposed in [33] and its performance is analyzed in [34] by finding bounds on the system BER.

4.0.3. Subcarrier Spacing

Subcarrier spacing, or number of subcarriers for a given bandwidth, is one of the most critical OFDM parameters. Subcarrier spacing may be changed adaptively depending on the Doppler spread, delay spread, and signal-to-noise ratio (SNR) of the channel. As subcarrier spacing increase, the inter-carrier interference (ICI) effect decrease as the spacing between subcarriers becomes larger compared to the Doppler shift. The raw BER performance of an OFDM system is shown in Fig. 5 as a function of FFT size. A center frequency of 5.8GHz, transmission bandwidth of 3.5MHz, and mobile speed of 100km/h are used in this figure. Note that since the bandwidth is fixed, x-axis can be interpreted as decreasing subcarrier spacing. Large subcarrier sizes cause higher peak-to-average-power ratio (PAPR) and the receiver becomes less immune to the frequency synchronization errors as the number of subcarriers increases.

While increasing subcarrier spacing cures the ICI problem, it means shorter symbol duration. Therefore, the relative size of required CP becomes larger, decreasing bandwidth efficiency. The effect of number of subcarriers on the system performance is investigated in [49]. The optimum number of subcarriers is found to be increasing with delay spread and the coherence time. The calculation of optimum subcarrier spacing for different channel conditions is studied in [27] by simulation and analytically in [56].

Variable FFT sizes is proposed for IEEE 802.16e standard to implement scalable OFDMA (see [62] and references therein) in which a constant subcarrier spacing is obtained by changing the FFT size and system bandwidth.

4.0.4. Adaptive Pilot Pattern

Pilot symbols may be inserted into the transmitted symbol stream regularly. These symbols can be used to find the channel values at their corresponding locations and these estimates may be filtered or interpolated to calculate the estimates of the channel parameters at other locations [35]. The rate of insertion of pilots in frequency and time direction (from OFDM symbol to OFDM symbol) can not be set arbitrarily. It depends on the variation of the channel parameters in time and frequency. It should be noted that an optimum pilot allocation is a trade-off between wasted energy in unnecessary pilot symbols, the fading process not being sampled sufficiently, and the spectral efficiency of the system. Hence, an optimum pilot allocation for a given channel might not be optimal for another channel as the fading process will be different.

Some other important elements for pilot arrangements can be the allocation of power to the pilots with respect to the data symbols, the transmitted modulation for the pilot tones *etc.* In most cases, the power for pilot tones and data symbols are equally distributed, however, it does not have to be uniform. For example, the channel estimation accuracy can be improved by transmitting more power at the pilot tones compared to the data symbols. However, for a given total power, this reduces the SNR over the data transmission.

It is clear from the above discussion about the pilot allocation that a better system performance can be obtained when the system is adaptive [23, 24, 30, 57]. In this case, the information about the channel statistics becomes very critical. The pilot allocation in the frequency direction requires the delay spread estimation, whereas the one in time direction requires Doppler spread estimation. If these estimates are available, then a pilot scheme using relatively less pilots but still providing an acceptable performance can be utilized. If this information is not available, then the pilot scheme can be designed based on the worst channel condition, *i.e.* the maximum expected delay spread and mobile speed.

In [48], the pilot patterns are chosen adaptively based on the prediction of the channel estimation error at the receiver. The transmitter tries to guarantee the minimum required SNR by using the minimum number of pilot subcarriers. Adaptation is performed over a block of OFDM symbols.

4.0.5. Dynamic Frequency Selection and Soft Spectrum Usage

Dynamic channel allocation and adaptive cell assignment algorithms have been studied since the early days of cellular systems. The main goal in assigning the available channel (or cell) to the users dynamically is to reduce the interference that the users experience. The adaptive system finds the most appropriate channel to be assigned to the users by considering the channel availability and the interference situations. Recent studies have been taking this concept beyond what is conventionally being done so far in cellular systems to include a soft spectrum usage not only within a specific network but also across the other wireless networks. The main motivation is the spectral crowding in some spectrum, especially in the unlicensed bands. Dynamic frequency selection (DFS) is an interference detection and avoidance, and channel sharing mechanism used in the unlicensed radio spectrum. Intended users are required to deploy a DFS mechanism to ensure that they do not invade an already occupied channel. Adaptive power control can be employed together with DFS mechanism where it can replace static power levels to increase the probability of selecting vacant channels while avoiding intrusion.

Further pushing the limit on the most efficient exploitation of all the available spectrum brings a new concept of soft spectrum usage. This involves the soft usage of the current licensed and unlicensed available spectrum. Even though some spectrum is licensed for a specific application, the usage of the spectrum for this application might not be to the full extend. Note that a recent study conducted by DARPA's XG initiative found that only 6% of the spectrum is fully utilized at all times. Currently, the major part of the spectrum is allocated (licensed) for a specific technology. Whether the spectrum is fully used or not, it is only available for a group of wireless devices. The static assignment of the spectrum to different technologies (or operators) causes the inefficient spectrum usage as quoted above. If a wireless device (or system) has the ability to operate over any carrier frequency with any bandwidth, and this device has the ability to identify the most suitable band (with less interference) to operate, then, the concept of the ideal soft spectrum usage will be a reality. However, with the current available technology, there are some limitations to achieve this goal, but, these limitations open new research areas to study. One of the major challenge is to be able to identify the most suitable band (carrier frequency and bandwidth) to operate. This requires scanning a wide spectrum and measuring a lot of parameters regarding to the channel and interference conditions. Once the channel behavior is accurately predicted, the transceivers can negotiate cognitive functions including channel coding, modulation rate, location, mobility *etc.* [29, 52, 58]. Another important challenge is to provide the transceiver the ability to operate over a wide range of spectrum with different bandwidths. The current transceivers include an analog front-end which is mostly fixed for a specific function to operate over a small range of frequencies. Such an analog front-end is not flexible and not programmable. This brings a new concept which is called software defined radio (SDR) where these fixed analog circuitry needs to be replaced with software programmable hardwares. The ideal SDR concept digitizes the received signal as soon as possible so that a flexible radio functionality can be obtained. As can be seen, this is a challenge with the current analog-to-digital-converter (ADC) capabilities and with the processing power available at this point. Therefore, currently, the new generation wireless systems are slowly integrating a version of this concept.

As far as OFDM is concerned, it has great ability to migrate to the soft spectrum usage. Its immunity to narrowband interference if designed properly and its ability to control different carriers independently are the features that can bring the soft spectrum usage into the reality for OFDM based systems. Note that this was one of the motivations why OFDM has become a strong option for Ultrawideband based wireless personal area networks. In the following, this will be described in more detail to show why OFDM is a good candidate for controlling interference and transmission spectrum.

An example for dynamically controlling interference in OFDM systems:

An OFDM approach can be one way of avoiding narrow-band interference (NBI) as briefly discussed above. NBI is especially a problem if the transmission bandwidth is very large. For example, in ultrawideband (UWB) communication, NBI is one of the major limiting factor. In OFDM based UWB, NBI can be avoided easily by an adaptive OFDM system design. First of all, the NBI will only affect some carriers in OFDM spectrum. Therefore, only the information that is transmitted over these frequencies will be affected by the interference. If the interfering carriers can be identified, transmission over these carriers can be avoided. In addition, by sufficient FEC and frequency interleaving, jamming resistance against NBI can be obtained easily. However, avoiding or adapting the transmission over the strongly interfered carriers can provide more spectrum and power efficiency, as they increase the immunity against NBI and hence relax the FEC power.

In OFDM, in order to identify the interfered carriers, the transmitter requires a feedback from the receiver. The receiver should have the ability to identify these interfered carriers. Once the receiver estimates these carriers, the relevant information will be sent back to the transmitter. The transmitter will then adjust the transmission accordingly. It can, for example, stop sending information on these carriers. Note that in such a scenario, the interference statistics need to be constant for a certain period of time. If the interference statistics change very fast, by the time the transmitter receives feedback and adjusts the transmission parameters, the receiver might observe different interference characteristics.

The feedback information can be various, including the interfered carrier index, the amount of interference on these carriers, the center frequency of NBI, the bandwidth of NBI *etc.* of the interfered carriers. One simple technique is to look at the average signal power in each carrier and compare it with a threshold. If the average signal power is greater than a threshold on a subcarrier at the receiver, that channel can be regarded as severely interfered by the NBI. Instead of making a hard decision on whether a carrier is interfered or not, soft estimation of the power of the interference can also be done [65].

4.1. Adaptation with Multiple Antennas

Adaptive antennas and adaptive beam forming techniques have also been studied extensively to increase the capacity and/or to improve the performance of wireless communication systems [42]. The adaptive antenna systems shape the radiation pattern in such a way that the information is transmitted (for example, from a base station) directly to the mobile user in narrow beams. This reduces the probability of another user experiencing interference in the network, resulting in improved link quality, which can also be translated into increased network capacity. Although, adaptive beam forming is an excellent way to utilize multiple antenna systems to enhance the link quality, recently different flavors of the usage of multi-antenna systems have gained significant interest. Space-time processing and multiple-input multiple-output (MIMO) antenna systems are some new developments that will allow further usage of multiple antenna systems in wireless communications. Adaptive implementation of these technologies is important for successful and efficient integration of them into wireless communication systems.

Combining adaptive antennas with OFDM for operation in a faded delay-spread channel is discussed in [55]. In this adaptive beamforming algorithm, a short training process first updates the weight vector and then a decision-directed technique is used for updating the weight vector. It has been demonstrated that this algorithm is able to extract desired signals while suppressing co-channel interferers.

MIMO and multi-antenna systems bring about a new dimension to wireless channel. The spatial dimension will be used in future communication systems for further improvement of the bandwidth and power efficiency. However, this dimension and the related parameter estimates need to be understood better. Research on parameter estimation for fast and accurate calculation of spatial selectivity, angular spread, and antenna correlation *etc.* is needed. Also, further research is required on the effect of mutual coupling between antenna elements, the effect of near-field scatterers on antenna patterns and antenna correlations, exploitation of pattern selectivity when the spatial selectivity is not enough, and adaptive generation of a desired pattern selectivity between antenna elements *etc.*

MIMO systems, which employ multiple transmit and multiple receive antennas, can provide huge capacity and/or improved performance gains by exploiting spatial selectivity of the channel. However, these gains, in reality, depend heavily on the statistical properties of the channel and the correlations between antenna elements. One of the factors that affects the antenna correlation is the characteristics of the scattering environment. Therefore, optimal way of using multiple antenna systems depends on the situation awareness. If the transmitter knows the instantaneous channel gains (the MIMO channel matrix), it can adapt the transmission to maximize the capacity of MIMO system [25, 53]. Similarly, the instantaneous antenna correlation values can be exploited to adapt the transmission. In many cases, estimation of the perfect instantaneous channel state and antenna correlation information, and feeding this information back to the transmitter might not be possible. This is the case especially when the mobility is high. Instead, other parameter measures like partial (statistical) channel information, average channel selectivity or angular spread would be useful for adapting the transmitter and receiver. Advanced signal processing techniques to calculate these partial channel and correlation information are needed.

By combining MIMO with OFDM systems, huge gains are achieved which are well suited for multimedia communication systems [51]. In MIMO-OFDM, the SNR not only varies over time and frequency but also depends on a number of parameters including the way the transmitted signals are mapped and weighed onto the transmit antennas, the processing technique used at the receiver, and the antenna polarization and propagation parameters such as mutual coupling between antennas [20]. Space-time adaptation aims to choose the best way of combining antennas either through space-time coding approach, beamforming or using the BLAST architecture. Those approaches can be coupled with adaptation of OFDM parameters, e.g., on each OFDM subcarrier, adaptive space-time coding, beamforming along with adaptive power and bit loading schemes can be employed yielding a space-time frequency adaptation scenario. The process can be generalized for multiuser MIMO-OFDM systems. In brief, MIMO-OFDM allows several domains of adaptation, however complexity in practical systems have to be taken into consideration.

5. Enabling Parameter Estimation

The methods studied in Section 4. requires some form of information about the dynamic radio channel. This information can be estimated at the physical layer as well as upper layers and can be fed to appropriate adaptation algorithms. The main requirements when employing an estimation scheme are accuracy, low complexity, robustness, and finally having low delay so that it can respond to system variations.

Some important parameters for OFDM adaptation are Doppler spread, Delay spread, channel estimate, noise plus interference power estimate, and SNR (or SINR). In this section, these important parameters and their estimation techniques will be discussed briefly.

5.1. Doppler Spread

Doppler shift is the frequency shift experienced by the radio signal when either the transmitter or receiver is in motion, and it is a measure of the spectral broadening caused by the temporal rate of change of the mobile radio channel. Therefore, time-selective fading and Doppler spread are directly related. The coherence time of the channel can be used to characterize the time variation of the time-selective channel. It represents the statistical measure of the time window over which the two signal components have strong correlation, and it is inversely proportional to the Doppler spread.

Knowing Doppler spread in mobile communication systems can improve detection and help to optimize transmission at the physical layer. Specifically, knowing Doppler spread can decrease unnecessary handoffs, adjust interleaving lengths to reduce reception delays, update rate of power control algorithms, *etc.* In addition, in OFDM systems, if the channel varies considerably within one OFDM symbol because of high mobility, orthogonality between subcarriers is lost, leading to ICI [39]. Doppler information can help in selection of appropriate transmission profiles that are immune to ICI and hence the overall system performance will be improved. More details on this topic can be found in Section 6..

Various methods based on the correlation of the channel and/or received signal have been used to estimate the Doppler spread in single carrier systems (see for example [54]). Similar approaches can also be employed in OFDM systems. For example, the autocorrelation between the repeated parts of the received signal due to CP is exploited in [19] to estimate the Doppler spread. This approach is simple and provides good estimate if the repeated parts are not disturbed by the multipath channel. However, the part of the CP that is undisturbed by the multipath channel may be small especially when the environment causes large delay spread. Similarly, the correlation of the channel can also be used in OFDM systems to obtain the Doppler information. Note that in OFDM systems the channel estimates are often obtained in frequency domain (i.e. frequency channel response). By obtaining the auto-correlation function (ACF) of a certain subcarrier over several symbols, Doppler spread can be estimated [47]. However, every subcarrier will have noise perturbation due to additive white Gaussian noise (AWGN) and ICI which appears to be fast varying over subcarriers. Therefore, by transforming the frequency domain channel components, the true channel and noise components (ICI and AWGN) may be separated. When IFFT is applied to the channel frequency response (CFR), the channel impulse response (CIR) will be concentrated on the first few taps, while noise and ICI terms will be spread over all IFFT size. This fact is used in [66] to overcome the bias due to ICI and noise by applying IFFT to the channel estimates and then using the time domain channel estimates to calculate Doppler spread.

5.2. Delay Spread

Delay spread describes the time dispersiveness of the channel, and it is related to frequency selectivity of the channel. The frequency selectivity can be described in terms of coherence bandwidth, which is a measure of range of frequencies over which the two frequency components have a strong correlation. The coherence bandwidth is inversely proportional to the delay spread [14].

The information about the frequency selectivity of the channel, and the corresponding time domain root-mean-squared (RMS) delay spread can be very useful for improving the performance of the wireless radio receivers through transmitter and receiver adaptation. Especially, in OFDM systems, this information will be very useful to adjust many transmission parameters. For example, one way to increase spectral efficiency is to adapt the length of the cyclic prefix depending on the radio environment. This adaptation requires the knowledge of the delay spread as discussed in Section 4.0.1.. Other OFDM parameters that could be changed adaptively using the knowledge of the dispersion are OFDM symbol duration and OFDM sub-carrier spacing. Dispersion estimation can also be used for receiver adaptation. A two-dimensional Wiener filter, which is implemented as a cascade of two one-dimensional filters, is used for channel estimation in [45]. The bandwidth of the second filter, which is in frequency direction, is changed depending on the estimated delay spread of the channel to keep the noise low and thus to improve the channel estimation.

Characterization of the frequency selectivity of the radio channel is studied in [59–61] using level crossing rate (LCR) of the channel in frequency domain. Frequency domain LCR gives the average number of crossings per Hz at which the measured amplitude crosses a threshold level. In [46], instantaneous RMS delay spread, which provides information about local channel dispersion, is obtained by estimating the channel impulse response (CIR) in time domain. The detected symbols in frequency domain are used to re-generate the time domain signal through inverse discrete Fourier transform (IDFT). This signal, then, is used to correlate the actual received signal to obtain CIR, which is then used for delay spread estimation. In [45], RMS delay spread is also calculated from the instantaneous time domain CIR, where in this case the CIR is obtained by taking IDFT of the frequency domain channel estimate. CIR is obtained by taking IDFT of the sampled CFR in [15] in order to decrease computational complexity. A technique based on the cyclic-prefix for delay spread estimation is proposed in [16]. This technique uses change of gradient of the correlation between cyclic prefix and the last part of the OFDM symbol as a strategy to detect the dispersion parameters. In [14], channel frequency correlation (CFC) estimation is used to get the average RMS delay spread of the channel by assuming an exponentially decaying power delay profile (PDP).

5.3. SNR and Noise Variance

The SNR is broadly defined as the ratio of the desired signal power to the noise power and has been accepted as a standard measure of signal quality. Adaptive system design requires the estimate of SNR in order to modify the transmission parameters to make efficient use of system resources. Poor channel conditions, reflected by low SNR values, require that the transmitter modify transmission parameters such as coding rate, modulation mode *etc.* to compensate channel and to satisfy certain application dependent constraints such as constant BER and throughput. Dynamic system parameter adaptation requires a real-time noise power estimator for continuous channel quality monitoring and corresponding compensation in order to maximize resource utilization. SNR knowledge also provides information about the channel quality which can be used by handoff algorithms, power control, channel estimation through interpolation, and optimal soft information generation for high performance decoding algorithms.

The SNR can be estimated using regularly transmitted training sequences, pilot data or data symbols (blind estimation). In OFDM systems, SNR is estimated in frequency domain since detection is done in this domain. In conventional SNR estimation techniques, the noise is usually assumed to be white and an SNR value is calculated.

Parallel with the possible data rates, the transmission bandwidth of OFDM systems is also large. UWB-OFDM [17] and IEEE 802.16 based WMAN [8] systems are examples of OFDM systems with large bandwidths. Because of these large bandwidths, noise can not be assumed to be white with flat spectrum across subcarriers. In [18], this assumption is removed by calculating SNR values for each subcarrier. However, the correlation of the noise variance across subcarriers is not used since noise variance is calculated for each subcarrier separately. In [13, 65], variation of the noise power across OFDM sub-carriers as well as across OFDM symbols is allowed. The noise variances at each subcarrier is estimated using a two dimensional sliding window whose size is calculated using local statistics of the noise. These estimates are specifically useful for adaptive modulation, and optimal soft value calculation for improving channel decoder performance.

5.4. Channel Quality Measures

Channel quality estimation is by far the most important measurement that can be used in adaptive receivers and transmitters [37]. Different ways of measuring the quality of radio channel are possible, and many of these measurements are done in the physical layer using baseband signal processing techniques. In most of the adaptation algorithms, the target quality measure is the FER or BER, as these are closely related to higher-level quality-of-service parameters like speech and video quality. However, reliable measurement of these qualities requires many measurements, and this causes delays in the adaptation as the process could be very long. Therefore, other types of channel quality measurements that are related to these might be preferred. When the received signal is impaired only by white Gaussian noise, analytical expressions can be found relating the BER to other measurements. For other impairment cases, like colored interferers, numerical calculations and computer simulations that relate these measurements to BER can be performed. Therefore, depending on the system, a channel quality is related to the BER. Then, for a target BER (or FER), a required signal quality threshold can be calculated to be used with the adaptation algorithm. The measurements can be performed at various points of a receiver, depending on the complexity, reliability, and delay requirements. There are trade-offs in achieving these requirements at the same time.

6. Adaptation in Mobile OFDM Systems

OFDM is a suitable solution for high data rate transmission because of its immunity to the time dispersion of the channel. However, frequency dispersion, or time-varying channel introduces another problem as OFDM systems loose their orthogonality when the channel varies over the OFDM symbol time, *i.e.* when the Doppler spread is a significant portion of the subcarrier spacing [39].

Assume that the transmitted signal is passed through a sample-spaced Rayleigh fading channel with maximum delay of LT_s, where T_s is the data symbol period. Then, the received time domain signal can be written as

$$y(n) = \sum_{l=0}^{L-1} h_{l,n} x(n) \tag{3}$$

where $h_{l,n}$ is the channel impulse response with L tabs. The output of the IFFT at the receiver can be calculated as [63]

$$\begin{aligned} Y(k) &= \sum_{m=0}^{N-1} \sum_{l=0}^{L-1} X(m) \frac{1}{N} \sum_{d=0}^{N-1} h_{l,d} e^{j2\pi d(m-k)/N} + W(k) &(4) \\ &= X(k)D(k,k) + \sum_{m=0, m\neq k}^{N-1} X(m)D(m,k) + W(k) &(5) \end{aligned}$$

The second term in the right hand size of (5) is the power leakage from the neighboring subcarriers and it is called ICI. Note that when there is no time variation in channel during a block period, *e.g.* $h_{l,n} = h_l$, the resulting ICI is zero and (5) reduces to (2).

For a fixed total bandwidth, the ICI contribution increases as the number of subcarriers increases. This happens since the OFDM symbol period increases and hence the variation of channel within the symbol duration. Analysis of the effects of Doppler spread for OFDM systems is given in [11, 40, 41]. Tight and general bounds for ICI due to Doppler spread are found in [36]. The ICI power is reported to be linearly proportional with the square of maximum Doppler frequency and symbol duration ($f_D T_s$). In other words, the ratio between the maximum Doppler frequency and subcarrier spacing determines the effect of ICI.

While increasing the subcarrier spacing may decrease the ICI, it increases the energy loss due to the cyclic prefix extension of OFDM systems. Hence the optimum window size is a function of frequency and time selectivities of the channel. This trade-off is shown in Fig. 6. The BER performance of an OFDM system with 3.5MHz bandwidth operating at 5.8GHz center frequency is shown for different FFT sizes and CP lengths. The SNR of the system is 30dB and mobile speed is 100km/h. As opposed to the Fig. 5, the affect of energy loss due to CP is taken into consideration in this figure. However, the length of the CP is chosen larger than the delay spread of the channel in order to only show the affect of CP and FFT sizes. As this figure shows the subcarrier spacing (or FFT size) that gives optimum performance is a function of the delay spread and Doppler spreads of the channel.

Transmitter adaptation may be used to overcome the mobility problem. The subcarrier spacing may be changed depending on the Doppler and delay spread of the channel in order

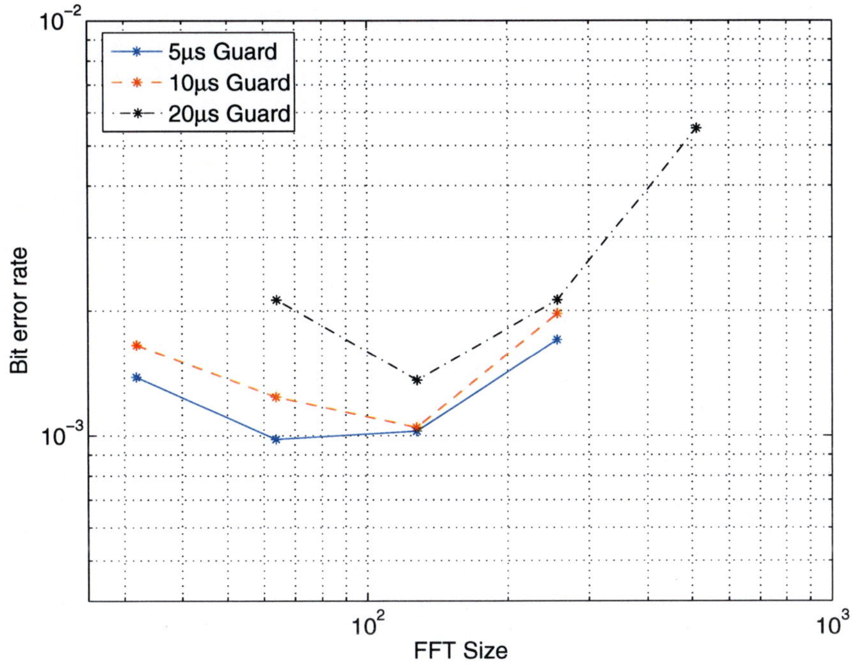

Figure 6. Bit error rate of an OFDM system in time-varying channel as a function of FFT size and guard interval length. The transmission bandwidth is 3.5MHz, the center frequency is 5.8GHz, the mobile speed is 100km/h and SNR is 30dB.

to reduce ICI while keeping the bandwidth usage maximum. Calculation of the optimum subcarrier spacing in doubly dispersive channels is investigated in [27,56]. The ICI problem can also be solved using advanced receiver algorithms such as frequency domain equalization or interference cancellation. However, these algorithms are computationally complex and they can only eliminate some of the disturbance due to ICI.

Another factor that needs to be considered in mobile OFDM systems is the estimation of fast varying channel. When the training symbols transmitted before the data symbols are used for estimation, an error floor for time-varying channels is introduced. To overcome this problem, pilot subcarriers or midambles may be used [44].

The ICI problem becomes more important as OFDM is used in applications that need to support mobility of the transceivers. One example for mobile OFDM systems is mobile WirelessMAN standard [31] which is currently being developed.

7. Case Study: IEEE 802.16 WirelessMAN

The IEEE 802.16d standard was published in 2004. It provides point to multipoint access for fixed subscribers. The draft standard IEEE 802.16e updates and expands this standard to allow for mobile subscriber stations traveling at vehicular speeds. While IEEE's role is to develop standards for the physical (PHY) and medium access control (MAC) layers, WiMAX forum ensures interoperability between vendors equipment through its certifica-

Table 1. Mandatory coding and modulation options for OFDM-256 physical layer of IEEE 802.16d.

Modulation	RS code	CC rate	Overall coding rate
BPSK	(12,12,0)	1/2	1/2
QPSK	(32,24,4)	2/3	1/2
QPSK	(40,36,2)	5/6	3/4
16-QAM	(64,48,8)	2/3	1/2
16-QAM	(80,72,4)	5/6	3/4
64-QAM	(108,96,6)	3/4	2/3
64-QAM	(120,108,6)	5/6	3/4

tion process and the technology is commonly referred as "WiMAX". Both IEEE 802.16d and IEEE 802.16e specifications make use of adaptive techniques for efficiently utilizing available resources. Hence, we have chosen these standards as a case study that ends this chapter. IEEE 802.16 defines multiple PHY layer options. The OFDM based PHYs employs different adaptive features. In this section, some of these features will be studied as an example of practical systems that use adaptive techniques discussed in this chapter.

7.1. CP Size

OFDM based PHY layer of 802.16 defines 4 different CP sizes of 8, 16, 32 and 64 samples for transmission in order to provide immunity to multipath channels. Although the size of CP can not be changed dynamically during transmission, it can be determined depending on the expected channel delay spread, larger delay spreads requiring larger CP sizes. The subscriber station (SS) should use the same CP size with base station (BS).

7.2. Modulation/Coding

In 802.16, channel coding is performed in three stages; randomization, FEC and interleaving. An FEC consisting of the concatenated Reed-Solomon outer coding and a rate compatible convolutional inner coding on both downlink and uplink are mandated for the 256 OFDM physical layer. The modulation and coding schemes may be adjusted individually for each SS on a frame-by-frame basis. These options for modulation and coding are shown in Table 1. The standard recommends that the SS use carrier-to-inteference-plus-noise ratio (CINR) as a metric to decide if a burst profile change is required.

7.3. Scalable OFDMA

Scalable OFDM concept is introduced in the IEEE 802.16e standard which will support users with vehicular speeds. The main idea of scalability is to keep the subcarrier spacing constant at different channel bandwidths by changing the FFT size. FFT sizes of 128, 256, 512, 1024 and 2048 will be supported. The SS may implement a scanning and search mechanism to detect the downlink signal when performing initial network entry and this may include dynamic detection of the FFT size and the channel bandwidth employed by the BS.

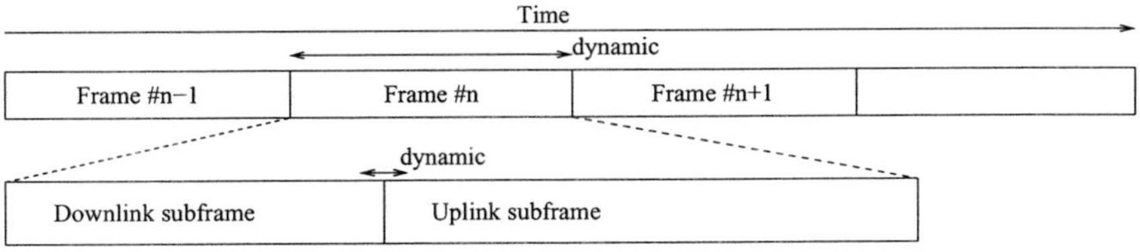

Figure 7. Example OFDM frame structure with TDD.

7.4. Dynamic Frequency Selection

DFS is defined as the ability to switch to different physical RF channels between transmit and receive activity based on channel measurement criteria and its implementation in license-exempt bands is mandatory as frequency planning is not practical in these bands. DFS can be used to avoid assigning a user to a channel occupied by another system. It is generally based on comparison of a CINR threshold against idle time receive signal strength indicator (RSSI) measurements. DFS is predominantly effective to combat interference from and to other systems operating in the same band.

DFS in WMAN consists of the following operations: scheduling for channel testing, testing channels for primary users, requesting and reporting of measurements, detecting primary users, discontinuing operations after detecting primary users, selecting and advertising a new channel.

7.5. Adaptive Power Control

The objective of the power control algorithm is to bring the received power density from a given subscriber to a desired level. The received power density is defined as total power received from a given subscriber divided by the number of active subcarriers. The base station shall be capable of providing accurate power measurements of the received burst signal. This value can then be compared against a reference level, and the resulting error can be fed back to the SS in a calibration message coming from the MAC. The implementation is left as a vendor-specific task.

7.6. Adaptive TDD

The 802.16 standard supports both FDD and TDD methods. In TDD mode, a frame consists of a downlink subframe and an uplink subframe which is illustrated in Fig. 7. The length of each subframe may be changed as well as the total frame length. This dynamic asymmetry between uplink and downlink channels matches bandwidth to the user demand.

8. Conclusion

As the demand for wireless access for voice and high data rate is increasing, new generation wireless systems needed to accommodate this demand with limited spectrum resources. This is only possible through better resource management and improved transmission technologies. Adaptive resource managements and adaptive transmission technologies allow improvements in spectral efficiency, power consumption, and overall system performance. Multi-carrier systems that change the transmission parameters adaptively has a potential for meeting this high demand.

Acknowledgments

The authors thank Francis E. Retnasothie for his help for the preparation of this document.

References

[1] IEEE 802.15 WPAN high rate alternative PHY task group 3a (tg3a) website.

[2] Asymmetric digital subscriber line (ADSL), 1995.

[3] Supplement to IEEE standard for information technology telecommunications and information exchange between systems - local and metropolitan area networks - specific requirements. part 11: wireless LAN Medium Access Control (MAC) and Physical Layer (PHY) specifications: high-speed physical layer in the 5 GHz band, Sept. 1999.

[4] Broadband radio access networks (BRAN); HIPERLAN type 2; physical (PHY) layer, Dec. 2001.

[5] Digital video broadcasting (DVB); framing structure, channel coding and modulation for digital terrestrial television, Jan. 2001.

[6] Radio broadcasting systems; digital audio broadcasting (dab) to mobile, portable and fixed receivers, May 2001.

[7] Broadband mobile access communication system (hiswana), Nov. 2002.

[8] IEEE standard for local and metropolitan area networks part 16: Air interface for fixed broadband wireless access systems, 2002.

[9] M.H. Ahmed, H. Yanikomeroglu, D. Falconer, and S. Mahmoud, Performance enhancement of joint adaptive modulation, coding and power control using cochannel-interferer assistance and channel reallocation, in *Proc. IEEE Wireless Commun. and Networking Conf.*, vol. 1, May 2003, pp. 306–310.

[10] M.H. Ahmed, H. Yanikomeroglu, and S. Mahmoud, Fairness enhancement of link adaptation techniques in wireless access networks, in *Proc. IEEE Veh. Technol. Conf.*, vol. 3, Oct. 2003, pp. 1554–1557.

[11] J. Armstrong, P. M. Grant, and G. Povey, Polynomial cancellation coding of OFDM to reduce intercarrier interference due to Doppler spread, in *Proc. IEEE Global Telecommunications Conf.*, vol. 5, Sydney, NSW, Australia, Nov. 1998, pp. 2771–2776.

[12] H. Arslan, Signal Processing Communications Handbook, CRC Press, 2004, ch. Adaptation Techniques and the Enabling Parameter Estimation Algorithms for Wireless Communication Systems.

[13] H. Arslan and S. B. Reddy, Noise variance and SNR estimation for OFDM based wireless communication systems, in *Proc. The 3rd Int. Conf. on Wireless and Optical Commun.*, Banff, Alberta, Canada, July 2003.

[14] H. Arslan and T. Yücek, Delay spread estimation for wireless communication systems, in *Proc. IEEE Symposium on Computers and Commun.*, Antalya, TURKEY, June/July 2003, pp. 282–287.

[15] ——, Estimation of frequency selectivity for OFDM based new generation wireless communication systems, in *Proc. World Wireless Congress,* San Francisco, CA, May 2003.

[16] C.R.N. Athaudage and A.D.S. Jayalath, Delay-spread estimation using cyclic-prefix in wireless OFDM systems, in *Proc. IEEE Int. Conf. on Acoust., Speech, and Signal Processing*, vol. 4, Apr. 2003, pp. 668–671.

[17] J. Balakrishnan, A. Batra, and A. Dabak, A multi-band OFDM system for UWB communication, in *Proc. IEEE Conference on Ultra Wideband Systems and Technologies*, Nov. 2003, pp. 354–358.

[18] S. Boumard, Novel noise variance and SNR estimation algorithm for wireless MIMO OFDM systems, in *Proc. IEEE Global Telecommunications Conf.*, vol. 3, Dec. 2003, pp. 1330–1334.

[19] J. Cai, W. Song, and Z. Li, Doppler spread estimation for mobile OFDM systems in rayleigh fading channels, *IEEE Trans. Consumer Electron.*, **49** (2003), pp. 973–977.

[20] S. Catreux, V. Erceg, D. Gesbert, and Jr. Heath, R.W., Adaptive modulation and MIMO coding for broadband wireless data networks, *IEEE Commun. Mag.*, **40** (2002), pp. 108–115.

[21] S. Celebi, Interblock interference (IBI) and time of reference (TOR) computation in OFDM systems, *IEEE Trans. Commun.,* **49** (2001), pp. 1895–1900.

[22] P. Dent, G.E. Bottomley, and T. Croft, Jakes fading model revisited, *IEE Electron. Lett.,* **29** (1993), pp. 1162–1163.

[23] A. Dowler, A. Doufexi, and A. Nix, Performance evaluation of channel estimation techniques for a mobile fourth generation wide area OFDM system, in *Proc. IEEE Veh. Technol. Conf.*, vol. 4, Vancouver, Canada, Sept. 2002, pp. 2036–2040.

[24] A. Dowler and A. Nix, Performance evaluation of channel estimation techniques in a multiple antenna OFDM system, in *Proc. IEEE Veh. Technol. Conf.*, vol. 2, Orlando, FL, Oct. 2003, pp. 1214–1218.

[25] G. J. Foschini and M. J. Gans, On limits of wireless communications in a fading environment when using multiple antennas, *Wireless Personal Commun.*, **6** (1998), pp. 311–335.

[26] S. Hara and R. Prasad, Overview of multicarrier CDMA, *IEEE Commun. Mag.*, **35** (1997), pp. 126–133.

[27] D. T. Harvatin and R. E. Ziemer, Orthogonal frequency division multiplexing performance in delay and Doppler spread channels, in Proc. *IEEE Veh. Technol. Conf.*, vol. 3, Phoenix, AZ, May 1997, pp. 1644–1647.

[28] W. Henkel, G. Taubock, P. Odling, P.O. Borjesson, and N. Petersson, The cyclic prefix of OFDM/DMT - an analysis, in *International Zurich Seminar on Broadband Communications.* Access, Transmission, Networking., Zurich, Switzerland, Feb. 2002, pp. 1–3.

[29] W. D. Horne, Adaptive spectrum access: Using the full spectrum space.

[30] C. Hou, S. Song, , and D. Cao, Channel estimation for adaptive OFDM system and effects of estimation error on system performance, in *Proc. IEEE Int. Workshop on Mobile and Wireless Commun. Network,* vol. 1, Stockholm, Sweden, Sept. 2002, pp. 200–204.

[31] IEEE 802.16e, mobile WirelessMAN.

[32] T. Keller and L. Hanzo, Adaptive modulation techiniques for duplex OFDM transmission, *IEEE Trans. Veh. Technol.*, **49** (2000), pp. 1893–1906.

[33] S.W. Lei and V.K.N. Lau, Adaptive interleaving for OFDM in TDD systems, *IEE Proc. Commun.*, **48** (2001), pp. 77–80.

[34] ——, Performance analysis of adaptive interleaving for OFDM systems, *IEEE Trans. Veh. Technol.*, **51** (2002), pp. 435–444.

[35] Y. Li, Pilot-symbol-aided channel estimation for OFDM in wireless systems, *IEEE Trans. Veh. Technol.,* **49** (2000), pp. 1207–1215.

[36] Y. (Geoffrey) Li and L. J. Cimini Jr., Bounds on the interchannel interference of OFDM in time-varying impairments, *IEEE Trans. Commun.,* **49** (2001), pp. 401–404.

[37] S. Nanda, K. Balachandran, and S. Kumar, Adaptation techniques in wireless packet data services, *IEEE Commun. Mag.*, **38** (2000), pp. 54–64.

[38] S. B. Reddy, T. Yücek, and H. Arslan, An efficient blind modulation detection algorithm for adaptive OFDM systems, in *Proc. IEEE Veh. Technol. Conf.*, Orlando, FL, Oct. 2003.

[39] P. Robertson and S. Kaiser, Analysis of the loss of orthogonality through Doppler spread in OFDM systems, in *Proc. IEEE Global Telecommunications Conf.*, vol. 1B, Rio de Janeireo, Brazil, 1999, pp. 701–706.

[40] P. Robertson and S. Kaiser, The effect of Doppler spreads in OFDM(A) mobile radio systems, in *Proc. IEEE Veh. Technol. Conf.*, vol. 1, Amsterdam, The Netherlands, Sept. 1999, pp. 329–333.

[41] M. Russell and G. L. Stüber, Interchannel interference analysis of OFDM in a mobile environment, in *Proc. IEEE Veh. Technol. Conf.*, vol. 2, Chicago, IL, July 1995, pp. 820–824.

[42] S. Anderson, H. Dam, U. Forssen, J. Karlsson, F. Kronestedt, S. Mazur, and K.J. Molnar, Adaptive antennas for GSM and TDMA systems, *IEEE Personal Commun. Mag.*, **6** (1999), pp. 74–86.

[43] B. Saltzberg, Performance of an efficient parallel data transmission system, *IEEE Trans. Commun.*, **15** (1967), pp. 805–811.

[44] M. Sandell and O. Edfors, A comparative study of pilot-based channel estimators for wireless OFDM, 1996.

[45] F. Sanzi and J. Speidel, An adaptive two-dimensional channel estimator for wireless OFDM with application to mobile DVB-T, *IEEE Trans. Broadcast.*, **46** (2000), pp. 128–133.

[46] H. Schober and F. Jondral, Delay spread estimation for OFDM based mobile communication systems, in *Proc. European Wireless Conf.*, Florence, Italy, Feb. 2002, pp. 625–628.

[47] ——, Velocity estimation for OFDM based communication systems, in *Proc. IEEE Veh. Technol. Conf.*, Kyoto, Japan, May 2002, pp. 715–718.

[48] O. Simeone and U. Spagnolini, Adaptive pilot pattern for OFDM systems, in *Proc. IEEE Int. Conf. Commun.*, vol. 2, June 2004, pp. 978–982.

[49] H. Steendam and M. Moeneclaey, Analysis and optimization of the performance of OFDM on frequency-selective time-selective fading channels, *IEEE Trans. Commun.*, **47** (1999), pp. 1811–1819.

[50] M. Sternad, T. Ottosson, A. Ahlén, and A. Svensson, Attaining both coverage and high spectral efficiency with adaptive OFDM downlinks, in *Proc. IEEE Veh. Technol. Conf.*, Orlando, FL, Oct. 2003.

[51] G.L. Stuber, J.R. Barry, S.W. McLaughlin, Ye Li, M.A. Ingram, and T.G. Pratt, Broadband MIMO-OFDM wireless communications, *Proc. IEEE*, **92** (2004), pp. 271–294.

[52] BBN Technologies, The XG vision.

[53] E. Telatar, Capacity of multiantenna Gaussian channels, tech. report, AT&T Bell Laboratories, June 1995.

[54] C. Tepedelenlioğlu, A. Abdi, G. Giannakis, and M. Kaveh, Estimation of Doppler spread and signal strength in mobile communications with applications to handoff and adaptive transmission, *Wireless Communications and Mobile Computing,* **1** (2001), pp. 221–242.

[55] L. M. Tuan, P. V. Su, J. Kim, and G. Yoon, A new RLS-based adaptive beamforming algorithm for smart antennas applied to an OFDM system, in *Proc. International Conference on Microwave and Millimeter Wave Technology*, Aug. 2002.

[56] F. Tufvesson and T. Maseng, Multiaccess, mobility and teletraffic - advances in wireless networks, Kluwer Academic Publishers, Dordrecht, The Netherlands, 1998, ch. Optimization of sub-channel bandwidth for mobile OFDM systems, pp. 103–114.

[57] X. Wang and K. J. R. Liu, OFDM channel estimation based on time-frequency polynomial model of fading multipath channel, in *Proc. IEEE Veh. Technol. Conf.*, vol. 1, Atlantic City, NJ, Oct. 2001, pp. 460–464.

[58] K. Werbach, Radio revolution.

[59] K. Witrisal and A. Bohdanowicz, Influence of noise on a novel RMS delay spread estimation method, in *Proc. IEEE Int. Symposium on Personal, Indoor and Mobile Radio Commun.*, vol. 1, London, U.K., Sept. 2000, pp. 560–566.

[60] K. Witrisal, Y.-H. Kim, and R. Prasad, RMS delay spread estimation technique using non-coherent channel measurements, *IEE Electron. Lett.*, **34** (1998), pp. 1918–1919.

[61] ——, A new method to measure parameters of frequency selective radio channel using power measurements, *IEEE Trans. Commun.*, **49** (2001), pp. 1788–1800.

[62] H. Yaghoobi, Scalable OFDMA physical layer in IEEE 802.16 WirelessMAN, *Intel Technology Journal*, **8** (2004).

[63] T. Yücek, Self-interference handling in OFDM based wireless communication systems, master's thesis, University of South Florida, Nov. 2003.

[64] T. Yücek and H. Arslan, A novel sub-optimum maximum-likelihood modulation classification algorithm for adaptive OFDM systems, in *Proc. IEEE Wireless Commun. and Networking Conf.*, Atlanta, GA, Mar. 2004.

[65] ——, Noise plus interference power estimation in adaptive OFDM systems, in *Proc. IEEE Veh. Technol. Conf.*, Stockholm, Sweden, May 2005.

[66] T. Yücek, R. A. Tannious, and H. Arslan, Doppler spread estimation for wireless OFDM systems, in *Proc. IEEE Sarnoff Symposium,* Princeton, NJ, Apr. 2005.

[67] Z.-Y. Zhang and L.-F. Lai, A novel OFDM transmission scheme with length-adaptive cyclic prefix, *Journal of Zhejing University Science,* **5** (2004), pp. 1336–1342.

In: Adaptation Techniques in Wireless Multimedia Networks ISBN 1-59454-883-8
Editors: Wei Li and Yang Xiao, pp. 241-264

Chapter 11

ONLINE STOCHASTIC OPTIMIZATION OF WIRELESS MULTIMEDIA NETWORKS*

***Sajal K. Das*[1], *M. A. Haleem*[2] *and R. Chandramouli*[1]**
[1]Department of Electrical and Computer Engineering,
Stevens Institute of Technology
[2]Department of Computer Science and Engineering,
University of Texas at Arlington, TX, USA

Abstract

Online optimization plays an important role in adaptive wireless multimedia networking. In adaptive networking the goal is to adapt some system parameters (e.g., source/channel coding rate, user schedule, and modulation/coding) to match the communication environment (e.g., wireless channel condition and remaining battery energy) to optimize the overall performance (e.g., throughput, delay, and received image quality). Online optimization techniques achieve this goal by "learning" the random, time-varying communication environment. In this chapter we discuss stochastic learning control based online optimization of wireless multimedia networks. Specifically, we consider the following two problems: (a) joint downlink adaptive rate selection and opportunistic user scheduling for a 3G multi-user network; and (b) wireless link state dependent video source coding rate control.

Keywords: stochastic learning automata, cross layer optimization, 3G wireless, user scheduling, video coding

1. Introduction

Adaptation in wireless networks can be implemented in several layers of the network protocol stack. Recently, there has been great interest in channel adaptive transmission techniques, with the availability of channel knowledge as side information - adaptive antenna arrays, space-time coding, and channel adaptive rate and power control to name a few.

*R. Chandramouli was supported in part by the grants: NSF ANI CAREER 0133761 and US Army/ARDEC.

These techniques rely on the estimation of channel parameters at the receiver and feedback of such information to the transmitter at the physical layer. An alternative approach would be "adaptive learning" of the best transmission parameters based on the success/failure of past communications. These approaches can make use of control information from layers other than physical layer and such information can be exchanged to physical layer if necessary. The performance of the class of such techniques depend on the desired time-scale of adaptation and other factors. In this chapter, we discuss adaptive online optimization of a wireless multimedia network in the following layers using stochastic techniques:

- **Physical (PHY) and Medium Access Control (MAC) Layer:** Joint optimization of rate control (adaptive modulation and coding) and multi-user scheduling to achieve desired quality of service (QoS).
- **Application Layer:** Wireless link state adaptive H.261 [8] based video rate control.

The quality of wireless communication channel in terms of signal attenuation is subject to periodic fluctuations over time due to multi-path fading and path loss variations [34, 65]. Thus with a fixed transmission power and rate, the channel bit error rate (BER) varies with time leading to excessive packet losses at deep fades, and under utilization of channel capacity at peaks. Adaptive rate selection [1, 2, 5, 24, 25, 60] has been of interest as a way to increase the throughput (successful bits/second) to achieve the capacity of wireless links.

Along with the adaptive rate control techniques for wireless links, optimum scheduling for multi user fading channel has been of interest for optimization of system throughput subject to quality of service (QoS) requirements of each user, in terms of the effective transmission rate. In a multi-user scenario, where a transmitter serves multiple receivers that are sufficiently far apart from each other having fading wireless channel with little or no correlations, transmission schedules can be done so as to exploit the inherent *multi-user diversity* to increase the system throughput. In [78], the authors present an opportunistic beamforming technique that randomly selects the antenna weights during each time frame, and schedules the user whose channel maximizes the received signal-to-noise plus interference ratio (SNIR) for the antenna weights.

Core to the success of above mentioned schemes is the channel state feedback that should take place at a sufficiently high rate. The *capacity regions* of multiuser wireless channel has been studied under the topic of *broadcast channels* [19]. The case with a single antenna at the transmitter and a single antenna at each receiver belongs to the category called *degraded broadcast channel*. The capacity in this case can be achieved by setting up a mechanism to cancel out the interference by the receiver with the stronger channel while the receiver with the weaker channel tolerates the interference. Let the total power at the transmitter normalized to the receiver noise is P_t, the attenuation factors at two receivers are $0 < k_1, k_2 < 1$. Let a fraction α of the total power is assigned to first receiver and the rest of it to the second receiver. The sum capacity achieved by the two receivers is $\log_2(1 + \alpha k_1 P_t) + \log_2(1 + \frac{(1-\alpha)k_2 P_t}{1+\alpha k_2 P_t}) = \log_2((1 + k_2 P_t)(\frac{1+\alpha k_1 P_t}{1+\alpha k_2 P_t}))$ which reduces to $\log_2(1 + k_2 P_t)$ when $k_1 = k_2$ *i.e.*, the channels are symmetrical. On the other hand for a system with time sharing of the channel, the sum rate is given by $\alpha \log_2(1 + k_1 P_t) + (1 - \alpha) \log_2(1 + k_2 P_t)$ (where α is the fraction of time allocated to receiver 1) which also reduces to $\log_2(1 + k_2 P_t)$ for symmetric channels. Thus it can be concluded that

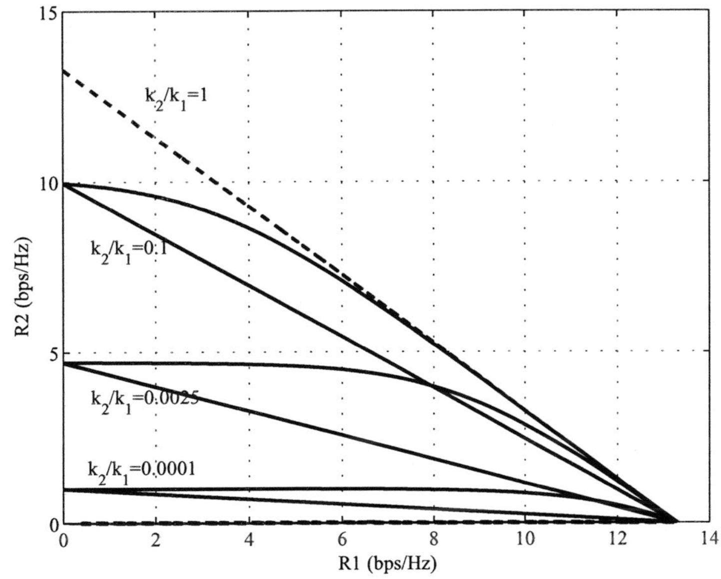

Figure 1. Capacity regions of degraded broadcast channel with two users

with symmetric channels the capacity regions of degraded broadcast channel is achieved by time sharing. However, when the channels are asymmetric, the capacity is more than that achievable by time sharing. This is exemplified in Fig. 1. In this illustration, the derivations of capacity regions is for a given channel instantiation under the assumption of *perfect channel knowledge* at the transmitter. There also are the underlying assumptions of continuous power and rate control. Nevertheless it was shown in [24] that adaptive rate control with fixed power achieves large performance gains compared to power control with fixed rate. Further it has also been reported that most of the capacity of continuous rate assignment can be achieved with a discrete set of rates of cardinality as small as 5.

When considering the time dimension however, a major challenge to achieving the aforementioned capacity arises. Time scales of the variations in the SNIR of a link varies based on the level of mobility. In the best known deterministic approach, the receivers estimate and feed back the channel state information to the transmitter prior to transmission of each data packet. As the rate of change in SNIR increases with increased speed, the performance becomes sensitive to feedback delays. Channel estimation errors at the receivers may degrade the performance further. As a result, the achieved throughput becomes significantly lower than that of a system with "perfect channel knowledge". In this light, stochastic adaptive techniques are found to provide simpler and efficient solutions which may result in performance comparable to that of best known deterministic approaches while resulting in savings of the bits used for feedback in the latter. These techniques can achieve performance close to that of methods with fast channel feedback, especially at low mobility scenarios.

There has been significant interest on the application of stochastic methods in the optimization of wireless networks. A stochastic learning automata based polling protocol was introduced in [51]. Successful use of stochastic approximation for power control has been demonstrated in [33]. Further, the use of a stochastic adaptive method namely, *stochastic learning automata* (SLA) to meet QoS requirements has been studied by Hall *et.al.* [28], and Haleem and Chandramouli [27]. The approach presented therein maintains a set of service probabilities which evolve with time in a way to converge to a probability distribution that satisfies the delay requirements of individual queues with different traffic intensities are guaranteed. It is seen in our studies that the SLA is a good choice to achieve QoS also in terms of the throughput requirements of users in the time varying wireless channel that uses adaptive transmission rate control. Thus both the adaptive rate assignment as well as the scheduling for QoS can be achieved with the algorithms derived from SLA. We note that selection of appropriate forms of SLA for each task is critical to the performance of the system. Readers interested in more information about the theoretical underpinnings of SLA may refer to the literature such as [50].

The stochastic adaptive learning and rate selection/scheduling methods we discuss in this chapter operate with minimal feedback requirements. In particular, instead of the multiple bits/frame feedback required at the physical/link (PHY/LINK) layer in the aforementioned schemes, the stochastic adaptive technique we discuss, use a single bit ACK/NACK feedback at the MAC layer for adaptive rate selection at the PHY/LINK layer. Thus it involves the exchange of control information from MAC layer to PHY/LINK layer and achieving savings in bit rate in the reverse channel and reduction in complexity at PHY/LINK layer. This can be considered a "cross layer" [64, 35, 18] implementation and "layer triggering" [18] in particuar.

Our discussion in this chapter also includes application of SLA techniques at application layer. Multimedia applications such as low bit rate wireless video transmission have lead to the study of issues in error-resilience. Wireless links not only suffer from limited bandwidth problems but are also highly vulnerable to channel errors. Video compression standards like H.261 [8] alleviate the bandwidth problem to a certain extent. The H.261 standard, also known as the $p \times 64$ standard was developed for video coding and decoding at the rate of $p \times 64$ kbits/s, where p is an integer from 1 to 30.

Most of the *state-of-the-art* video codecs treat source and channel coding separately. Bandwidth reduction is achieved by the source coder by removing the redundancy in the source statistics. Error protection against channel error is taken care of by the channel coder through the addition of redundancy in the transmitted data. However, this separation is justifiable only in the limit of an arbitrary encoding/decoding complexity. But, we know that in practice complexity and delay are the main constraints for communication systems. Therefore, the separation of source and channel coding is no longer optimal. This implies that source and channel coding should depend on each other leading to *joint source-channel coding* [38].

JSCC tries to design the source coder and channel coder in some joint way, which can provide better error protection and bandwidth utilization. JSCC schemes can be broadly classified into three different categories, *joint source-channel encoding* (JSCE) [21]-[58], *joint source-channel decoding* (JSCD) [62]-[39], [3]-[11], [66]-[15] and rate allocation strategies [32]-[16]. As the names suggest, these deal with the joint design of encoders, decoders and the rate allocation between the channel and source codes respectively. Channel-matched source rate control or quantization has been shown to be an effective way to add error-resilience to the transmission of compressed images and video over noisy channels

[13, 38].

Channel-matched source quantization and adaptive source rate control are effective ways to improve the received video quality. However, the performance depends on how fast and reliably the channel parameters (such as the bit error probability p_e) can be estimated. In many applications p_e is computed using pilot symbol aided techniques. This causes large delays which may not be acceptable for real-time applications. It has also been observed that up to a 14% loss in capacity can be incurred due to pilot symboling [12]. Therefore, it is desirable to reliably estimate the channel statistics and also achieve rate control through adaptive quantization *on the fly* with minimal overhead. In this chapter, we discuss such a mechanism [14] using stochastic learning automata.

The rest of the chapter is organized as follows. We discuss the stochastic behavior of wireless channel in Section 2.. A brief introduction to stochastic learning automata is presented in Section 3.. Section 4. deals with online joint randomized user scheduling and rate selection in the downlink. Channel adaptive stochastic video rate control is discussed in Section 5.. Section 6. contains a brief summary.

2. Stochastic Behavior of Wireless Channel

Wireless channel is characterized by the time varying attenuation and the additive random noise [34, 65, 61, 45]. The attenuation of the radio signals is due to the diverse mechanisms by which electromagnetic wave propagation takes place. These can be attributed to reflection, diffraction, and scattering. The average path loss (depends on the distance) follows the "power law" where the received power decays as a function of the distance raised to some power. In addition, due to different sizes and nature of objects crossing the path between transmitter and receiver, there could be a large numbers of random attenuations of the transmitted signal. The effect of the collection of such random objects is modeled using a "log-normal" distribution. The distance dependent path loss and the log-normal shadow fading contribute to the "large scale" variations in the signal attenuation. There are differences in the nature of large scale "path loss" for indoor and outdoor propagation environments. Widely popular outdoor empirical models for large scale path loss in different kinds of terrestrial environments have been formulated by Hata [30]. For a survey of indoor propagation models see [29]. It is seen that as the receiver moves with respect to the transmitter, the slow large scale variations are observed in the received signal and have a deterministic and a random component.

Superimposed on large scale variations are small scale fluctuations called *small scale fading*. Factors influencing small scale fading are multi-path propagation, speed of the mobile, speed of surrounding objects, and the transmission bandwidth of the signal. The parameters used in the qualitative and quantitative assessment of the small scale fading are *delay spread*, *coherence bandwidth*, *doppler spread*, and *coherence time*. The delay spread describes the extent to which the multi-path environment disperses a signal. The coherence bandwidth indicates the range of frequencies over which the channel can be considered fixed *i.e.*, have constant attenuation and linear phase shift. The rms delay spread and the coherence bandwidth have approximately an inverse relationship. While delay spread and coherence bandwidth describe the time dispersive nature of the channel in a local area,

Doppler spread and coherence time is used to explain the time varying nature of it. The latter two parameters also have an approximate inverse relationship to each other.

Based on the level of multipath time delay spread, the small scale fading can be "flat" or "frequency selective". Similarly, based on the Doppler spread it can be "fast" or "slow" in nature. In flat fading the bandwidth of the signal is smaller than the bandwidth of the channel, or equivalently the delay spread is less than the symbol period. The opposites apply for frequency selective fading. Fast fading results due to high Doppler spread, coherence time smaller than the symbol period, or in other words when the channel variations are faster than base band signal variations.

In the analysis and simulation of wireless systems and networks, the time varying nature of the received envelope of a flat fading signal or an individual multipath component is modeled with Rayleigh distribution. Ricean distribution is used when there is a line-of-sight path between the transmitter and receiver on top of which the reflected components are superimposed. Clarke's model [17] describes the statistical characteristics in terms of the superposition of scattered components. In our simulations studies, we use the Jakes' simulator [34] which uses a variation of this model. In essence, the time variations of attenuation of wireless channels is a stochastic process and there has been tremendous amount of work to model this behavior deterministically and statistically. Systems are being designed to perform optimally based such models. The alternative approaches are to use measurement and feedback on the fly, of the channel quality, and adaptive learning. The focuss of this chapter is on adaptive learning techniques and stochastic learning automata in particular.

3. Stochastic Learning Automaton

Stochastic learning automaton is a sub class of the general class of learning automaton (LA). In the quest to solve the learning problem, Tsetlin [75] created a new model of machine learning which is now called a learning automaton. The goal of such an automaton is to determine the optimal action out of a set of allowable actions in an unknown (random) environment. The optimal action for an automaton is defined as the action that maximizes the probability of getting a reward. The functionality of the learning automaton can be described in terms of a sequence of repetitive feedback cycles in which the automaton interacts with the environment. During a cycle, the automaton chooses an action, which triggers a response from the environment. The response can be either a reward or a penalty. The automaton uses this response and the knowledge acquired in the past actions to choose the next action. By learning to choose the optimal action the automaton discovers the behavior of the unknown environment.

In the first LA designs, the transition and the output functions were time invariant, and for this reason these LA were considered fixed structure stochastic automata. Tsetlin, Krylov, and Krinsky [75, 76] presented notable examples of this type of automata. Later, Varshavskii and Vorontsova in [77] introduced variable structure stochastic learning automaton (VSLA). These automata are characterized by the fact that the state transition probabilities or the action selection probabilities are updated with time. For such an automaton, there is a bi-directional correspondence between the set of states and the set of actions of the automaton, each state corresponding to a particular action. Consequently, the set of states becomes redundant in the definition of a SLA, and is completely defined by a set of actions (which is the output of the automaton), a set of inputs (which is usually the response of the environment) and a learning algorithm T. The learning algorithm operates on an *action probability vector*, $p(n) = [p_1(n), ..., p_r(n)]'$, where $p_i(n), i = 1, ..., r$ is the probability that the automaton will select the action α_i at the time step n: $p_i(n) = P[\alpha(n) = \alpha_i], i = 1, ..., r$, and it satisfies, $\sum_{i=1}^{r} p_i(n) = 1$ for all $n = 1, 2, \ldots$ VSLA are completely defined by a set of action probability updating rules operating on $p(n)$ [41, 42, 50]. Formally SLA can be defined as follows.

Definition. A stochastic learning automaton is a 5-tuple $\{\alpha, \beta, p, T, C\}$ where α is the set of actions, β is the set of input to the automaton (the set of output from the environment), $T : [0, 1]^r \times \alpha \times \beta \rightarrow [0, 1]^r$ is an updating scheme such that $p(n+1) = T[\alpha(n), \beta(n), p(n)]$ where $p(n) = [p_1(n), ..., p_r(n)]'$ is the action probability vector, $\alpha(n)$ is the action chosen at time n, and $\beta(n)(= \{0, 1\})$, is the response it has obtained. $\beta(n) = 0$ is a reward and $\beta(n) = 1$ corresponds to a penalty. $C = \{c_1, c_2, ..., c_r\}$ is the set of penalty probabilities conditioned on the chosen action, where $c_i = P(\beta(n) = 1|\alpha(n) = \alpha_i), i = 1, 2, ..., r$.

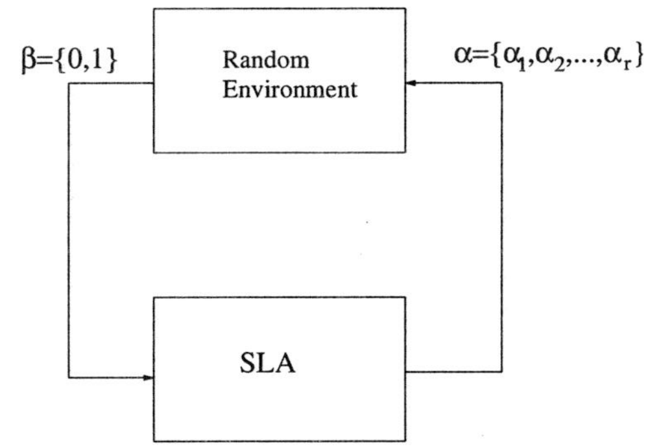

Figure 2. Stochastic learning automata.

The values of c_i are unknown and it is assumed that C has a unique minimum element. The working of the learning automaton can be described as follows. One has the flexibility to choose an initial value for the action probability vector $p(1)$. In the absence of any a priori information $p_i(n) = \frac{1}{r}, \forall i$ is the default choice. Initially at $n = 1$, one of the actions is chosen by the automaton at random following the initial probability distribution. This action is then applied to the system and the response from the environment is observed. If the response is favorable ($\beta(1) = 0$) then the probability of choosing that action for the next period of time is updated according to the updating rule T. The next action is chosen according to the updated action probability distribution. This process is repeated until a stopping criterion is reached. When the learning process stops the algorithm has learnt to choose the optimal action and the characteristics of the unknown random environment. In fact, ergodic SLA [41, 49, 52] converge in distribution and the asymptotic distribution of the action probability vector is independent of the initialization. Because of this independence property they are suitable for non-stationary environments.

The concept of discretized probability space was introduced to LA in [72]. This increases the speed of convergence. If the values allowed are equally spaced in the interval

[0,1] the discretization is said to be linear, otherwise non-linear. Various discrete automata have been presented in the literature[52, 54].

Another important characteristic of learning automata is the philosophy of the learning paradigm. Giving more importance to rewards than to penalties in the probability updating rules, can considerably improve the convergence properties of LA. Thus we have *Linear reward-penaly* (LRP) schemes and *Linear reward-inaction* (LRI) schemes based on whether we penalize or not an action α when $\beta(n) = 1$.

3.1. Estimator and Pursuit Algorithms

In attempting to design faster converging learning algorithms, Thathachar and Sastry [73] opened another avenue by introducing a new class of algorithms, called *estimator* algorithms. They maintain running estimates for the reward probability of each possible action, and use them in probability updating equations. Estimator algorithms make changes on the action probability vector based only on the running estimates of the reward probabilities—they ignore the feedback received from the environment, except to update the reward estimates themselves. A detailed description of the estimator algorithms can be found in [43, 44, 55, 73, 74]. *Pursuit* algorithms are a subset of the estimator algorithms. The action probability vector in these algorithms pursues the action that is currently estimated to be the optimal action. This is achieved by increasing the probability of the action whose current estimate of being rewarded is maximal [55, 74]. Estimator algorithms can be designed to update the probability vector $p(n)$ based on either "long term" or "short term" properties of the environment. The pursuit algorithms introduced in [74] adapts to only the long term properties. Later, Oommen and Lanctot [55] presented a discretized version of a pursuit algorithm which considered both the short-term and the long-term perspectives and embraced the *reward-inaction* learning paradigm. In [53], Oommen and Agache mention that the information on short term behavior of the environment resides in the most recent responses that are obtained by the automaton. They presented schemes in which the long term information crystallized in term of the running reward-probability estimates, and the short term information used by considering whether the most recent response was a reward or a penalty. They formulated and compared four different pursuit algorithms namely *Continuous Pursuit Reward-Penalty* (CPRP) algorithm as introduced by Thathachar and Sastry [74], *Discrete Pursuit Reward-Inaction* (DPRI) algorithm as presented by Oommen and Lanctot in [55], *Continuous Pursuit Reward-Inaction* (CPRI) algorithm, and *Discrete Pursuit Reward-Penalty* (DPRP) algorithm. They presented numerical comparison and concluded that the DPRI surpasses all other forms of the pursuit algorithms. The relative performances are in the order DPRI $>$ DPRP $>$ CPRI $>$ CPRP. In general, all the pursuit algorithms maintain two vectors $W(n)$ and $Z(n)$ where $Z(n)$ is the number of times the ith action α_i has been chosen and $W(n)$ is the number of times the action α_i has been rewarded. The reward estimate $\hat{d}(n)$ for α_i, is computed to be $\hat{d}(n) = \frac{W(n)}{Z(n)}$. Following an update of the reward estimates, the probability of the action with the current highest reward estimate is increased and that of other actions are decreased.

4. Joint Randomized Scheduling and Adaptive Rate Selection

In this section we discuss in details of an online technique based on SLA for joint user scheduling and MCS selection.

4.1. Randomized User Scheduling

The user scheduler assumes a *best effort* queue and a set of K users each with a throughput requirement. The term best effort queue here refer to a pool of data (possibly from more than one user) which does not have a strict short term throughput requirement. Users to be served in each time slot is selected according to a service probability distribution, $p^s(n) = [p_0^s(n), p_1^s(n), \cdots, p_K^s(n)]$ where $p_0^s(n)$ is the probability associated with the best effort queue. Service probability of a user is updated following transmission of P packets to "the user" or when the time elapsed since the last update of the user is Q slots, whichever occurs first. The update is an increment/decrement in the probability of assignment of the user in concern. The increment/decrement of the probability of assignment is a function of the current value and the difference between the achieved short term throughput and the requested throughput. The increment/decrement of the probability is compensated for, by a corresponding decrement/increment in the probability of the best effort queue. The user scheduling algorithm is summarized as follows.

Parameters

P	number of packets transmitted before an update.
Q	maximum time elapse (slots) between updates.
R^{req}	the vector of requested throughput values.
$R^{ave}(n)$	the vector of achieved throughput values during the averaging window
a, b, η	scaling parameters ($0 < \{a, b\} \leq 1$ and $\eta > 0$)

Pseudo-code

Initialize $p_i^s(n) = 1/(K+1)$, for $0 \leq i \leq K$.
Repeat

1. Take control from PHY/LINK layer *rate adaptation* algorithm.
2. Select a user to serve in the next time slot with probabilities
$$p^s(n) = [p_0^s(n), p_1^s(n), \cdots, p_K^s(n)]$$
3. If $Q_0 = \min\{\text{time for } P, Q\}$ achieved for the selected user i, goto step 4; else goto step 7.
4. Compute $\nu = \frac{R_i^{req} - R_i^{ave}(n)}{\eta R_i^{req}}$.
5. Let
$$\beta_i(n) = \begin{cases} 1, & \textit{if} \quad \nu > 1 \\ \nu, & \textit{if}\, 1 > \nu > -1 \\ -1, & \textit{if} \quad \nu < -1 \end{cases} \tag{1}$$

6. Update $p^s(n)$ according to the following.
   for $\beta_i(n) > 0$,

$$p_i^s(n+1) = p_i^s(n) + ap_i^s(n)\beta_i(n) \tag{2}$$
$$p_0^s(n+1) = p_0^s(n) - ap_i^s(n)\beta_i(n) \tag{3}$$

   for $\beta_i(n) < 0$,

$$p_i^s(n+1) = p_i^s(n) - bp_i^s(n)|\beta_i(n)| \tag{4}$$
$$p_0^s(n+1) = p_0^s(n) + bp_i^s(n)|\beta_i(n)| \tag{5}$$

7. Pass control to PHY/LINK layer *rate adaptation* algorithm for the scheduled user.

End Repeat

4.2. Adaptive Rate Selection

Once the scheduler selects a user i based on the current probability vector $p^s(n)$, the rate adaptation algorithm randomly selects a rate (corresponding to an MCS) from the set of r rates, following the probability distribution $p_i^R(n) = [p_{i1}^R(n), \cdots, p_{ir}^R(n)]$ of user i. The number of bits in the packet is selected such that the data frame can be completely transmitted within a transmission time interval (TTI) with the selected MCS. In a typical 3G wireless system such as high speed down-link packet access (HSDPA), a data frame may extend to more than one TTI. The method presented here is readily applicable to such scenario as well. In the formulation of the problem and the algorithm to follow, n is the index of the sequence of TTIs and the SINR of the channel during n^{th} TTI is expressed by $\gamma(n)$. The probability of frame error with a given channel SINR, and j^{th} rate (MCS) is expressed as $P_{ij}^e(\gamma(n))$. The set of rates available are $\{R_j : j = 1, 2, \cdots, r\}$ (bits/s). Thus the throughput achieved with a rate R_j is given by

$$D_{ij}(n) = R_j(1 - P_{ij}^e(\gamma(n))),\ i = 0, \cdots, K;\ j = 1, \cdots, r \tag{6}$$

Ideally, the transmitter is required to find the index of the best MCS m_i of user i such that

$$m_i = \arg\max_j D_{ij}(n) \tag{7}$$

While this is computable in an straight forward manner with the knowledge of channel state during each TTI, the stochastic learning and rate selection algorithm achieves this via an iterative process. The rate selection probability vector is altered by an iterative updating process such that the probability of assigning the best rate (MCS) approaches "one" while the remainint entries in the vector approaches zero. At $n = 1$, the probabilities $p_{ij}^R(n)$, $j = 1, \cdots, r$ are assigned equal values of $1/r$. Then the rate selection and transmission proceeds with the fixed $p_i^R(n)$ until every rate is selected at least M (a tunable parameter) number of times after which $p_i^R(n)$ is augmented at each n. Following each transmission, the transmitter receives an ACK/NACK signal indicating the successful reception/failure of the data packet. The current and the past ACK/NACK signals are used in augmenting the probability vector $p_i^R(n)$ toward the optimum. This is done by maintaining a time varying estimate of throughput values, $\hat{D}_{ij}(n)$ for each rate $R_j, j = 1, \cdots, r$. Following each TTI, an update of $\hat{D}_{ij}$ and $p_i^R(n)$ are carried out considering the last M ACK/NACK signals of each rate. Thus the length of the moving window used in the estimation (in terms of number of TTIs) could vary over the time. We may write,

$$\hat{D}_{ij}(n) = \frac{R_j}{M} \sum_{k=L_{ij}(n)-M+1}^{L_{ij}(n)} J_{ij}(k) \tag{8}$$

where $J_{ij}(k)$ is an indicator function such that $J_{ij}(k) = 1 \; or \; 0$ depending on whether the feedback following k^{th} use of rate R_j is an ACK or NACK. $L_{ij}(n)$ is the number of TTIs for which the rate R_j is selected during the time from the start till the nth TTI. Following the transmission of each data frame, the index $\hat{m}_i$ of the best rate is decided, the probabilities, $p_{ij}^R(n)$, $j \neq \hat{m}_i$ are decreased by Δ $(0 < \Delta < 1)$, and the probability of the estimated best rate, $p_{i\hat{m}_i}^R(n)$ is increased by $(r-1) \times \Delta$ where $\Delta = \frac{1}{N}$ is the smallest step size. N here is a tunable resolution parameter. If the channel state remains fixed for sufficiently long time, the algorithm can increase the selection probability $p_{im}^R(n)$ of the best rate to unity (and set $p_{ij}^R(n) = 0$ for all $i \neq m$). While this could maximize the throughput in a stationary channel, adaptivity to time varying channel requires us to maintain non-zero values of $p_{ij}^R(n)$ for all j and for all n to avoid "absorbing states". Therefore we maintain a minimum probability of B called "bias" for all rates. The proposed rate selection algorithm can be summarized as follows.

Parameters

M	number of transmitted packets required for the estimation of throughput $\hat{D}_{ij}$.
N	resolution parameter in the probability step size.
Δ	$(=\frac{1}{N})$ is the adaptation step size.
m	the rate index of the largest element in $\hat{D}_i(n) = [\hat{D}_{i1}(n), \hat{D}_{i2}(n), \cdots, \hat{D}_{ir}(n)]$.
$L_{ij}(n)$	number of times the rate R_j is selected from time 0 till n for user i.
$J_{ij}(k)$	"1" or "0" on receiving an ACK/NACK following the *kth* use $(1 \leq k \leq L_{ij}(n))$ of rate R_j.
B	a bias to prevent $p_{ij}^R(n) = 0 \; \forall i$ and $\forall j$ to facilitate tracking of time varying channel.

Pseudo-code

Initialize $p_{ij}^R(n) = 1/r$, for all i and for all j.
Repeat

1. Take control from *user scheduler* at time n.

2. Pick a rate $R_j (1 \leq j \leq r)$ according to probability distribution $p_i^R(n)$.

3. Update $I_{ij}(k)$ and $L_{ij}(n)$ on receiving an ACK/NACK signal.

4. Update $\hat{D}_{ij}(n)$according to (8).

5. If $L_{ij}(n) \geq M$ for all j (initialization phase completed) goto step 6; else goto step 7.

6. Detect the index $\hat{m}_i$ of the estimated best rate $R_{\hat{m}_i}$ and update $p_i^R(n)$ according to the following equations:

$$p_{ij}^R(n+1) = max\{p_{ij}^R(n) - \Delta, B\}, \forall j \neq \hat{m}_i \tag{9}$$

$$p_{i\hat{m}_i}^R(n+1) = 1 - \sum_{j \neq \hat{m}_i} p_{ij}^R(n+1). \tag{10}$$

7. Pass control to *user scheduler* along with R_i^{ave} information.

End Repeat

4.3. Simulation Results

Numerical computations and simulations were carried out with parameters of a 3G wireless system namely HSDPA operating at 2.0GHz. A frequency flat fading radio link was assumed. The transmitter and the receiver were assumed to have single antennas. The set of six transmission rates, $\{0.12, 0.24, 0.36, 0.48, 0.60, 0.72\}$ (Mb/s) corresponding to a set of MCS was included used. The ACK/NACK signals to follow the transmission of each data frame were simulated using a set of pre-derived frame error probability versus SINR curves. These curves were derived for the performance in additive white Gaussian Noise (AWGN) channel with an interleaver/deinterleaver and turbo-coder/decoder in the system. The set includes one curve for each MCS for the range of SINR of interest. The frame duration was taken to be one TTI which is 0.667ms. Instantiations of the fading channel were generated using Jakes' model [34] with 13 taps.

4.3.1. Throughput Performance

It was found from the system simulations that $M = 1$ results in the best performance except at very low speeds. This observation is consistent with the intuitive fact that when the channel variations take place at time scales comparable to the TTI, the estimate $\hat{D}_j(n)$ would not improve by increasing M. Thus the best estimate is achieved with minimum M. Next, we compare the performance of the proposed method to that of an ideal scheme where the channel state in each TTI is known to the transmitter.

Shown in Fig. 3 are the average throughput of the proposed algorithm as a percentage of the throughput of ideal scheme, at a set of speeds. With each set of parameters, the simulation was performed for a sufficient length of time (in the order of 60,000 frames) and the average throughput values were computed for each such parameter setting. The optimum values of parameters M, N, and B maximizing the average throughput at each speed were found by repeating the simulation for a range of values of these parameters. At zero speed (stationary channel), the proposed method achieves 100% of the throughput of ideal scheme. A 71.6% throughput is achieved at a speed of 3 km/h. As speed increases, the value of N achieving best throughput decreases and becomes $N = 1$ around 1km/h. Further, it is seen that as speed increases, the optimum bias B increases. Note that timely detection of state changes requires testing of every rate at sufficiently small time intervals, which in turn requires sufficiently large probabilities of selection for every rate. An increase in the value of B fulfils this. With smaller than optimum values of B, the penalty arising out

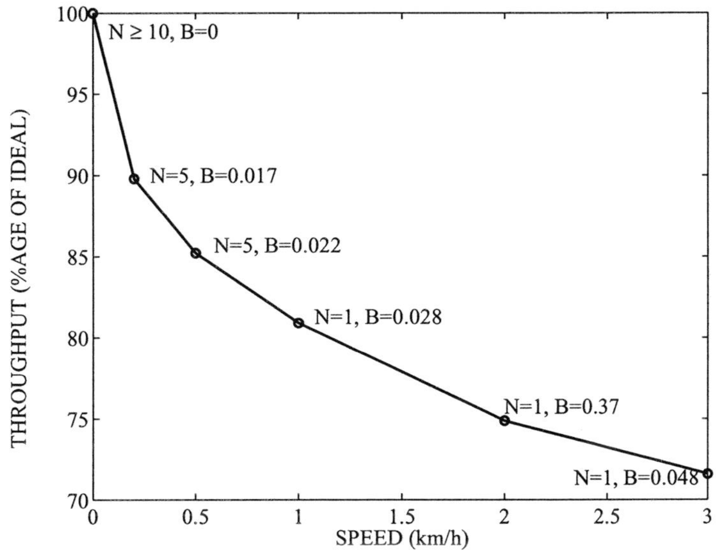

Figure 3. Throughput performance of stochastic adaptive algorithm with increasing speed. Optimum N and B are shown at each data point. Other settings are $M = 1$ and average SINR=0 dB.

of delayed detections becomes more severe than the loss due to the drop in the maximum probability of selecting the best rate.

The performance of the joint scheduling and rate selection is illustrated next for time varying channels. The realizations of channels were generated for 5000 time slots (frames). The parameter settings of user selection automaton were $P = 10, Q = 50, a = 0.2, b = 0.1, \textit{and}\ \eta = 0.2$. Those of rate selection automaton were $M = 1, N = 5$, and $B = 0.022$ for time varying channels. The simulation proceeds as follows. At the start all the users are assigned equal probability of selection of $1/(K + 1)$ and all MCS are assigned equal probability of selection of $1/r$ for all users. The MCS selection probabilities of a user is kept fixed until all the values are selected at least for transmission of M packets. Then the estimations of throughput corresponding to each rate, and update of MCS selection probabilities start and continue in each time slot. User selection probabilities are updated at the end of each P packets or when Q time slots elapsed since last update.

The average SINRs of the users were set to 15, 12, and 9dB. Fig. 4 shows the signal envelops. Fig. 5 gives the evolution of user selection probabilities. When the SINR of a user increases/decreases, the probability of selection adaptively decreases/increases to maintain the average transmission rate to the requested value. Note that as in Fig. 6, the short term average transmission rates achieves the requested rates within first few iterations and remains at the requested values. There is no effort by the algorithm to regulate the transmission rate of the best effort queue. Shown in Fig. 7 is the evolution of rate assignment probability vector, $p_1^R(n)$ of QoS user 2. The correspondence to the variation in channel SINR is observable.

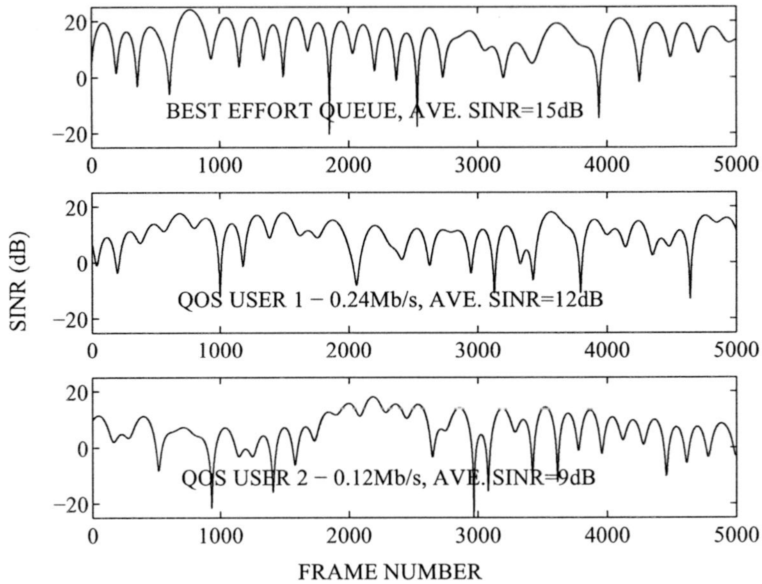

Figure 4. SINR variation against time for the set of users at a speed of 0.5km/h (based on Jakes' model). Frame duration = 1 TTI (0.667ms).

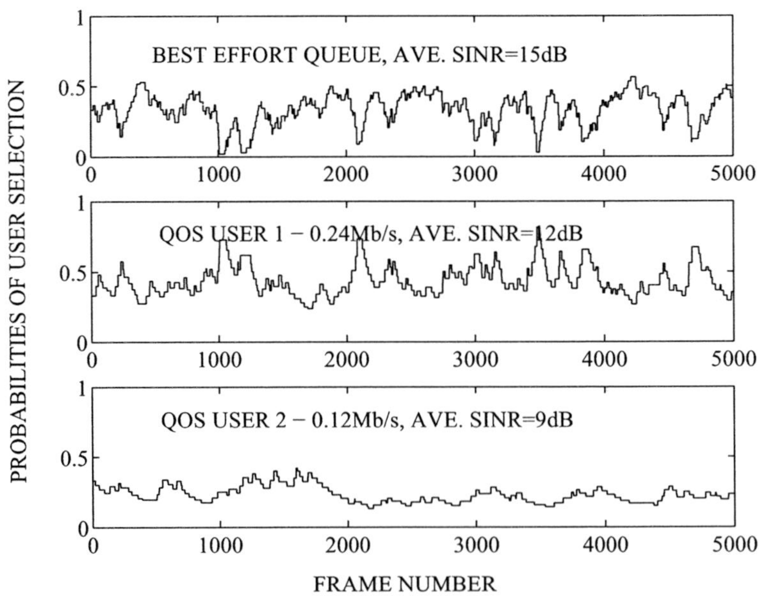

Figure 5. Time evolution of user selection probabilities. Frame duration = 1 TTI (0.667ms).

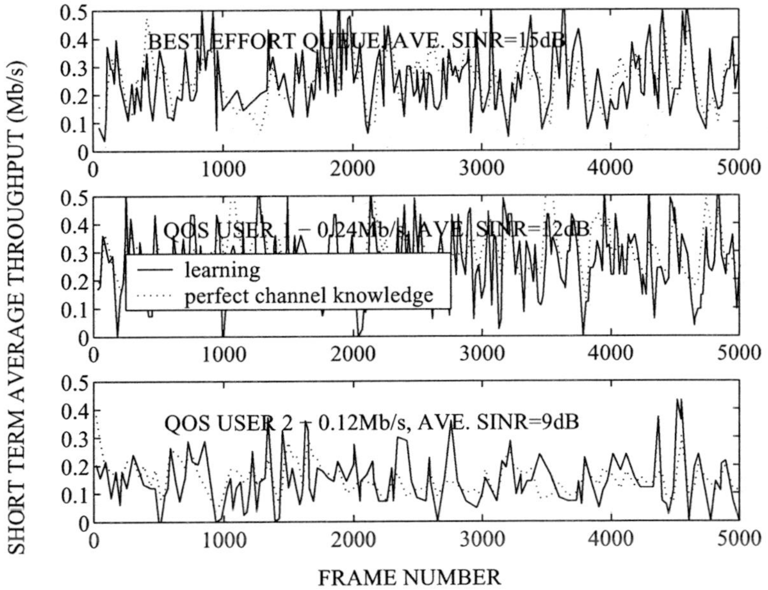

Figure 6. Short term throughput of DPRI and rate assignment with perfect channel knowledge. Frame duration = 1 TTI (0.667ms).

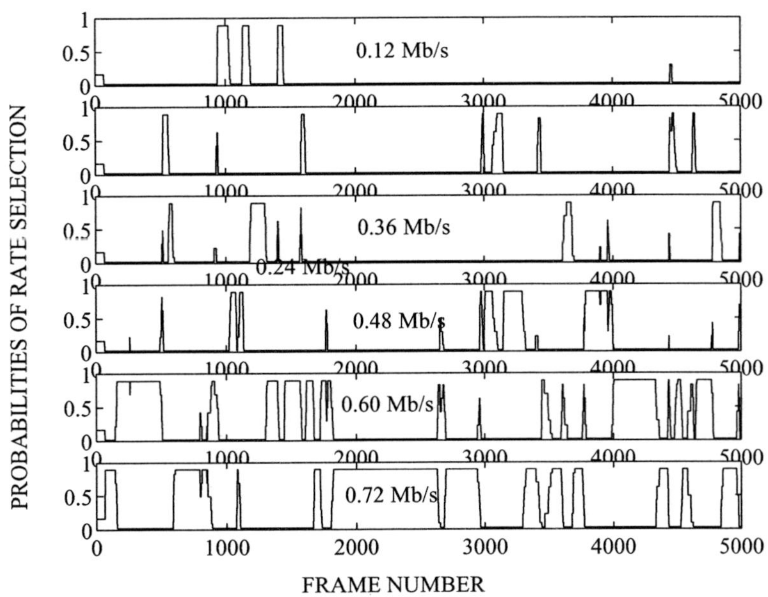

Figure 7. Time evolution of rate selection probabilities for QoS user 2. Frame duration = 1 TTI (0.667ms).

5. Joint Adaptive Video Coding Rate Control and Channel Tracking

Adaptive rate control is one way to achieve error-resilience. For channel matched source rate control the wireless channel bit error probability (p_e) needs to be estimated first. Transmitting pilot symbols to estimate p_e is a common solution. But, this introduces delay and costs bandwidth. Use of VSLA alleviates this problem by enabling quantization rate control and estimation of p_e jointly. We consider the H.261 video coder throughout this section.

Without loss of generality let us assume that p_e can take one of the three values, 10^{-1}, 10^{-2}, or 10^{-3}. At the beginning of the transmission the p_e value for the channel is unknown to the source coder. Now, let the set of actions of the learning automaton, $\alpha = \{\alpha_1, \alpha_2, \alpha_3\}$ correspond to choosing the three channel matched optimal quantizers [14] for channel bit error rates 10^{-1}, 10^{-2}, and 10^{-3}, respectively. Since there is a one-to-one correspondence between channel bit error probability and the quantizer choice, by learning the optimal quantization parameter the VSLA implicitly estimates the current p_e value. Let a favorable response ($\beta = 0$) from the receiver imply that the peak signal to noise ratio (PSNR) of the received video frame is greater than or equal to that of the previous frame and let 1 correspond to an unfavorable response. We assume here that PSNR and the number of blocks in error in the received frame are inversely proportional. The penalty probability for the choice of each action is unknown and defines the environment (channel's conditions). The goal is to maximize the PSNR of the received video frames by learning the channel condition and choosing the corresponding optimal rate for the H.261 quantizer. This corresponds to learning the action with the minimum penalty probability. We update the quantizer choice probabilities using linear reward inaction (LRI) [48] control. The typical steps involved in the proposed channel-adaptive video rate control algorithm are as follows:

Step 1: I-frame is quantized using a high rate quantizer and transmitted with sufficient protection.
Step 2: The video decoder, after reconstructing this frame, quantizes the frame using the quantization tables that have been optimized for the channel bit error probabilities and stores them in a buffer.
Step 3: The VSLA chooses a quantizer randomly for the nth frame with the probability distribution $p(n) = \{p_1(n), p_2(n), p_3(n)\}$.
Step 4: The quantized frame is transmitted along with the the quantizer information as protected side information.
Step 5: A response β based on the PSNR of the decoded frame is transmitted as a feedback information to the encoder.
Step 6: Based on the β value the probability distribution $p(n)$ of choosing the quantizers for the $n + 1$st frame is updated using LRI.
Step 7: Steps 1-6 are repeated until the the probability of choosing the optimal action for the unknown p_e converges to 1.

When the algorithm converges it has learnt the unknown p_e value of the channel and the corresponding optimal quantizer. We note that this method could result in the first few

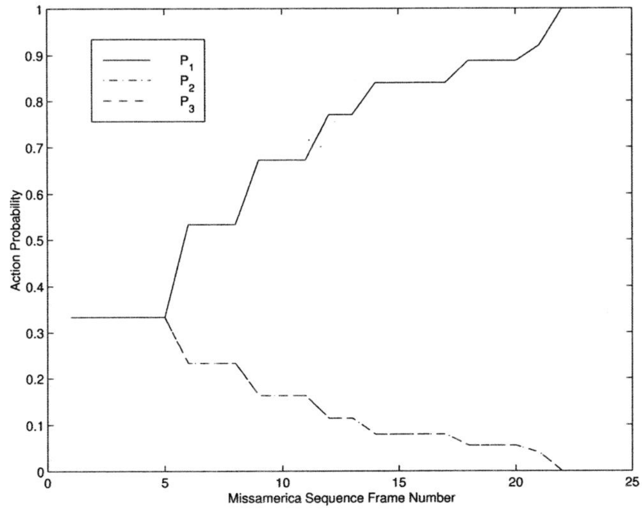

Figure 8. Convergence of proposed stochastic learning control to the channel-optimized quantizer corresponding when $p_e = 10^{-1}$ for the Missamerica sequence; $a = 0.3$.

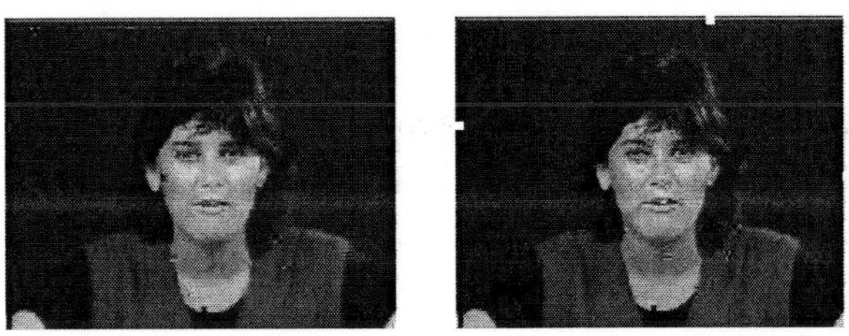

(a) Frame 25, M = 1.3 (b) Frame 25, M = 2.0

Figure 9. Received video frames for two different quantization rates when $p_e = 10^{-2}$.

frames to be sub-optimally quantized. This is the cost incurred in on-line channel estimation. However, the learning delay can be controlled by the value of the reward parameter in LRI control. Depending on the value of the error tolerance a suitable reward parameter can be chosen. A higher reward parameter results in faster convergence of the LRI learning algorithm at the expense of convergence to a sub-optimal solution.

Since $p(n)$ converge w.p.1 [14] let $\lim_{n\to\infty} \Psi(n) = \Psi^*$ w.p. 1. Then, the average rate of convergence of LRI learning is [48] given by $\rho_{av} \approx \frac{E[\Psi(n+1)-\Psi^*]}{E[\Psi(n)-\Psi^*]}$. After every frame n, $E[\Psi(n) - \Psi^*]$ decreases by a factor of ρ_{av}. If $\tilde{T}$ is the time taken for $E[\Psi(0) - \Psi^*]$ to decrease to d times its value, then $\tilde{T} = logd/log\rho_{av}$. For slowly changing channels, if the channel bit error probability changes then the encoder will receive a series of penalties. Then the learning process is started again. Therefore, the encoder is in sync with the channel again.

We present experimental results for gray level, qcif format, Miss America video sequence at 10 frames/second. The uncompressed bit rate is about 2Mbps and the video

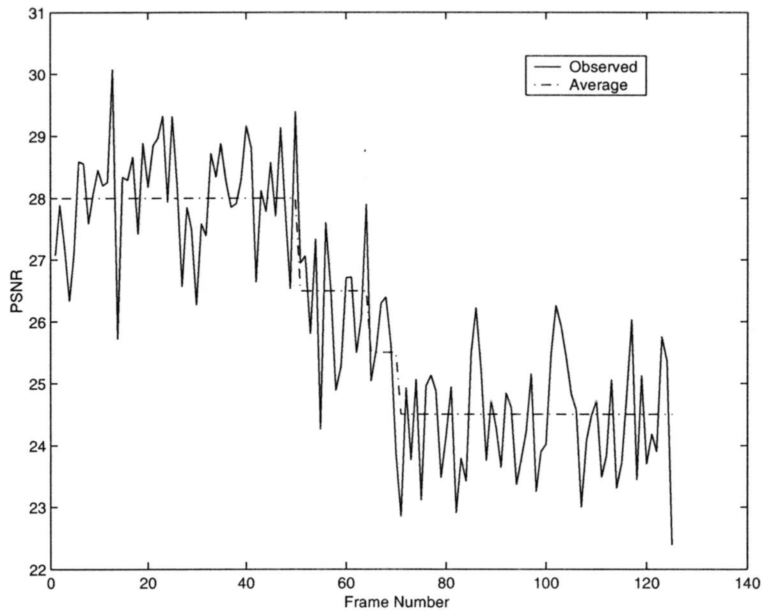

Figure 10. Wireless channel tracking capability when $a = 0.3$.

sequence is compressed to 35Kbps by a suitable of choice quantization parameter (M). Figure 8 shows the convergence of the linear LRI learning control algorithm to the optimal quantizer when $p_e = 10^{-1}$. Here, $p_1(n)$ corresponding to the optimal quantizer converges to 1 while $p_2(n)$ and $p_3(n)$ converge to zero. Both $p_2(n)$ and $p_3(n)$ have the same behavior as n varies.

Figure 9 shows the decoded video frames for different values of M (quantization rate) when $p_e = 10^{-2}$. We can see that the visual quality of Figure 9(a) corresponding to the optimal rate is better. Figure 10 shows the channel tracking capability of the proposed algorithm. Here, 125 frames of the Miss America sequence was used and the channel bit error rate was allowed to change from 10^{-2} to 10^{-1}. We see from the figure that during the channel transition period (at the 50th frame) the rate controller senses the change and starts searching again in the space of channel optimized quantizers. This search continues for nearly 14 frames before the correct quantizer is learned with a high probability. The quality at the decoder during these 14 frames fluctuates as seen in Figure 10.

Clearly, there is an important related question here: how often do we trigger the learning algorithm in a dynamically changing wireless environment? Unfortunately this question does not have a straightforward answer. However, there is a clear trade-off here. Consider the two cases: (a) the channel has changed *significantly* and the learning algorithm has not been triggered and (b) learning algorithm has been triggered erroneously. In the first case, suppose the channel has changed from a higher bit error rate to a lower one, then, by not triggering the learning algorithm to compute the new (higher) source coding rate, the source coder induced distortion at the receiver will be high. On the other hand, if the channel has changed from a lower bit error rate to higher, then the channel induced distortion will be larger. In case (b), the loss will be in terms of PSNR during the learning period when the quantizer randomly tries different quantization parameters. One way to alleviate this problem would be not to rely only upon physical layer information but also use information from higher layers of the protocol stack.

6. Summary

Using VSLA for online optimization of wireless multimedia networks seems to be a promising research direction. Both the examples discussed in this chapter seem to support this conclusion. VSLA's lend themselves to rigorous mathematical analysis. Simulation results show that VLSA based control of multimedia networks is practically viable. However, one issue in using VLSA, just like any other iterative control algorithm, is the choice of learning rate. Currently there is no rigorous method to choose this value. Therefore, we have to resort to some trial-and-error. The choice has to be made based on the rate, delay, channel estimation accuracy and other factors.

References

[1] Alternative to 8 psk modulation/coding scheme for hsdpa, TSGR1#20(01)0471 Texas Instruments, (2001).

[2] Physical layer aspects of utra high speed down-link packet access (release 4) 3gpp, TR25.848 v4.00, (2001).

[3] F. Alajaji, N. Farvardin, and T. Fuja, Detection of binary Markov sources over channels with additive Markov noise, *IEEE Transactions on Information Theory,* **42** (1996), pp. 230–239.

[4] F. Alajaji and T. Fuja, A communication channel modeled on contagion, *IEEE Transactions on Information Theory,* **40** (1994), pp. 2035–2041.

[5] K. Balachandran, S. R. Kadaba, and S. Nanda, Channel quality estimation and rate adaptation for cellular mobile radio, *IEEE JSAC,* **17** (1999), pp. 1244–1256.

[6] R. Bauer and J. Hagenauer, Iterative source/channel-decoding using reversible variable length codes, *IEEE Data Compression Conference,* (2000), pp. 93–102.

[7] ——, Symbol by symbol MAP decoding of variable length codes, 3rd ITG Conference Source and Channel Coding, (2000).

[8] V. Bhaskaran and K. Konstantinides, Image and video compression standard, Kluwer Academic Publishers, 1995.

[9] P. Burlina and F. Alajaji, An error resilient scheme for image transmission over noisy channels with memory, *IEEE Transactions on Image Processing,* **7** (1998), pp. 593–600.

[10] M. Bystrom and J. W. Modestino, Combined source-channel coding for transmission of H.263 coded video with trellis-coded modulation over a slow fading Rician channel, *IEEE International Symposium on Information Theory,* (1998), p. 12.

[11] M. Bystrom, S.Kaiser, and A.Kopansky, Soft source decoding with applications, *IEEE Transactions on Circuits and Systems for Video Technology,* **11** (2001), pp. 1108–1120.

[12] J. Cavers, An analysis of pilot symbol assisted modulation for rayleigh fading channels, *IEEE Trans. on Vehicular Technology,* **40** (1991), pp. 686–693.

[13] R. Chandramouli, S. Kumar, and N. Ranganathan, Joint optimization of quantization and on-line channel estimation for low bit-rate video transmission, *Proc. of IEEE International Conf. on Image Processing,* **1** (1998), pp. 649–653.

[14] R. Chandramouli, K.P. Subbalakshmi, and N. Ranganathan, Stochastic channel-adaptive rate control for wireless video transmission, *Pattern Recognition Letters,* **25** (2004), pp. 793–806.

[15] Q. Chen and K. P. Subbalakshmi, Trellis decoding for MPEG-4 streams over wireless channels, *SPIE Electronic Imaging: Image and Video Communications and Processing,* (2003), pp. 810–819.

[16] G. Cheung and A. Zakhor, Bit allocation for joint source/channel coding of scalable video, *IEEE transactions on Image Processing*, **9** (2000), pp. 340–356.

[17] R. H. Clarke, A statistical theory of mobile-radio reception, *Bell Systems Technical Journal,* **47** (1968), pp. 957–1000.

[18] M. Conti, G. Maselli, G. Turi, and S. Giordano, Cross-layering in mobile ad hoc network design, *IEEE Comp. Mag.*, (2004), pp. 48–51.

[19] T. M. Cover, Comments on broadcast channels, *IEEE Trans. Info. Theory,* **44** (1998), pp. 2524–2530.

[20] N. Demir and K. Sayood, Joint source/channel coding for variable length codes, *IEEE Data Compression Conference,* (1998), pp. 139–148.

[21] J. G. Dunham and R. M. Gray, Joint source and noisy channel trellis encoding, *IEEE Transactions on Information Theory,* **27** (1981), pp. 516–519.

[22] R.M. Fano, A heuristic discussion of probabilistic decoding, *IEEE Transactions on Information Theory*, **9** (1963).

[23] N. Farvardin, A study of vector quantization for noisy channels, *IEEE Transactions on Information Theory,* **36** (1990), pp. 799–809.

[24] A. J. Goldsmith and S-G.Chua, Variable-rate variable-power mqam for fading channels, *IEEE Trans. Info. Theory,* **45** (1997), pp. 1218–1230.

[25] A. J. Goldsmith and P. P. Varaiya, Capacity of fading channels with channel side information, *IEEE Trans. Info. Theory,* **43** (1997), pp. 1986–1992.

[26] L. Guivarch, J.-C. Carlach, and P. Siohan, Joint source-channel soft decoding of Huffman codes with turbo-codes, *Data Compression Conference,* (2000), pp. 83–92.

[27] M. A. Haleem and R. Chandramouli, Adaptive downlink scheduling and rate selection: A cross layer design, *IEEE JSAC Spec. Issue on Mobile Computing and Networking,* (2005).

[28] J. Hall and P. Mars, Satisfying qos with a learning based scheduling algorithm, *Proc. of IEEE IWQoS,* (1998).

[29] H. Hashemi, The indoor radio propagation channel, *Proc. IEEE,* **81** (1993), pp. 943–968.

[30] M. Hata, *IEEE Trans. Veh. Tech.,* **VT-29** (1980), pp. 317–325.

[31] A. Hedayat and A. Nosratinia, On joint iterative decoding of variable-length codes and channel codes, *IEEE Conference on Communications, Control, and Computing,* (2002).

[32] B. Hochwald and K. Zeger, Tradeoff between source and channel coding, *IEEE Transactions on Information Theory,* (1997), pp. 1412–1424.

[33] T. Holliday, A. Goldsmith, P. Glynn, and N. Bambos, Distributed power and admission control for time varying wireless networks, *IEEE Globecom,* (2004), pp. 768–774.

[34] W. C. Jakes, ed., phMicrowave Mobile Communications, *IEEE*, New York, 1974.

[35] V. Kawadia and P. R. Kumar, *A cautionary perspective on cross layer design.*

[36] J. Kliewer and R. Thobaben, Combining FEC and optimal soft-input source decoding for the reliable transmission of correlated variable-length encoded signals, *IEEE Data Compression Conference,* (2002), pp. 83–91.

[37] A. Kopansky and M. Bystrom, Sequential decoding of MPEG-4 coded bitstreams for error resilience, *The 33rd Annual Conference on Information Sciences and Systems*, (1999).

[38] A. Kurtenbach and P. Wintz, Quantization for noisy channels, *IEEE Trans. on Communication Tech.,* **17** (1969), pp. 291–302.

[39] K. Lakovic and J. Villasenor, Combining variable length codes and turbo codes, *IEEE Vehicular Technology Conference,* (2002), pp. 1719–1723.

[40] K. Lakovic, J. Villasenor, and R. Wesel, Robust joint Huffman and convolutional decoding, *IEEE Vehicular Technology Conferenc*, (1999), pp. 2551–2555.

[41] S. Lakshmivarahan, Learning Algorithms Theory and Applications, Springer-Verlag, New York, 1981.

[42] S. Lakshmivarahan and M. A. L. Thathachar, Absolutely expedient algorithms for stochastic automata, *IEEE Trans. man. Cybern.*, **SMC-3** (1973), pp. 281–286.

[43] J. K. Lanctot, Discrete estimator algorithms: A mathematical model of computer learning, master's thesis, Dept, Math. Statistics, Carleton Univ., Ottawa, Canada, 1989.

[44] J. K. Lanctot and B. J. Oommen, Discretized estimator learning automata, *IEEE Trans. on Syst. Man and Cybern.*, (1992), pp. 1473–1483.

[45] William C. Y. Lee, Mobile Communications Design Fundamentals, John Wiley and Sons Inc., NY, 1993.

[46] A. H. Murad and T. E. Fuja, Exploiting the residual redundancy in motion estimation vectors to improve the quality of compressed video transmitted over noisy channels, *Proc. Int. Conf. Image Processing,* (1998).

[47] ——, Joint source-channel decoding of variable-length encoded sources, *IEEE Information Theory Workshop*, (1998).

[48] K. Narendra and M.A.L. Thathachar, Learning automata : An introduction, Prentice Hall, 1989.

[49] K. S. Narendra and M. A. L. Thathachar, Learning automata - a survey, *IEEE Trans. on Syst. Man and Cybern.*, **SMC-4** (1974), pp. 323–334.

[50] K. S. Narendra and M. A. L. Thathachar, phLearning Automata:An introduction, Englwood Cliffs, NJ:Prentice-Hall, 1989.

[51] P. Nicopolitidis, G. I. Papadimitriou, and A. S. Pomportis, *Learning automata-based polling protocol for wireless lans*, **51** (2003), pp. 453–463.

[52] B. J. Oommen, Absorbing and ergodic discretized two-action learning automata, *IEEE Trans. Syst., Man, Cybern.,* **SMC-18** (1986), pp. 282–296.

[53] B. J. Oommen and M. Agache, Continuous and discretized pursuit learning schemes: Various algorithms and their comparison, *IEEE Trans. syst., Man, and Cybern.,* **31** (2001), pp. 277–287.

[54] B. J. Oommen and E. R. Hansen, The asymptotic optimality of discretized linear reward-inaction learning automata, *IEEE Trans. Syst., Man, Cybern.*, **14** (1984), pp. 542–545.

[55] B. J. Oommen and J. K. Lanctot, Discretized pursuit learning automata, *IEEE Trans. Syst., Man, Cybern.*, **20** (1990), pp. 931–938.

[56] M. Park and D. J. Miller, Low-delay optimal MAP state estimation in HMMS with application to symbol decoding, *IEEE Signal Processing Letters,* **4** (1997), pp. 289–292.

[57] ——, Joint source-channel decoding for variable-length encoded data by exact and approximate MAP sequence estimation, *IEEE Transactions on Communications,* **48** (2000), pp. 1–6.

[58] N. Phamdo, F. Alajaji, and N. Farvardin, Quantization of memoryless and Gauss-Markov sources over binary Markov channels, *IEEE Transactions on Communication,* **45** (1997), pp. 668–674.

[59] N. Phamdo and N. Farvardin, Optimal detection of discrete Markov sources over discrete memoryless channels, *IEEE Transactions on Information Theory,* **40** (1994), pp. 186–193.

[60] X. Qiu and K. Chawla, On the performance of adaptive modulation in cellular systems, *IEEE Trans. Comm.*, **47** (1999), pp. 884–895.

[61] Theodore S. Rappaport, *Wireless Communications:Principles and Practice* , Prentice Hall, NJ, 1996.

[62] K. Sayood and J. C. Borkenhagen, Use of residual redundancy in the design of joint source channel coders, *IEEE Transactions on Communications,* **39** (1991), pp. 839–846.

[63] K. Sayood, F. Liu, and J. D.Gibson, A constrained joint source/channel coder design, *IEEE Transactions on Communications,* **12** (1994), pp. 1584–1593.

[64] S. Shakkottai, T. S. Rappaport, and P. C. Karlsson, Cross-layer design for wireless networks, *IEEE Comm. Mag.*, (2003), pp. 74–80.

[65] S. Stein, Fading channel issues in system engineering, *IEEE JSAC,* **5** (1987), pp. 68–89.

[66] K. P. Subbalakshmi and Q. Chen, Joint source-channel decoding for MPEG-4 coded video over wireless channels, IASTED International Conference on Wireless and Optical Communications, (2002), pp. 617–622.

[67] K. P. Subbalakshmi and J. Vaisey, On the joint source-channel decoding of variable-length encoded sources: The additive markov channel case, *IEEE Transactions on Communications,* (To Appear).

[68] ——, Optimal decoding of entropy coded memoryless sources over binary symmetric channels, *IEEE Data Compression Conference,* (1998), p. 573.

[69] ——, Joint source-channel decoding of entropy coded Markov sources over binary symmetric channels, *IEEE International Conference on Communications* , **1** (1999), pp. 446–450.

[70] ——, Optimal decoding of entropy coded Markov sources over channels with memory, *The 33rd Annual Conference on Information Sciences and Systems,* **2** (1999), pp. 624–629.

[71] ——, On the joint source-channel decoding of variable-length encoded sources: the BSC case, *IEEE Transactions on Communications,* **49** (2001), pp. 2052–2055.

[72] M. A. L. Thathachar and B. J. Oommen, Discretized reward-inaction learning automata, *J. Cybern. Infrom. Sci.,* (1979), pp. 24–29.

[73] M. A. L. Thathachar and P. S. Sastry, A class of rapidly converging algorithms for learning automata, in *Proc. IEEE Int. Conf. Cybern. Soc.*, Bombay, India, Jan. 1984, pp. 602–606.

[74] ——, Estimator algorithms for learning automata, in *Proc. Platinum Jubilee Conf. Syst. Signal Process.* Bangalore, India, Dec. 1986, pp. 29–32.

[75] M.L. Tsetlin, On the behavior of finite automata in random media, *Automat. Telemek. (USSR),* **22** (1961), pp. 1345–1354.

[76] M. L. Tsetlin, Automaton Theory and the Modeling of Biological Systems, Academic, New York, 1973.

[77] V. I. Varshavskii and I. P. Vorontsova, On the behavior of stochastic automata with variable structure, *Automat. Telemek. (USSR),* **24** (1963), pp. 327–333.

[78] P. Viswanath, D. N. C. Tse, and R. Laroia, Opportunistic beamforming using dumb antennas, *IEEE Trans. Info. Theory,* **48** (2002), pp. 1277–1294.

[79] J. Wen and J. D. Villasenor, Utilizing soft information in decoding of variable length codes, *Data Compression Conference,* (1999), pp. 131–139.

In: Adaptation Techniques in Wireless Multimedia Networks ISBN 1-59454-883-8
Editors: Wei Li and Yang Xiao, pp. 265-284

Chapter 12

ECONOMIC AND ADAPTIVE ADMISSION CONTROL OF NON-STATIONARY TRAFFIC IN WIRELESS MULTIMEDIA NETWORKS*

Ki-Dong Lee[1 †] ***and Antonio Iera***[2 ‡]
[1]Department of Electrical and Computer Engineering,
the University of British Columbia (UBC),
2332 Main Mall, Vancouver, B.C., Canada V6T 1Z4
[2]Dipartimento di Informatica, Matematica,
Elettronica e Transporti (DIMET), the University "Mediterranea"
of Reggio Calabria, 89100 Reggio Calabria, Italy

Abstract

As the major target of mobile service shifts from voice to real-time multimedia data, handover management is becoming more important issue. In radio link layer aspects, it has been one of the main issues how many resources to reserve for handover calls while avoiding a critical sacrifice (degradation) in new call blocking performance. This issue has been extensively studied using the M/M/c/c queue model and its variants. However, the blocking performance of handover traffic in the emerging wireless network is not easily formulated with an acceptable level of accuracy because of the non-stationary nature of handover traffic, which causes difficulty in quantitatively making optimal decisions for channel assignment to meet a given quality of service (QoS) requirement. Even though many challenging schemes have been proposed to dynamically determine the number of guard channels, most schemes either use heuristic methods or are based on simplifying assumptions, causing invalid decision results. As an alternative, we suggest a quantified method to minimize the *fraction of the number of blocked calls* out of the number of total calls under non-stationary handover traffic. We develop new mathematical formulations of that fraction and optimization models with efficient exact solution algorithms. Performance analysis shows that the proposed method improves the blocking probabilities of handovers and new calls; this

---

*Date of submission: March 28, 2005; September 25, 2005.
†E-mail address: kidongl@ece.ubc.ca
‡E-mail address: iera@ing.unirc.it

improvement results from a highly adaptive and reactive characteristic to the fluctuating handover traffic condition.

Keywords: resource management, handover, admission control

AMS: 15A15, 15A09, 15A23

1. Introduction

To achieve high spectral efficiency and high utilization of radio resources has been the main fucus of existing and future broadband mobile networks [1, 2]. As the major wireless communication service shifts from voice to multimedia applications, it has become one of the hot issues to design and develop advanced broadband mobile networks supporting multiple classes of traffic. To efficiently accommodate multimedia traffic through a limited available bandwidth, many challenging schemes for admission control have been proposed [1, 12, 13, 19]. The major concern of these studies focus on adaptive technique for both maximizing the utilization of valuable radio resources and maximizing the quality-of-service that the users experience. Because the traffic load is expected to be even more fluctuating in the emerging wireless network, developing adaptation techniques is imperative to achieve the aforementioned purposes. This fluctuation phenomenon usually results from the highly dynamic moving behavior of mobile users caused by cells getting smaller in size. This gave rise to an important issue of economic management of handovers and the prioritized channel assignment was extensively studied over the years as a solution to the issue.

Most previous studies using quantitative optimization are based on the call blocking probability which is formulated under the assumption of the classical stationary Poisson process. The modeled blocking probability provides an acceptable level of accuracy in land mobile cellular systems. However, the accuracy of the blocking probability formula is not guaranteed in the emerging wireless network because of the intrinsic non-stationary nature of handover traffic. As a solution to making optimal channel assignment decisions for prioritized handover calls, we develop a new mathematical formulation for the call blocking fraction under non-stationary traffic conditions. We think that the proposed method needs a few more collaborative modules for field applications. However, we believe that the proposed method is a new guideline challenging the prioritized channel assignment in the emerging wireless network.

To design a personal communication system with a higher transmission rate, many transmission techniques and network design concepts have been proposed for wireless network evolution. In the emerging wireless network, subscribers can make real-time voice call and higher rate of multimedia connections. However, it is necessary to solve handover and mobility management problems in order to provide a higher transmission rate using the emerging wireless network. There are challenging schemes for the purpose of improving traffic performance, and they are classified into dynamic channel assignment (DCA) and fixed channel assignment (FCA) [8, 10]. In DCA, the total channels are pooled and then dynamically assigned to each cell with a certain reuse distance and returned to the pool after use. In FCA, each cell is assigned a fixed number of channels. DCA has a flexible assignment opportunity and guarantees better blocking performance than FCA, but DCA requires centralized real-time control, which results in operational complexity and performance degradation under heavy traffic conditions. Although several hybrid schemes of DCA and FCA have also been proposed, FCA is still more useful in practical systems, and improving the real-time control efficiency in DCA is still an open problem to study.

Several prioritized admission schemes have been proposed to accommodate handover requests at a smaller dropping probability [3, 8, 10]. In FCA, it is imperative to dynamically determine the number of guard channels for handovers out of the fixed number of channels under non-stationary handover processes. Under stationary traffic conditions, the number of guard channels [3] is simply optimized using blocking probability (BP) formulated by a Birth-Death queueing model [11]. However, it is not guaranteed that this method provides valid decision results under non-stationary traffic conditions as observed in the present and emerging wireless networks.

Rather than heuristic methods [12, 13, 14], we suggest a quantitative method in searching for a variable target number of guard channels according to the fluctuation of handover traffic. To minimize the *fraction of lost calls* out of the total calls, we develop a mathematical formulation of this quantity. This quantity is usually regarded as the blocking probability in the literature [3], where the formula indicates the *time average*, i.e., the fraction of system saturation time out of the whole time span [11]. Under stationary traffic conditions, the blocking probability is an unbiased estimator of this quantity. Under non-stationary traffic conditions, however, it is not an unbiased estimator and cannot be simply formulated. As an alternative, we define a new measure called the *blocking fraction* (BF), which denotes the fraction of the number of blocked calls out of the number of total calls, and mathematically formulate the BF. With this, we formulate optimization problems to optimize the number of guard channels and develop exact solution algorithms. We evaluate the performance of the proposed method with extensive simulations. According to the performance evaluation results, the proposed method provides better performance in both handover blocking and new call blocking than previous methods and shows a highly adaptive and reactive characteristic to the variation of handover traffic conditions.

The rest of this paper is organized as follows. The system model, problem definition, and the proposed procedure are presented in Section 2.. Mathematical formulations and solution procedures for optimal resource allocation are presented in Sections 3. and 4.. The admission control is presented in Section 5.. Following performance evaluation, we conclude our work in Section 7..

2. The Proposed Access Control

2.1. System Model

We consider a resource allocation problem in a mobile multimedia system with multiple cells. Each cell has a resource allocation module and a traffic rate prediction module as shown in the reference model of Fig. 1. There is handover among cells, and handovers have higher priority than new calls. The traffic rate prediction modules of adjacent base stations share traffic and mobility information, if any, with each other. With this collaborative function, each traffic rate prediction module estimates future traffic parameters and reports the information to the resource allocation module.

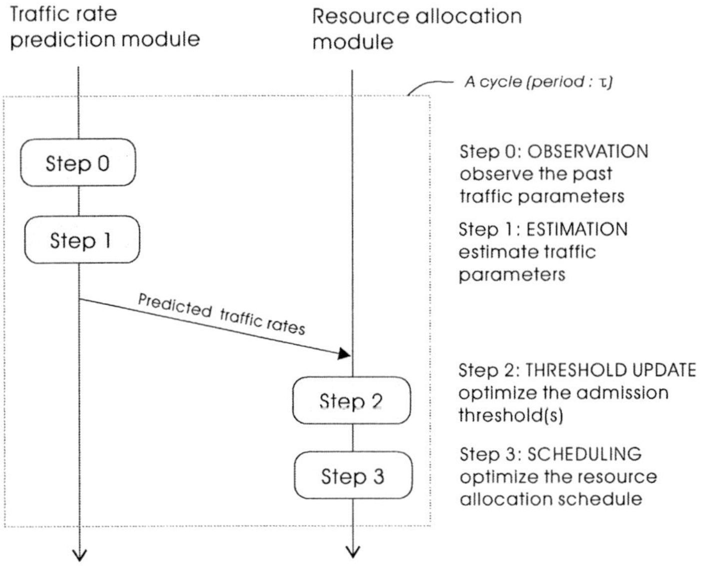

Figure 1. A reference model for admission control of non-stationary traffic.

2.2. Problem Definition

A counting process is called stationary if the arrival rate $\lambda(t)$ is constant, and is called non-stationary if $\lambda(t)$ varies over time. Fig. 2 shows sample stochastic paths of stationary traffic and non-stationary traffic. Under a stationary process, the blocking probability can be formulated by a Birth-Death process with sufficient accuracy [15]. With that formulation, we can simply find an optimal number of guard channels for prioritized handover admission [3].

However, it is not simple to formulate the blocking probability itself under a non-stationary process. Thus, it is necessary to investigate an alternative to make an optimal channel assignment plan for every time t. Accurate estimation of the future traffic rate is a prerequisite for optimal planning of channel assignment. Developing a useful estimation technique for field applications is an interesting further work, but it is not the focus of this paper. Mathematical models for traffic load estimation can be found in [16, 17]. In [16], each base station estimates future handover traffic load using *local* information alone, i.e., the estimation uses the traffic information of ongoing calls in the local cell but does not use the traffic information of ongoing calls in the neighboring cells. Since the incoming handover traffic to a base station is dependent upon the number of ongoing calls in the neighboring cells, this local method may seem to be unsuitable for estimation accuracy, but the method has sufficient accuracy according to the test results. In [17], several location-based estimation models are developed to estimate offered traffic. Since those models have practical explanatory variables, such as user density at location x at time t, the methods are expected to have an excellent accuracy and to be practical in such systems with the explanatory variables given (or simply estimatable as in highway cellular systems). In this paper, we focus on developing a new formulation for optimal planning of channel assignment in each resource allocation module (scheduler) given that traffic rate prediction modules pro-

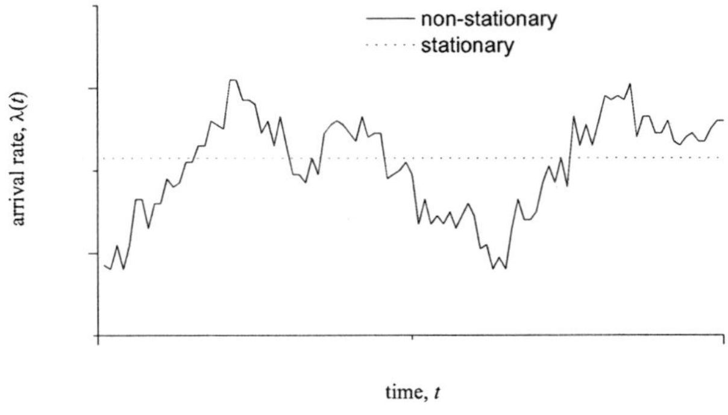

Figure 2. Sample paths of non-stationary traffic and stationary traffic.

vide future traffic information to resource allocation modules. With the new formulation, the admission control threshold value is dynamically updated so that the BF is minimized.

2.3. Adaptive Access Control with BF

Under a stationary process and FCA [3], the target number of guard channels need not be updated if it is optimized once. However, the target number should be dynamically updated according to traffic conditions under a non-stationary process. Thus, resource allocation modules should update the target number in order to maintain an acceptable level of quality of service (QoS). Fig. 1 shows the proposed procedure, where the resource allocation module is our major concern. In the figure, the traffic rate prediction module estimates the probability mass function (PMF) of the number of arriving call requests and reports that PMF to the resource allocation module, and the resource allocation module updates the target threshold value for admission control. Details on the proposed method are presented in the following sections.

3. Mathematical Formulation

3.1. Notation and Underlying Assumptions

For each cell, we define the following notation.

- $c(t)$ is the total resource units (RUs) allocated to a cell. RUs could be time slots (in TDMA), frequency channels (in FDMA), code channels (in CDMA), or a combination of them (in CDMA/TDD) according to the multiple access scheme employed in the reference system. We call an RU a channel.
- $z(t)$ is the number of available channels at time t.
- $r(t)$ is the number of channels reserved for prioritized handover admission. If there are available channels less than or equal to $r(t)$, only handovers are accepted.
- $x(t) \triangleq z(t) - r(t)$ is the number of channels shared by both new calls and handovers (time index t may be dropped if doing so does not create confusion).

Table 1. Comparison of BP and BF.

metric	description
BP	average fraction of time of system saturation $= \lim_{t\to\infty} \frac{\int_0^t I(u)du}{t}$, where $I(t) = \begin{cases} 1, & \text{system is saturated at } t \\ 0, & \text{otherwise} \end{cases}$
BF	average fraction of the number of arriving calls during system saturation out of the total arriving calls $= \lim_{t\to\infty} \frac{\int_0^t \lambda(u)\cdot I(u)du}{\int_0^t \lambda(u)du}$, where $\lambda(t)$ denotes the arrival rate at t

We have BP = BF, given that $\lambda(t)$ is a constant, i.e., the process is stationary

- Random variables Y_H and Y_N denote the number of handover arrivals and the number of new call arrivals during a given time window τ, respectively.
- Random variable $K(x)$ denotes the number of new calls that arrive until the number of total calls (whether handover or new calls) that arrive is x during τ. That is, $K(x)$ denotes how many new calls arrive till x-th call (whether handover or new call) arrives during a given time window τ.
- $\mathbf{R}(z) \triangleq \{0, 1, \cdots, z\}$.

3.2. Formulation of Blocking Fraction

The concept of blocking probability widely used in the literature [3] is *time average*, i.e., the fraction of time during which all channels are busy. By theorem PASTA (Poisson Arrivals See Time Average) [11], a random arrival from a Poisson stream sees that the system is saturated with a probability equal to the fraction.

However, it is difficult to formulate the time average of system saturation under a non-stationary process. Thus, we introduce a concept of *number average*, i.e., the fraction of the number of blocked calls out of the number of total calls. Given that there are z available channels, the average number of blocked handovers when r $(= z - x)$ channels are reserved for handover admission is given by

$$\begin{aligned} D_H(r:z) &= \sum_{i=0}^{z-r} D_H(r:z|K(z-r)=i)\mathbb{P}(K(z-r)=i) \\ &= \sum_{i=0}^{z-r}\sum_{y=z-i}^{\infty} (y-(z-i))\mathbb{P}(Y_H=y)\mathbb{P}(K(z-r)=i), \end{aligned} \quad (1)$$

where $D_H(r:z)$ denotes the average number of blocked handovers because of system saturation when r channels are reserved for prioritized handover admission. If we have one more channel for handovers, $D_H(r:z)$ decreases. This is simply because more handovers could be admitted with the additional channel. Thus, it is obvious that $D_H(r:z)$ is a strictly decreasing function of r.

Similarly, the penalty due to new call blocking is defined as

$$\begin{aligned} D_N(r:z) &= \sum_{i=0}^{z-r} D_N(r:z|K(z-r)=i)\mathbb{P}(K(z-r)=i) \\ &= \sum_{i=0}^{z-r}\sum_{y=i}^{\infty}(y-i)\mathbb{P}(Y_N=y)\mathbb{P}(K(z-r)=i). \end{aligned} \tag{2}$$

In (1) and (2), $\mathbb{P}(K(z-r)=i)$ is given by

$$\begin{aligned} \mathbb{P}(K(z-r)=i) &= \mathbb{P}[A(t+\tau)-A(t)=(z-r)-i] \\ &\quad \times\mathbb{P}[B(t+\tau)-B(t)=i], \end{aligned} \tag{3}$$

where random variables $A(t)$ and $B(t)$ denote the number of handover requests and the number of new call requests, respectively, during the interval $[0,t)$. The PMFs of $[A(t+\tau)-A(t)]$, $[B(t+\tau)-B(t)]$, Y_H, and Y_N are given by traffic rate prediction modules.

For a finite number of computations, (1) and (2) are rewritten as

$$\begin{aligned} D_H(r:z) &= \sum_{i=0}^{z-r}\Big(\mathbb{E}(Y_H)-(z-i)+\sum_{y=0}^{(z-i)-1}((z-i)-y)\mathbb{P}(Y_H=y)\Big) \\ &\quad \times\mathbb{P}(K(z-r)=i), \end{aligned} \tag{4}$$

$$D_N(r:z) = \sum_{i=0}^{z-r}\Big(\mathbb{E}(Y_N)-i+\sum_{y=0}^{i-1}(i-y)\mathbb{P}(Y_N=y)\Big)\mathbb{P}(K(z-r)=i), \tag{5}$$

respectively. Thus, the blocking fraction of handovers and the blocking fraction of new calls are defined by

$$\mathrm{BF}_H(r:z) = \frac{D_H(r:z)}{\mathbb{E}(Y_H)}, \tag{6}$$

$$\mathrm{BF}_N(r:z) = \frac{D_N(r:z)}{\mathbb{E}(Y_N)}, \tag{7}$$

respectively.

3.3. Examples of BF

We consider a simple case where the stochastic processes of handover arrival and new call arrival are mutually independent. We assume that

$$\begin{aligned} \mathbb{P}(Y_H=y) &= e^{-\mathbb{E}(Y_H)}\frac{\mathbb{E}(Y_H)^y}{y!}, \\ \mathbb{P}(Y_N=y) &= e^{-\mathbb{E}(Y_N)}\frac{\mathbb{E}(Y_N)^y}{y!}. \end{aligned}$$

The meaning of this assumption is that the handover arrival process is stationary and has independent increment during the reference time window τ [1]. From the above assumption,

[1]Note that the stochastic processes of handovers and new calls can be modelled with sufficient accuracy by this assumption under the conditions that τ is sufficiently small and the arrival rate and the departure rate at the beginning of the time window and those at the end of the time window are equally likely.

we have a conclusion that random variable $K(z-r)$ follows a binomial distribution with parameters $(z-r, p)$, where

$$p = \frac{\mathbb{E}(Y_N)}{\mathbb{E}(Y_N) + \mathbb{E}(Y_H)}.$$

That is, we have

$$\mathbb{P}(K(z-r) = i) = \binom{z-r}{i} p^i (1-p)^{(z-r)-i}. \tag{8}$$

As a result, we have

$$D_H(r:z) = \sum_{i=0}^{z-r} \left(\mathbb{E}(Y_H) - (z-i) + \sum_{y=0}^{(z-i)-1} ((z-i)-y) e^{-\mathbb{E}(Y_H)} \frac{\mathbb{E}(Y_H)^y}{y!} \right) \tag{9}$$
$$\times \binom{z-r}{i} p^i (1-p)^{(z-r)-i},$$

$$D_N(r:z) = \sum_{i=0}^{z-r} \left(\mathbb{E}(Y_N) - i + \sum_{y=0}^{i-1} (i-y) e^{-\mathbb{E}(Y_N)} \frac{\mathbb{E}(Y_N)^y}{y!} \right) \tag{10}$$
$$\times \binom{z-r}{i} p^i (1-p)^{(z-r)-i}.$$

3.4. Blocking Fraction Minimization Problem

The threshold value of control variable r should be dynamically updated according to traffic conditions so that the BF is minimized. We mathematically formulate BF minimization problems $(PH(z))$ and $(PN(z))$ as non-convex combinatorial optimization problems [18], respectively, for each cell. The role of $(PH(z))$ is to minimize the average fraction of blocked new calls out of the total new calls subject to the constraint that the average fraction of blocked handovers out of the total handovers is not greater than a threshold value ξ_H.

$(PH(z))$
Minimize $\mathrm{BF}_N(r:z)$
Subject to $\mathrm{BF}_H(r:z) \leq \xi_H,\ r \in \mathbf{R}(z).$

Similarly, the role of $(PN(z))$ is to minimize the average fraction of blocked handovers out of the total handovers subject to the constraint that the average fraction of blocked new calls out of the total new calls is not greater than ξ_N.

$(PN(z))$
Minimize $\mathrm{BF}_H(r:z)$
Subject to $\mathrm{BF}_N(r:z) \leq \xi_N,\ r \in \mathbf{R}(z).$

3.5. Model Extension to Multiclass System using the Goal Programming Concept

One of the most important features in the challenging wireless networks is the so-called 'multiclass service' [19]. Recently, many resource allocation techniques in multiclass wireless networks with different priority levels have been studied [12, 14, 19, 4]. Using the economic concept of our penalty function, the PMP in a multiclass system can be formulated by the goal programming (GP) technique [20]. The proposed formulation is a simple extension of $(PH(z))$ which is an example of a two priority-class system (handover and new call).

The basic idea of our GP formulation is what we refer to as *highest priority first served* (HPFS). An incoming request of class i is accepted, if there is available capacity for the class:

$$z(k) \geq \phi_i + \sum_{j=1}^{i-1} \phi_j \cdot r_j(k), \tag{11}$$

where ϕ_i is necessary for the incoming request and $\left[\sum_{j=1}^{i-1} \phi_j \cdot r_j(k)\right]$ denotes the reserved capacity for classes $j \in \{1, \cdots, i-1\}$. Let $\mathbf{r}(k) = (r_1(k), \cdots, r_{M-1}(k))$, where $r_n(k)$ is a threshold (in numbers of calls) at step k in a cutoff priority-based admission policy, where incoming requests of classes $i \in \{n+1, \cdots, M\}$ are rejected if the amount of available capacity is less than $\left[\phi_i + \sum_{j=1}^{n} \phi_j \cdot r_j(k)\right]$ [2]. In HPFS, the constraint of priority class 1 must be satisfied first, that of priority class 2 next, and so on [20].

$(GP(z(k)))$

minimize $\quad D_M(\mathbf{r}(k) : z(k))$

subject to

$$E[Y_1]^{-1} \cdot D_1(\mathbf{r}(k) : z(k)) \leq \xi_1 \tag{12}$$

$$E[Y_2]^{-1} \cdot D_2(\mathbf{r}(k) : z(k)) \leq \xi_2 \tag{13}$$

$$\vdots$$

$$E[Y_{M-1}]^{-1} \cdot D_{M-1}(\mathbf{r}(k) : z(k)) \leq \xi_{M-1} \tag{14}$$

$$\sum_{j=1}^{M-1} \phi_j \cdot r_j(k) \leq z(k). \tag{15}$$

In $(GP(z(k)))$, Y_i is the number of class i arrivals and ξ_i is the target threshold of the fraction of blocked calls out of the total calls in class i. The basic idea of $D_i(\mathbf{r}(k) : z(k))$ is the same as that of $D_H(r : z)$ and it can be derived accordingly. We label constraints (12)-(14) as QoS requirements[3], and constraint (15) as a budget constraint during τ. In the GP,

[2] In $(PH(z))$ and $(PN(z))$, we express r RUs rather than $\phi_i \cdot r$ RUs under an assumption that ϕ_i's for the two classes are unity because it is generally accepted that the handovers and new calls of the same traffic class require the equivalent effective bandwidth.

[3] Each QoS requirement is called a *goal*. There are $(M-1)$ goals in $(GP(z(k)))$.

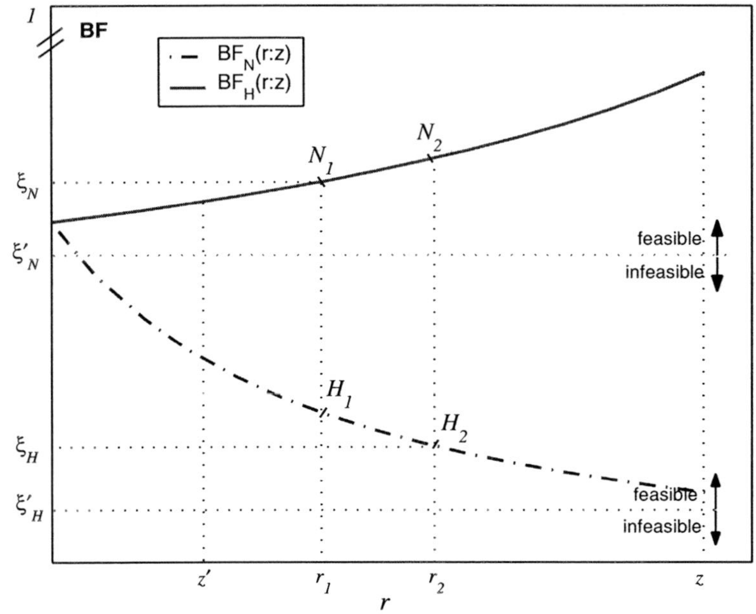

Figure 3. Graphical insight of $(PH(z))$ and $(PN(z))$.

constraint (12) has the first priority, constraint (13) the second priority, and so on. As long as it is possible, constraint (12) must be satisfied independent of the other QoS requirements of lower priority, and then constraint (13) must be satisfied independent of the other QoS requirements of lower priority, and so on. Therefore, the $(GP(z(k)))$ can be solved by the solution algorithm proposed in Section 4. with simple modification.

4. Solution Procedure

Without loss of generality, we present our algorithm in a two-priority class system.

4.1. Graphical Illustration and Property Analysis of $(PH(z))$ and $(PN(z))$

The constraint of $(PH(z))$ can be rewritten as

$$f_H(r : z, \xi_H) \leq 0, \; r \in \mathbf{R}(z), \tag{16}$$

where

$$f_H(r : z, \xi_H) = \mathrm{BF}_H(r : z) - \xi_H. \tag{17}$$

Similarly, we define

$$f_N(r : z, \xi_N) = \mathrm{BF}_N(r : z) - \xi_N \tag{18}$$

for $(PN(z))$.

Fig. 3 gives graphical illustrations of the two problems under consideration. In $(PH(z))$, $\mathrm{BF}_N(r : z)$ should be minimized over the feasible set

$$A_H(z, \xi_H) = \{r | f_H(r : z, \xi_H) \leq 0\}. \tag{19}$$

In the figure, we obtain $A_H(z, \xi_H) = \{r_2, \cdots, z\}$ and the minimum is $\mathrm{BF}_N(r_2 : z)$ (point N_2). Suppose that we require better QoS for handovers, i.e., we have a new target threshold $\xi'_H < \xi_H$ such that $A_H(z, \xi'_H) = \emptyset$. In this case, the problem is infeasible (violation of constraint). However, $r^* = z$ is recommended given that handovers have higher priority. Suppose that the number of available channels is z', then $A_H(z', \xi_H) = \emptyset$ and the problem is infeasible. In this case, $r^* = z'$ is recommended. The recommendation is based on the idea of minimizing the degree of violation (note that $f_H(r : z, \xi'_H)$ and $f_H(r : z', \xi_H)$ are minimal at $r^* = z$ and at $r^* = z'$, respectively). Suppose that there are z_1 available channels, i.e., $z = z_1$ and $(PH(z))$ is infeasible at $z = z_1$. If there are more channels than z_1, then the graphs in Fig. 3 shift down. There is a resolving point z^* such that $(PH(z^* - 1))$ is infeasible and $(PH(z^*))$ is feasible (resolving infeasibility). If $z^* > z_1$, we say that the cell is *congested*. If $z^* > c$, we say that the cell is *perfectly congested*. The resolving point for an infeasible $(PH(z))$ can be found

$$z^* = \min\{z | f_H(z : z, \xi_H) \leq 0\}. \tag{20}$$

In $(PN(z))$, $\mathrm{BF}_H(r : z)$ should be minimized over the feasible set

$$A_N(z, \xi_N) = \{r | f_N(r : z, \xi_N) \leq 0\}. \tag{21}$$

Similarly, we obtain $A_N(z, \xi_N) = \{0, \cdots, r_1\}$ and the minimum is $\mathrm{BF}_H(r_1 : z)$ (point H_1). Suppose that we require better QoS for new calls, i.e., we have a new target threshold $\xi'_N < \xi_N$ such that $A_N(z, \xi_N) = \emptyset$. In this case, $r^* = 0$ is recommended. According to the idea introduced in the previous example, $f_N(r : z, \xi'_N)$ is minimal at $r^* = 0$. The resolving point for an infeasible $(PN(z))$ can be found

$$z^* = \min\{z | f_N(0 : z, \xi_N) \leq 0\}. \tag{22}$$

Proposition. [Optimum of $(PH(z))$] If there exists an r^* such that

$$r^* = \arg\max\{f_H(r : z, \xi_H) \leq 0\} \text{ for all } r \in \mathbf{R}(z), \tag{23}$$

then r^* is the optimal solution of $(PH(z))$.

Proof. Suppose that there exists an r^* such that $r^* = \arg\max\{f_H(r : z, \xi_H) \leq 0\}$ for all $r \in \mathbf{R}(z)$. Then the feasible set $A_H(z, \xi_H)$ can be expressed as

$$A_H(z, \xi_H) = \{r^*, \cdots, z\}. \tag{24}$$

Since $D_N(r : z)$ is a monotonically increasing function of r, $\min A_H(z, \xi_H)$ minimizes $D_N(r : z)$. The smallest element of the feasible set is r^*. This completes the proof. □

Step 1 (Feasibility Check)[4]
If $f_H(z : z, \xi_H) > 0$, then $(PH(z))$ is infeasible.
Find the resolving point z^*
Return z^*, and $r^* := z$ as a solution of recommendation (Stop).
Else, Set $z^* := 0$.
Step 2 ($\min\{g_H(r)\}$ using Golden Section Method)[5]
Step 2.0 (Initialization)
Set $k := 0$ and Take $[r_k, \overline{r}_k] := [0, z]$.
Calculate $b_k := \lfloor r_k + s(\overline{r}_k - r_k) \rfloor$, $\overline{b}_k := \lceil \overline{r}_k - s(\overline{r}_k - r_k) \rceil$ where $s = \frac{3-\sqrt{5}}{2}$.
Step 2.1 (Iteration)
If $g_H(b_k) < g_H(\overline{b}_k)$, Set
$r_{k+1} := r_k, \overline{r}_{k+1} := b_k$ if $g_H(r_k) \leq g_H(b_k)$
$r_{k+1} := r_k, \overline{r}_{k+1} := \overline{b}_k$ if $g_H(r_k) > g_H(b_k)$.
If $g_H(b_k) > g_H(\overline{b}_k)$, Set
$r_{k+1} := \overline{b}_k, \overline{r}_{k+1} := \overline{r}_k$ if $g_H(\overline{b}_k) \geq g_H(\overline{r}_k)$
$r_{k+1} := b_k, \overline{r}_{k+1} := \overline{r}_k$ if $g_H(\overline{b}_k) < g_H(\overline{r}_k)$.
If $g_H(b_k) = g_H(\overline{b}_k)$, Set
$r_{k+1} := b_k, \overline{r}_{k+1} := \overline{b}_k$.
Step 2.2 (Optimality Check)
If $|r_{k+1} - \overline{r}_{k+1}| \leq 1$, Return z^*, and $r^* := \arg\min\{g_H(r_{k+1}), g_H(\overline{r}_{k+1})\}$ (Stop).
Else, Set $k := k + 1$ and Go to Step 2.1.

Figure 4. Solution procedure. [4] For $(PN(z))$, $(PN(z))$ is infeasible, if $f_N(0 : z, \xi_N) > 0$. Return $r^* := 0$ as a solution of recommendation. [5] Given $[r_k, \overline{r}_k]$, determine $[r_{k+1}, \overline{r}_{k+1}]$ so that $r^* \in [r_{k+1}, \overline{r}_{k+1}]$.

Proposition. [Optimum of $(PN(z))$] If there exists an r^* such that

$$r^* = \arg\max\{f_N(r : z, \xi_N) \leq 0\} \text{ for all } r \in \mathbf{R}(z), \tag{25}$$

then r^* is the optimal solution of $(PN(z))$.

Proof. Suppose that there exists an r^* such that $r^* = \arg\max\{f_N(r : z, \xi_N) \leq 0\}$ for all $r \in \mathbf{R}(z)$. Then the feasible set $A_N(z, \xi_N)$ can be expressed as

$$A_N(z, \xi_N) = \{0, \cdots, r^*\}. \tag{26}$$

Since $D_H(r : z)$ is a monotonically decreasing function of r, $\max A_N(z, \xi_N)$ minimizes $D_N(r : z)$. The largest element of the feasible set is r^*. This completes the proof. □

4.2. Exact Solution Algorithm

We propose an efficient and exact algorithm to solve $(PH(z))$ and $(PN(z))$. With simple substitution, $(PN(z))$ can be solved by the same procedure for $(PH(z))$ thanks to its similarity to $(PH(z))$. Thus, we present our algorithm for $(PH(z))$ only.

By Proposition 4.1., we only have to find r^* such that (23) is satisfied. In order to find r^* with a fast convergence rate, we employ the well-known Golden Section Method.

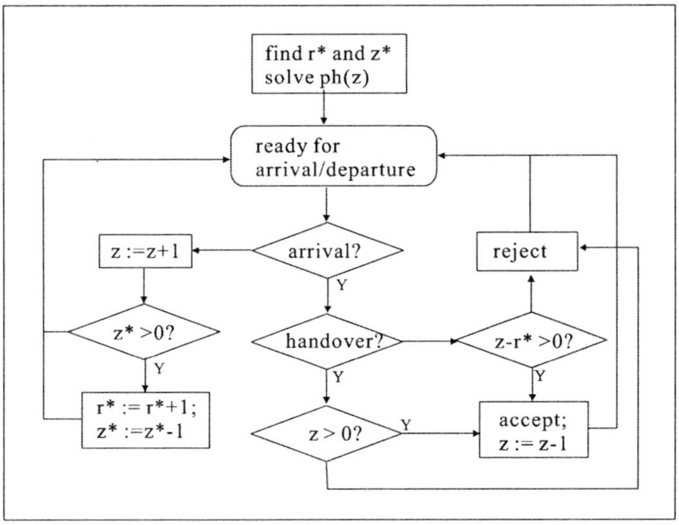

Figure 5. Resource allocation for $(PH(z))$ during a step.

The Golden Section Method iteratively finds the optimum of a *unimodal function* with an excellent rate of convergence. In the Golden Section Method, the error range within which the optimum is included *converges geometrically* [18].

Let us define a unimodal function of $r \in \mathbf{R}(z)$ to find r^* in (23) as follows:

$$g_H(r) = \begin{cases} -f_H(r : z, \xi_H), & f_H(r : z, \xi_H) \leq 0 \\ c_H, & f_H(r : z, \xi_H) > 0 \end{cases}, \tag{27}$$

where c_H is a constant such that $g_H(r) \geq \sup\{-f_H(r : z, \xi_H)\}$ for all $r \in \{r | f_H(r : z, \xi_H) > 0\}$ (We use ξ_H as a feasible candidate for c_H). (Similarly, a unimodal function

$$g_N(r) = \begin{cases} -f_N(r : z, \xi_N), & f_N(r : z, \xi_N) \leq 0 \\ c_N, & f_N(r : z, \xi_N) > 0 \end{cases} \tag{28}$$

could be used to solve $(PN(z))$, where $c_N = \xi_N$.) It is obvious that the minimum of $g_H(r)$ is the optimum of $(PH(z))$. The proposed algorithm is presented in Fig. 4. The resulting solution does not always mean that it is the optimal solution. In cases where the problem is infeasible (violation of constraint), the algorithm returns a solution of recommendation that minimizes the degree of violation as discussed in the previous section. Otherwise, the algorithm always returns the optimal solution.

In our algorithm, the concept of a *resolving point* is used for the allocation of released channels due to handover departure and call completion. With the value of z^*, we know the state of congestion: if $z^* = 0$, no congestion; if $z^* > 0$, congestion.

5. Admission Control

The procedure of the proposed resource allocation algorithm is shown in Fig. 5. Consider a reference cell i at step k. Before starting step $k + 1$, r^* and z^* should be determined by

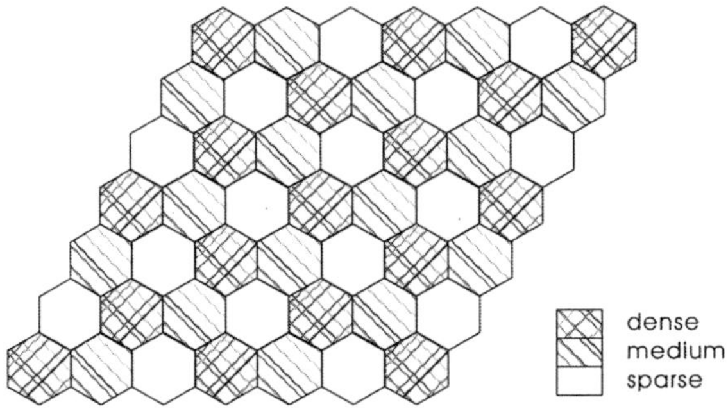

Figure 6. The cell layout with three types of user densities. S: sparse; D: dense; M: medium

Table 2. Simulation parameters.

param.	description	value
$c(t)$	no. of channels per cell	60
T_c	call duration (s)	unif(80,150)
T_H	channel holding time (s)	unif(50,150)
$\lambda_N(t)$	new call rate (calls/hour/cell)	
	in dense area	unif(150,250)
	in medium area	unif(100,200)
	in sparse area	unif(50,150)
$\lambda_H(t)$	handover rate (calls/hour/cell)	depends on T_c and $\lambda_N(t)$
τ	time window (s)	100

t: $0 \leq t < 72000$ (in *simulated* seconds)

solving $(PH(z))$. If an event occurs, the type is checked to determine whether it is traffic arrival or traffic departure. For traffic arrival, the type is checked to determine whether the type is handover or new call. Handovers are accepted only if the number of available channels is greater than zero, i.e., $z > 0$, whereas new calls are accepted only if the number of available channels is greater than r^*, i.e., $z > r^*$. If the request is accepted, one available channel disappears ($z := z - 1$) in the cell. If the event is traffic departure, an additional available channel appears ($z := z + 1$). If the present $(PH(z))$ is infeasible ($z^* > 0$), it is reasonable that the additional opportunity be used for handovers ($r^* := r^* + 1$), which results in reducing the resolving point value ($z^* := z^* - 1$).

6. Performance Evaluation

6.1. Simulation Environments

We evaluate the performance of the proposed admission control using simulation. The simulation model has a group of 49 cells with three types of user densities (S: sparse; D: dense; M: medium) as shown in Fig. 6 In practice, each cell has a varying distribution (density) of users. This phenomenon depends upon the geographic and environmental reasons that gather users time to time. However, for the sake of simplicity, we assume that the types of user distribution of the respective cells are given and that a group of randomly chosen users pass through the three kinds of regions (cells) at random. That is, each user passes heavy

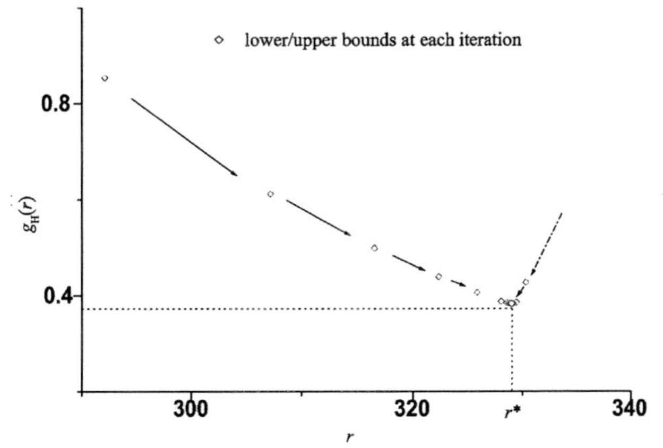

Figure 7. Golden section search for r.

traffic areas (dense user distributions), medium traffic areas, and light traffic areas (sparse user distributions).

The traffic rate of new calls is randomly chosen from a uniform distribution every τ according to the user density type as shown in Table 2.

6.2. Results and Discussions

Fig. 7 shows the rate of convergence of the Golden section search used in searching for r. Due to the fast convergence rate, the search produces a suboptimal solution with error range less than 10^{-4} in about 15 iterations.

We use simulations to analyze the performance of our algorithm, considering two performance measures: the average fraction of blocked handovers, and the average fraction of blocked new calls. In addition, we observe the mean and the standard deviation of $r(t)$, denoted by $\mathbb{E}(r(t))$ and $\sigma(r(t))$, respectively, during our simulation period. These two statistics are important in understanding the highly adaptive feature of our resource allocation algorithm in a system with highly non-stationary traffic arrival streams. To compare the performance, we consider two previous schemes which have been proposed for a land mobile cellular system and considered admission schemes based on traffic estimation: scheme1 [14] and scheme2 [19].

6.2.1. Scheme 1

The admission control scheme proposed in [14] can be used for multiple traffic class systems. The key characteristic of that algorithm consists in reserving a certain amount of bandwidth in neighboring cells for any priority connection.

For non-priority connections, there is no reservation of bandwidth regardless of the type (i.e., handover or new call). Since this algorithm reserves bandwidth in (all) neighboring cells for priority connections, the handover dropping probability can be remarkably improved for those connections at the cost of a good amount of bandwidth to reserve. A higher bandwidth reservation ratio may give rise to decrease of admission opportunity for

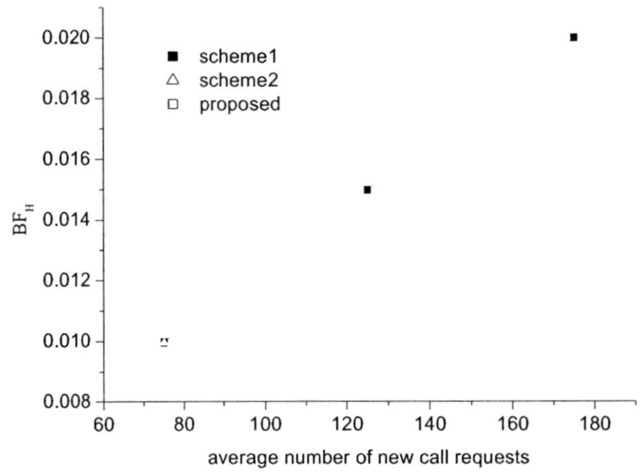

Figure 8. The blocking fraction of handovers for $PH(z)$ (in dense, medium, and sparse area). The average number of handover requests depends upon the average number of new call requests, the average channel holding time, and so on.

other incoming connections.

6.2.2. Scheme 2

In [19], `ExpectedMax` Strategy was proposed. For all possible call arrival/departure event sequences during next time interval, a base station computes the maximum net bandwidths that will be needed to deal with handover calls and the expected value of them. This value is used for the guard threshold value of handover call admission. The base station may update the guard threshold value once every $K \geq 1$ new call requests.

Figs. 8 and 9 show the the average fraction of blocked handovers and the average fraction of blocked new calls when we solve $(PH(z))$ problems at $\xi_H = 0.013$ (in dense traffic area), $\xi_H = 0.012$ (in medium traffic area), and $\xi_H = 0.010$ (in sparse traffic area). In the figure, it is observed that the blocking fractions for scheme 1 and scheme 2 are over the target values ξ_H's in dense and medium areas. However, this does not mean that the proposed scheme is dominant over these two schemes. This only means that a targeted blocking fraction ξ_H may be realized by using a hard constraint shown in the formulation of $(PH(z))$. That is, the proposed access control scheme is just designed to facilitate satisfying a target level of QoS but those two previous schemes were not.

In these figures, it is observed that both BF_H and BF_N obtained by the proposed scheme are lower at the same time than those by the two competing schemes. This is because the proposed scheme is more highly adaptive to the non-stationary traffic variation than the other two schemes, which is observed in Fig. 11 where the standard deviation of the guard thresholds $\sigma(r(t))$ is greater than those of the other schemes. Also, it can be inferred that the fraction of idle bandwidth in using the proposed scheme is smaller than those fractions in using the other two schemes from the fact that the number of accommodated (non-blocked) calls is improved.

In Figs. 8 and 9, it is observed that scheme 1 produces the highest blocking fractions in all cases. However, this fact does not always mean that scheme 1 is inferior to the other two

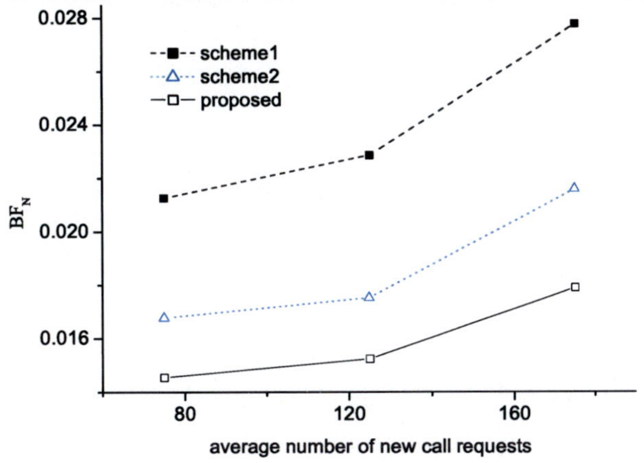

Figure 9. The blocking fraction for $PN(z)$: BF of new call (in dense, medium, and sparse area).

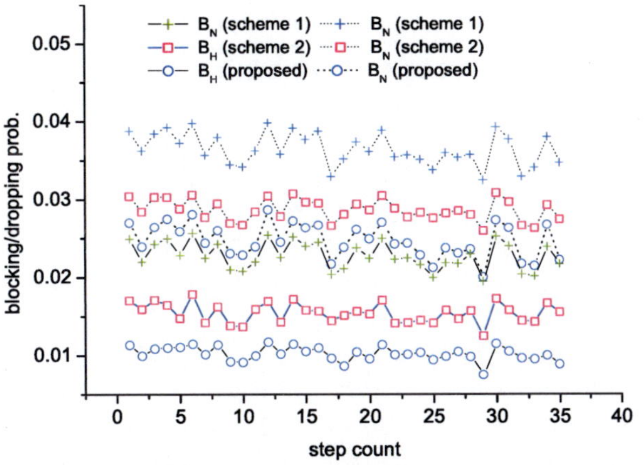

Figure 10. The trend of blocking fractions of handover and new calls for $PH(z)$ at $\xi_H = 0.012$.

schemes. Inspecting the algorithm in [14], we can find that all neighboring base stations are responsible for reserving bandwidth for every single call which seems to be handed over. As a result, a lot of bandwidth may be reserved at the cost of a lot of bandwidth waste. In practice, it is difficult to estimate the next cell for a call (or active user) residing in a cell in land mobile cellular systems. Thus, scheme 1 may be more practical in land mobile systems although there is a little bit more waste of bandwidth.

Fig. 10 shows the trends of the fraction of blocked handovers and that of new calls for each scheme, where we solve $(PH(z))$ at $\xi_H = 0.012$ for the proposed scheme. The simulation was performed for about 3500s (i.e., 35τ, $\tau = 100$s), where we assume that the traffic distribution is non-uniform. It is observed that the blocking fraction of the proposed scheme is dominant over those of the other schemes on average.

According to the results, the proposed scheme improves both handover blocking performance and new call blocking performance. In previous work, the improvement in handover blocking performance caused degradation in new call blocking performance in prioritized handover admission. This phenomenon has been explained by trade-off. We stress that the amount of bandwidth reserved for handover can be dynamically updated by a "smart" control scheme so that the negative effect of the reserved bandwidth on new call blocking performance can be minimized. Such a smart function is even more important in non-stationary traffic conditions. If a control scheme estimates varying traffic conditions with excellent accuracy and reflects the information in admission control, it is called "smart" (cf. smart antenna).

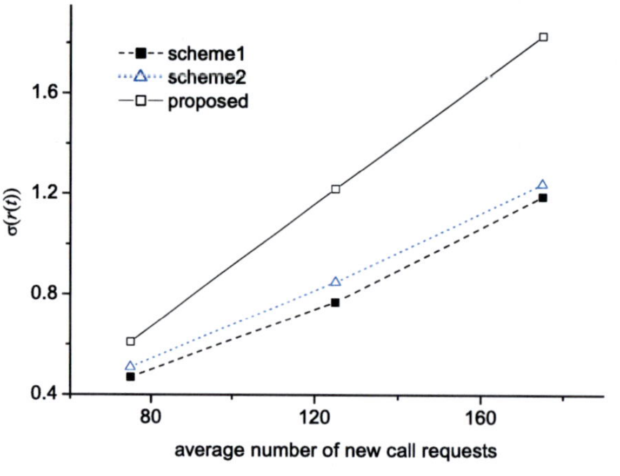

Figure 11. The standard deviation of $r(t)$ during simulation, $\sigma(r(t))$.

To support the results in Figs. 8 and 9, we show the trends of admission thresholds $r(t)$ of the respective schemes. The standard deviation of $r(t)$ shows the degree of fluctuation. As Fig. 11 shows, the proposed scheme offers the most highly adaptive feature in determining $r(t)$.

7. Conclusions

This paper has focused on adaptive admission control with dynamically determining the amount of bandwidth to reserve for prioritized handover admission under non-stationary traffic arrival processes. Since the handover traffic has a non-stationary characteristic, the classical blocking probability formulation under the assumption of stationary traffic causes invalid results in emerging wireless networks with a lot of cells small in size. We have suggested a novel concept for blocking performance measurement and provided a new mathematical formulation. With this, we can dynamically determine an admission threshold for accommodating non-stationary handover calls. Simulation results show that the proposed scheme is highly adaptive to fluctuating traffic conditions, which results in improvement of both handover blocking performance and new call blocking performance.

References

[1] A. N. Dudin and V. I. Klimenok, "Optimal admission control in a queueing system with heterogeneous traffic," *Operations Research Letters*, vol. 31, no. 2, Mar. 2003.

[2] Z. Lei, and S.B. Slimane, "A multicarrier allocation (MCA) scheme for variable-rate 3Gwireless systems," *IEEE Commun. Mag.*, vol. 38, no. 10, Oct. 2000.

[3] S.-H. Oh and D. W. Tcha, "Prioritized channel assignment in a cellular radio network," *IEEE Trans. Commun.*, vol. 40, no. 7, Jul. 1992.

[4] K.-D. Lee and S. Kim, "Optimization for adaptive bandwidth reservation in wireless multimedia networks," *Computer Networks*, vol. 38, no. 5, 2002.

[5] I.F. Akyildiz, H. Uzunalioglu, and M. D. Bender, "Handover management in Low Erarth Orbit (LEO) satellite networks," *Mobile Net. and Applic.*, vol. 4, pp.301-310, 1999.

[6] F. Dosiere, T. Zein, G. Maral, J. P. Bortes, "A model for the handover traffic in low earth orbiting (LEO) satellite networks for personal communications," *IEEE Globecom*, vol. 1, pp. 574–578, Houston, 1993.

[7] E. Chlebus, "Nonstationary traffic model for mobile satellite communications systems," *Proc. IEEE PIMRC'96*, vol. 2, pp. 442-446 , Oct. 1996.

[8] E. Del Re, R. Fantacci, G. Giambene, "Efficient dynamic channel allocation techniques with handover queuing for mobile satellite networks," *IEEE J. Select. Areas Commun.*, vol. 13, no. 2, Feb. 1995.

[9] E. Del Re, R. Fantacci, G. Giambene, "Handover queuing strategies with dynamic and fixed channel allocation techniques in low Earth orbit mobile satellite systems," *IEEE Trans. Commun.*, vol. 47, no. 1, Jan. 1999.

[10] S. Tekinay and B. Jabbari, "Handover and channel assignment in mobile cellular networks," *IEEE Commun. Mag.*, vol. 29, no. 11, pp. 42– 46, Nov. 1991.

[11] R. W. Wolff, *Stochasti Modeling and the Theory of Queues*, Prentice Hall, 1989.

[12] D. A. Levine, I. F. Akyildiz, and M. Naghshineh, "A resource estimation and call admission algorithm for wireless multimedia networks using the shadow cluster concept," *IEEE/ACM Trans. Networking*, vol. 5, no. 1, pp. 1-12, Feb. 1997.

[13] M. Naghshineh and M. Schwartz, "Distributed call admission control in mobile/wireless networks," *IEEE J. Selec. Areas Commun.*, vol. 14, May 1996.

[14] C. Oliveria, J. B. Kim, and T. Suda, "An adaptive bandwidth reservation scheme for high-speed multimedia wireless networks," *IEEE J. Selec. Areas Commun.*, vol. 16, no. 6, pp. 858-874, Aug. 1998.

[15] E. Cinlar, *Introduction to Stochastic Processes*, Prentice-Hall, New Jersy, 1975.

[16] T. Zhang, E. Berg, J. Chennikara, P. Agrawal, J.-C. Chen, and T. Kodama , "Local predictive resource reservation for handoff in multimedia wireless IP networks," *IEEE J. Selec. Areas Commun.*, vol. 19, no. 10, Oct 2001.

[17] K. K. Leung, W. A. Massey, and W. Whitt, "Traffic models for wireless communication networks," *IEEE J. Selec. Areas Commun.*, vol. 12, no. 8, Oct 1994.

[18] D. P. Bertsekas, *Nonlinear Programming*, ISBN 1-886529-14-0, Athena Scientific, 1995.

[19] P. Ramanathan, K.M. Sivalingam, P. Agrawal, and S. Kishore, "Dynamic resource allocation schemes during handoff for mobile multimedia wireless networks," *IEEE J. Selec. Areas Commun.*, vol. 17, no. 7, pp. 1270-1283, Jul. 1999.

[20] F. S. Hillier, and G. J. Lieberman, *Introduction to Operations Research*, 5ed, McGraw-Hill, 1990.

In: Adaptation Techniques in Wireless Multimedia Networks ISBN 1-59454-883-8
Editors: Wei Li and Yang Xiao, pp. 285-306

Chapter 13

MODELING THE EFFECTS OF MOBILITY ON QOS IN WIRELESS MULTIMEDIA NETWORKS

***K. Mitchell* and K. Sohraby*†**
Dept. of Computer Science and Electrical Engineering,
University of Missouri – Kansas City,
5100 Rockhill, Kansas City, MO 64110, USA

Abstract

A multi-cell analytic model for cellular networks is presented for multimedia applications having diverse bandwidth requirements. We introduce control strategies from wireline networks and examine them in a mobile environment where the network supports multiple classes of subscribers having different bandwidth requirements. The expressions for call blocking using these control strategies have product form solutions. We summarize our previous work, which is a generalization of the Erlang fixed point approximation using multi-rate, multi-class, Erlang loss formulas for each class of traffic. We produce expressions for originating calls lost, handoff calls lost, forced-termination, and mean channel occupancy for each class of traffic in each cell for different control strategies. The multi-cell analytic model allows us to investigate the effects of asymmetric loads and mobility patterns in the network where non-random mobility is modeled by assigning a "mobility" class to each traffic stream. Due to the Erlang insensitivity property, the performance measures of interest rely only on the mean of the service time distribution, allowing general dwell times and general call holding times to be incorporated into the model. We present an iterative scheme for determining the residual time of the call holding time distribution that allows us to investigate the effects of variance in both the call holding time distribution and the dwell time distribution on network performance. For traffic exhibiting non-stationary behavior, we propose an adaptive control strategy that adjusts the amount of bandwidth assigned to each traffic class. The analytic results are supported by simulation.

Keywords: cellular networks, multi-class, mobility, bandwidth sharing

---

*E-mail address: mitchellke@umkc.edu
†E-mail address: sohrabyk@umkc.edu

1. Introduction

Future generations of mobile cellular networks are expected to support a variety of multimedia applications having diverse bandwidth requirements. These networks will serve multiple classes of subscribers with each class having distinctively different quality of service (QoS) requirements. Unlike networks with stationary subscribers, multiple requests for resources may be made during the duration of a single call as subscribers move from cell to cell. Forced termination of calls already in progress occurs if resources are not available when a subscriber experiences a handoff. Forced termination is considered to be more serious than blocking of new call attempts, so it is essential to avoid or minimize forced termination provided that it does not negatively impact other classes of traffic.

In order to minimize blocking, an increase in capacity can be achieved by increasing the number of channels in a cell, reducing the cell size, or introducing efficient power control algorithms. Networks supporting multimedia traffic are expected to have small cell sizes, resulting in higher handoff rates between cells than those of traditional cellular networks. It is therefore important to ascertain the impact of mobility on the performance of these networks.

There have been many proposals for minimizing forced termination in cellular networks serving a single class of traffic [1, 2, 5, 7, 11, 13]. A non-preemptive cut-off priority model with guard channels for which the original calls are buffered was presented by Daigle and Jain [4] and by Guerin [6]. A dual buffer system in which both new calls and handoff calls were buffered was proposed by McMillan [13] and a model using hysteresis control was introduced by Choi and Sohraby [2]. These models assumed traffic where all originating calls required a single channel.

Bandwidth allocation strategies fall into one of two categories: fixed channel allocation (FCA), where each cell has a fixed number of channels, and dynamic channel allocation (DCA) where channels are dynamically assigned to cells based on the offered load to each cell.

The cut-off priority strategies introduced in the papers cited above have been proven effective for FCA and have also been employed in DCA for load balancing between cells. The effectiveness of these policies for multi-class traffic having different bandwidth requirements in a multi-cell environment was investigated in our earlier work [14] and is summarized and expanded in this work.

Models which provide a multiple cell analysis with an arbitrary mobility pattern must be developed. Unfortunately, no product form solution for a multi-cell analysis exists, and the resulting state space explosion prohibits an exact analysis [9, 19]. Clearly numerical approximation techniques must be employed. The Erlang fixed-point approximation (FPA) used by Kelly [9] for blocking provides a good approximation for large networks if the mobility rate is not extremely high. Here, we apply the FPA to a multi-class model.

In this work we present a summary of our multi-cell, multi-class, analytic model in order to investigate the effects of mobility on bandwidth allocation strategies based on cut-off priority schemes. We also introduce a cut-off priority strategy in which only a portion of traffic is selectively blocked above a certain threshold and we investigate the effects of mobility on the *fairness* that different bandwidth allocation strategies have on different classes of traffic.

The contribution of this work is the generalization of the Erlang fixed point approximation to a multi-class model so that a multi-cell mobile network can be studied. We study cut-off priority bandwidth management strategies from wireline networks which result in product form solutions [3, 10], and apply them to cellular networks [14]. The combination of strategies which result in product form solutions and the Erlang fixed point approximation allows the use of the multi-rate Erlang B formulas to derive performance parameters of interest. The solution to these formulas are well known, stable, efficient, and give accurate results. Additionally, we develop a model where the mobility of subscribers in each cell is broken down into streams [15]. Each stream is modeled as a separate class of traffic that is characterized by its own dwell time distribution. We incorporate non-exponential call durations by introducing an iterative FPA scheme to approximate the vector for the residual call duration. Given a dwell time distribution $F_w(t)$ and a call holding time distribution $F_s(t)$, only the mean of min $(F_w(t), F_s(t))$ is needed to derive the performance parameters of interest, allowing us to use the fixed point approximation methodology for Erlang loss systems.

The rest of this work is organized as follows: In section 2 of this work the system description is given. In section 3, we present solutions for new call blocking probabilities, handoff blocking probabilities, forced termination probabilities, and mean channel occupancy. In section 4 we produce numerical results. In section 5 we introduce an enhancements to the mobility model which include breaking mobility into separate streams "classes" and the inclusion of non-exponential dwell and call service times. The conclusion can be found in section 6.

2. System Description

Consider a network of M cells, where each cell numbered from $m = 0, 1, \ldots, M$ has C_m channels. Traffic arriving at each cell is partitioned into r separate classes based on bandwidth requirements. Each class i of subscriber requires bandwidth c_i where c is an integral number of channels (i.e., time-slots for TDMA systems, frequency assignments for FDMA, and code sequences for CDMA systems). Additionally, for each class of traffic, handoffs have priority over new call arrivals. Therefore, each class is separated into two types of calls, namely, new calls (designated as A) and handoff calls (designated as H). New calls and handoff calls are themselves treated as separate classes, resulting in a total of $2r$ classes for the system.

It is assumed that the arrival process for each class i into cell m is Poisson with rate λ_{i_m}. Since it is conceivable that different classes of applications may have different holding times, each class of call has an exponential call holding time distribution [1, 5, 7] with mean of $1/\mu_i$. Mobility of subscribers may cause handoff before a call is terminated. The rate of handoff from cell j to cell k for class i traffic is denoted as $\alpha_{i_{jk}}$. Hence, the time a subscriber of class i spends at cell j is exponentially distributed with mean $1/\sigma_{i_j}$ where

$$\sigma_{i_j} = \mu_i + \sum_{k \neq j} \alpha_{i_{jk}}. \tag{1}$$

Bandwidth allocation policies using cut-off priority schemes can be implemented by assigning thresholds to each class of traffic. A Markov state diagram for a small two class

1(4,0) 2(3,0) 3(3,1) 4(3,2) 5(2,0) 6(2,1) 7(2,2) 8(2,3) 9(2,4) 10(1,0) 11(1,1) 12(1,2) 13(1,3) 14(1,4) 15(1,5) 16(1,6) 17(0,0) 18(0,1) 19(0,2) 20(0,3) 21(0,4) 22(0,5) 23(0,6) 24(0,7) 25(0,8) D1Th_1 D2Th_2

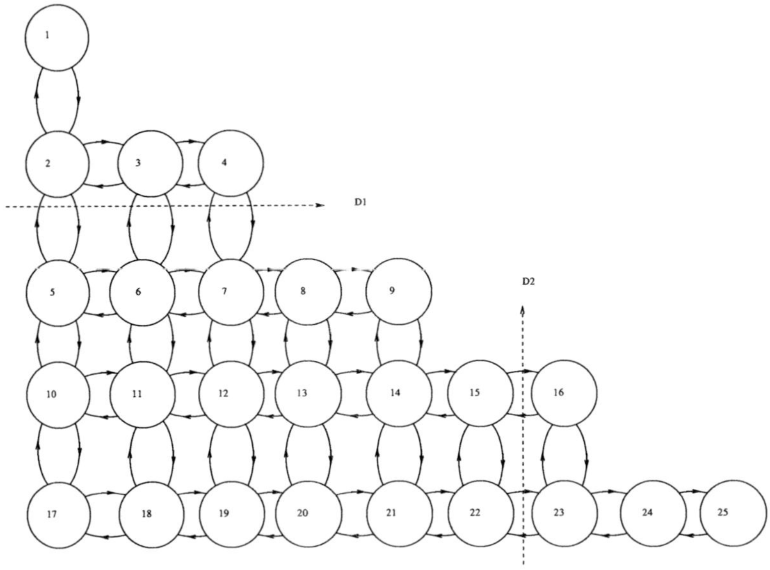

Figure 1. Markov state diagram of a two class cell with eight channels

cell with eight channels is shown in Fig. 1 where $c_1 = 2$ and $c_2 = 1$. We denote the number of channels that class i occupies in cell j before cut-off policies are implemented as Th_{i_j}. If the number of subscribers of type i in cell j, designated as n_j, exceeds Th_{i_j}, then successive arrivals of that class are accepted with probability w_{i_j}. Multiple thresholds with multiple weights can be defined for each class in each cell resulting in complex bandwidth management policies. Lea and Alyatama [10] generalized on Kelly's analysis [8] and demonstrated that these policies resulted in product form solutions which we describe in the following section.

3. System Analysis

In this section we start with describing expressions for steady state probabilities, blocking probabilities, and forced-termination probabilities for a single cell. We then generalize to a single cell with control policies and finally to a multi-cell network.

3.1. Multi-class Single Cell with No Control Policies

For clarity reasons, we drop all subscripts referring to cell m, as we refer to single cell in this section. Let the steady state vector for the number of subscribers n_i of each type i in a cell be $\boldsymbol{\pi}(\boldsymbol{n})$, where $\boldsymbol{n} = (n_1, n_2, \ldots, n_{2r})$. The state space for each cell Ω consists of all allowable states such that $\sum_{i=0}^{2r} n_i c_i \leq C$. The expression for the offered load for each class i to the cell is denoted as

$$\rho_i = \frac{\lambda_i}{\sigma_i}, \tag{2}$$

where σ_i is σ_{i_j} from equation (1) with the subscript j dropped for clarity.

The steady-state probability mass function has the simple product form

$$\pi(\boldsymbol{n}) = \frac{1}{G}\prod_{i=1}^{2r}\frac{\rho_i^{n_i}}{n_i!}, \tag{3}$$

where G is the normalization constant

$$G = \sum_{\boldsymbol{n}\in\Omega}\prod_{i=1}^{2r}\frac{\rho_i^{n_i}}{n_i!}. \tag{4}$$

Choudhury, Leung, and Whitt [3], and Nilsson [16], both present stable and efficient algorithms for calculating the blocking probabilities for large loss systems.

Since arrivals are assumed to be Poisson, the probability of blocking for class i, denoted as P_{B_i}, is the probability of being in any state vector $\boldsymbol{n}$ in the state space Ω such that not enough channels are available to support the arrival of another subscriber of type i.

Define θ_i as the set of population vectors $\boldsymbol{n}$ such that

$$\theta_i = \{\boldsymbol{n}|(n_i+1)c_i + \sum_{l\neq i} n_l c_l > C \text{ and } \boldsymbol{n}\in\Omega\}. \tag{5}$$

The expression for the blocking probability P_{B_i} is

$$P_{B_i} = \frac{1}{G}\sum_{\boldsymbol{n}\in\theta_i}\prod_{i=1}^{2r}\frac{\rho_i^{n_i}}{n_i!}. \tag{6}$$

The expression for the mean channel occupancy $\bar{C}$ is

$$C = \frac{1}{G}\sum_{\boldsymbol{n}\in\Omega}\prod_{i=1}^{2r}\frac{\rho_i^{n_i}}{n_i!}n_i c_i. \tag{7}$$

3.2. Multi-class Single Cell with Control Policies

Thresholds are defined for each class Th_i in each cell of the network. For states above each threshold defined in section 2, traffic from each class is blocked with probability $1 - w_i$, where $0 \leq w_i \leq 1$. We define ρ_i^+ as

$$\begin{aligned} \rho_i^+ &= w_i\rho_i \quad \text{if} \quad n_i c_i > Th_i, \\ \rho_i^+ &= \rho_i \quad \text{otherwise.} \end{aligned} \tag{8}$$

Substituting ρ_i^+ from (8) into (3) gives the steady-state probability mass function

$$\pi(\boldsymbol{n}) = \frac{1}{G}\prod_{i=1}^{2r}\frac{\rho_i^{+n_i}}{n_i!}, \tag{9}$$

where the normalization constant is

$$G = \sum_{\boldsymbol{n}\in\Omega} \prod_{i=1}^{2r} \frac{\rho_i^{+n_i}}{n_i!} \tag{10}$$

In order to derive blocking probabilities for class i, we define θ_i^+ as the set of population vectors $\boldsymbol{n}$ such that

$$\theta_i^+ = \{\boldsymbol{n}|Th_i \leq n_i c_i < C,\ \boldsymbol{n} \in \Omega \text{ and } \boldsymbol{n} \notin \theta_i\}. \tag{11}$$

The expression for the blocking probability P_{B_i} is now

$$P_{B_i} = \frac{1}{G} \sum_{\boldsymbol{n}\in\theta_i} \prod_{i=1}^{2r} \frac{\rho_i^{+n_i}}{n_i!} + \frac{1}{G} \sum_{\boldsymbol{n}\in\theta_i^+} \prod_{i=1}^{2r} \frac{\rho_i^{+n_i}}{n_i!}(1 - w_i) \tag{12}$$

and the expression for the mean channel occupancy $\bar{C}$ is

$$\bar{C} = \frac{1}{G} \sum_{\boldsymbol{n}\in\Omega} \prod_{i=1}^{2r} \frac{\rho_i^{+n_i}}{n_i!} n_i c_i. \tag{13}$$

3.3. Generalization to a Multi-cell Network

In this section we use the Erlang fixed point approximation to effectively model a multi-cell network and we generalize the approximation to include multiple classes. The FPA is used to estimate the handoff traffic load for each class of subscriber to each cell in the network based on the subscriber's mobility pattern. We define mobility probability matrices $\boldsymbol{\Phi}_i$ for each class i, where each element

$$[\boldsymbol{\Phi}_i]_{jk} = \frac{\alpha_{i_{jk}}}{\sigma_{i_j}} \tag{14}$$

describes the probability of traffic of class i going from cell j to cell k. Next define the matrix $\boldsymbol{M}_i(n)$ for class i at iteration n where each element on the diagonal is

$$[\boldsymbol{M}_i(n)]_{mm} = 1 - P_{B_{im}}(n). \tag{15}$$

The expression $P_{B_{im}}(n)$ is the blocking calculated for class i at cell $m = 1, 2, \ldots, M$ at iteration n using expression (6) or (12).

For each class i we define two arrival rate vectors. One for the originating calls denoted as $\boldsymbol{\Lambda}_{Ai}$ where

$$\boldsymbol{\Lambda}_{Ai} = [\lambda_{i_1}, \lambda_{i_2}, \ldots, \lambda_{i_M}], \tag{16}$$

and one for the handoff calls at the nth iteration, denoted as $\boldsymbol{\Lambda}_{Hi}(n)$ where

$$\boldsymbol{\Lambda}_{Hi}(0) = [0, 0, \ldots, 0]. \tag{17}$$

From the above equations it can be seen that we assume the starting point for the Erlang fixed point approximation to be original calls only and that the initial rate of handoff calls into each cell is zero.

For each iteration of the FPA, the rate matrix $\boldsymbol{\Lambda}_{Hi}(n)$ is updated as follows:

$$\boldsymbol{\Lambda}_{Hi}(n) = \boldsymbol{\Lambda}_{Ai}\boldsymbol{M}_i(n-1)\boldsymbol{\Phi}_i + \boldsymbol{\Lambda}_{Hi}(n-1)\boldsymbol{M}_i(n-1)\boldsymbol{\Phi}_i. \quad (18)$$

At each iteration n, the matrix $\boldsymbol{M}_i(n)$ is recalculated based on $\boldsymbol{\Lambda}_{Ai}$ and $\boldsymbol{\Lambda}_{Hi}(n)$ obtained by the FPA. The FPA terminates when the differences between successive normalization constants at each iteration fall below a given threshold ϵ

$$|G(n)/G(n-1) - 1.0| < \epsilon. \quad (19)$$

3.4. Forced Termination Probabilities

Forced termination probabilities are calculated separately for each class i so the subscript i is dropped for purposes of clarity. Definitions should be apparent from the context. Once a new call has been successfully initiated in cell j there are three possibilities for a transition out of cell j. The first possibility is that the call completes successfully within cell j with probability

$$\text{Pr[success]} = \frac{\mu}{\sigma_j}. \quad (20)$$

The call may also successfully complete a handoff to cell k with probability

$$\text{Pr}[jk] = \frac{\alpha_{jk}\left(1 - P_{B_k}\right)}{\sigma_j}, \quad (21)$$

or the call can be dropped when a handoff occurs between cell j and k with probability

$$\text{Pr[dropped]} = \frac{\alpha_{jk}P_{B_k}}{\sigma_j}. \quad (22)$$

Define the single step probability transition matrix $\boldsymbol{P}$ with dimension $M+2$ with two absorbing states. The absorbing state representing the probability of a call being dropped is denoted as d and the state representing the probability of a successful completion of a call is denoted as s. Transition probabilities are defined by

$$[\boldsymbol{P}]_{jk} = \frac{\alpha_{jk}\left(1 - P_{B_k}\right)}{\sigma_j}, \quad (23)$$

$$[\boldsymbol{P}]_{jd} = \frac{\sum_{k\neq j}\alpha_{jk}P_{B_k}}{\sigma_j}, \quad (24)$$

$$[\boldsymbol{P}]_{js} = \frac{\mu_j}{\sigma_j}, \quad (25)$$

$$[\boldsymbol{P}]_{dd} = [\boldsymbol{P}]_{ss} = 1. \quad (26)$$

The initial vector $\boldsymbol{\pi}$ is the distribution of successful originating call arrivals such that the jth element of $\boldsymbol{\pi}$ is

$$[\boldsymbol{\pi}]_j = \frac{\lambda_j\left(1 - P_{B_j}\right)}{\sum_j \lambda_j\left(1 - P_{B_j}\right)}, \quad (27)$$

$$[\boldsymbol{\pi}]_d = [\boldsymbol{\pi}]_s = 0. \quad (28)$$

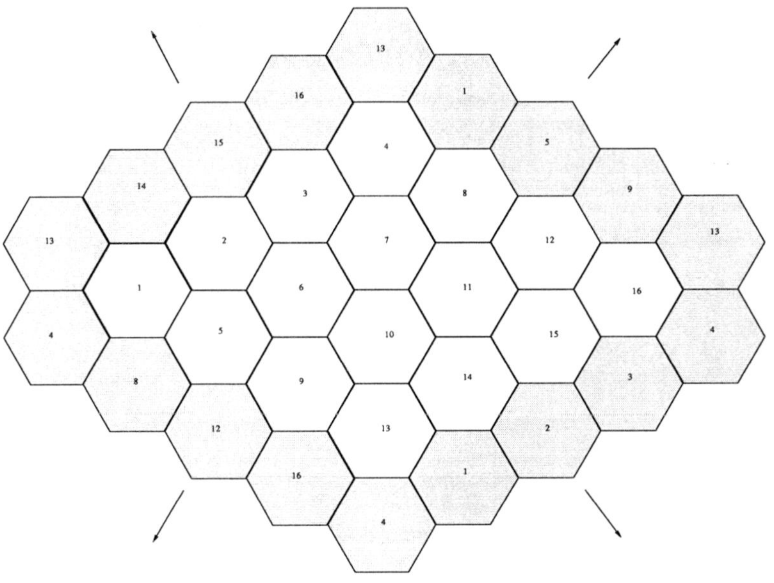

Figure 2. Sixteen cell model with wraparound

The expression for the call dropout probability in the network becomes

$$\Pr[\text{dropout}] = \lim_{i\to\infty} \left[\pi \boldsymbol{P}^i\right]_d . \tag{29}$$

4. Numerical Results

We consider a network of 16 hexagon shaped cells which are wrapped around in order to model a large network as depicted in Fig. 2. For hexagon shaped cells, we use the following formula taken from [7] for the arrival rate λ_{i_m} for each class i in cell m.

$$\lambda_{i_m} = 3\sqrt{3/2}\lambda'_{i_m} R_m^2 \tag{30}$$

where R_m is the cell radius for cell m and λ'_{i_m} is the originating call arrival rate per unit area for class i.

Each cell consists of 120 channels. We assume each channel to have bandwidth in the range of present day digitized voice in PCS systems (about 16 Kbps). Class 1 traffic consists of low bit rate voice channels and requires one channel ($c_1 = 1$). Class 2 traffic consists of medium bandwidth Internet applications and requires four channels ($c_2 = 4$), and class 3 traffic is composed of high bandwidth multi-media applications and requires ten channels ($c_3 = 10$). These bandwidths were chosen for illustrative purposes as the model can accept any integer value depending on the application.

We first show results for a symmetrically loaded network. We later show an example where the load is different for some cells in the network and study the effect mobility of one class has on loss probabilities for the other classes. We allow the mobility rate to vary from 0 to 16 times μ_i for each class of traffic and study the following strategies:

• Strategy 1: No bandwidth allocation control. Bandwidth is allocated on a first come first serve basis for each type of traffic requesting it.
• Strategy 2: Fixed partition. Each type of traffic has a fixed number of channels it can access. The network in this case is decomposable as the performance measures for each class of traffic is independent of other classes of traffic.
• Strategy 3: Variable partitions. Partitions can be set up for cut-off for certain classes only. Usually employed for priority of handoffs over new call arrivals.
• Strategy 4: Selective blocking beyond a certain threshold. Traffic of class i is admitted with a certain probability w_i above a threshold Th_i.

4.1. Comparison to Simulation Results

The fixed-point approximation proposed by Kelly [9] produces accurate results for relatively low mobility. Since the state space explosion makes an exact analysis impossible, we compare loss probabilities with simulation results for a network with relatively high load in order to define the bounds for which this method can give accurate results. Since it is expected that the high bandwidth traffic is at a disadvantage, the load to the system is scaled proportionally for each class. Each class in each cell has load $\rho_1 = 30.0$, $\rho_2 = 7.5$, and $\rho_3 = 3.0$. Fig. 3 shows a comparison of blocking for the three classes compared with simulation for mobility ranging from 0 to 16 times μ_i. The figure shows that the analytic model is very accurate.

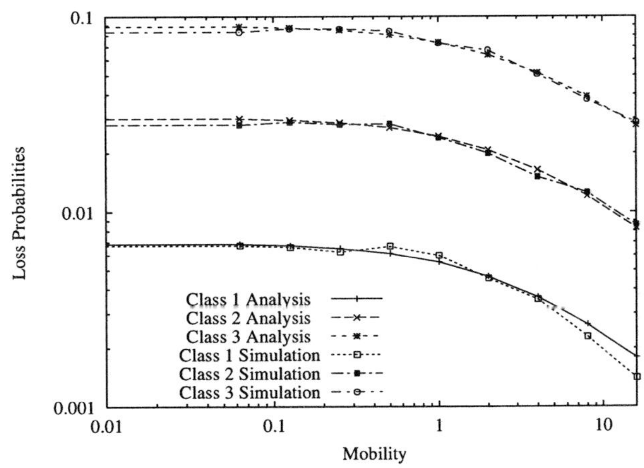

Figure 3. Comparison of analytic with simulation results

4.2. Strategy 1: No Control Policies

Fig. 4 shows originating call loss probabilities and forced termination probabilities for the three classes of traffic for varying mobility. Load to the system was moderate with $\rho_1 = 20.0$, $\rho_2 = 5.0$, and $\rho_3 = 2.0$. Although the offered load was balanced, class 3 traffic was still at a distinct disadvantage. Due to the PASTA (Poisson arrivals see time averages) property, originating call loss probabilities and handoff loss probabilities are the same for

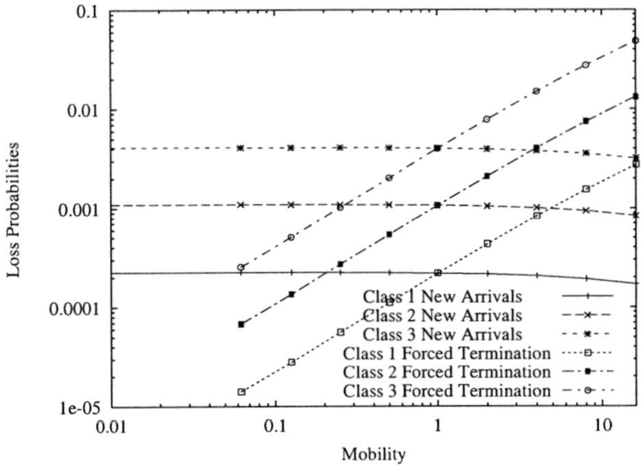

Figure 4. Call blocking with no control policies

each class [17], but since forced termination is a function of mobility, increased mobility rates result in an increase in forced termination.

Fig. 5 shows originating call loss probabilities and forced termination probabilities for the three classes of traffic where the mobility for classes 1 and 2 are fixed at sixteen times μ_i. Load to the system was moderate with $\rho_1 = 20.0$, $\rho_2 = 5.0$, and $\rho_3 = 2.0$. Mobility for class 3 traffic is varied from 0 to 16 times μ_3. The mobility of class 3 traffic did not affect the other traffic appreciably. However, this observation is not true if the network is not symmetrically loaded, or if each class has a different mobility pattern in the network.

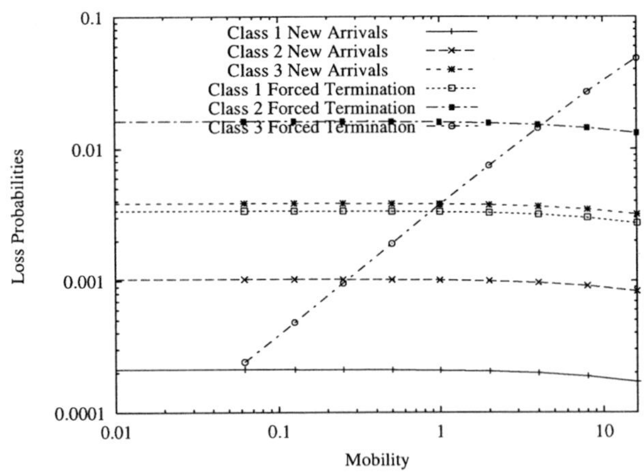

Figure 5. Mobility effects of class 3 traffic on other classes

4.3. Strategy 2: Fixed Partition of Bandwidth

Fig. 6 shows originating call loss probabilities and forced termination probabilities for the three classes of traffic for varying mobility. Load to the system was light but not balanced with $\rho_1 = 1.0$, $\rho_2 = 1.0$, and $\rho_3 = 1.0$. The number of channels in the cell was strictly

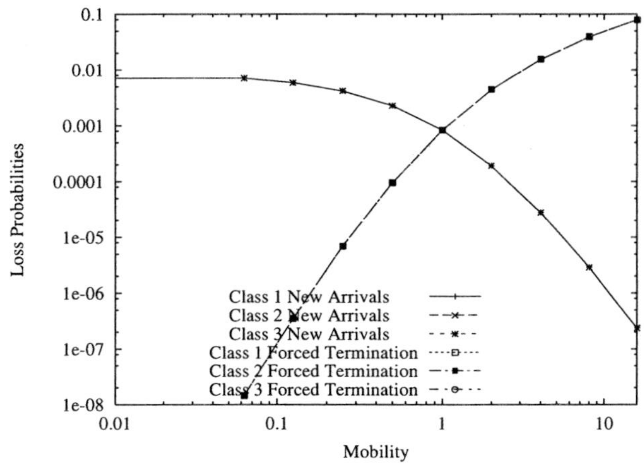

Figure 6. Call blocking for completely partitioned bandwidth

partitioned proportionally based on the number of channels each class of traffic requires. Threshold values were set at $Th_1 = 8$, $Th_2 = 32$, and $Th_3 = 80$. Since fairness issues between offered calls and handoff calls have been well studied and are not a part of our studies here, one half of each partition is reserved for originating calls, the other half for handoffs. The ratio of the two halves can be changed to implement cut-off priority policies for handoff traffic based on mobility, but we elect to leave this ratio fixed. Fig. 6 shows that strict fairness between classes can easily be achieved but at the expense of higher overall blocking probabilities and a drop in mean channel occupancy.

4.4. Strategy 3: Partial Sharing

Fig. 7 shows originating call loss probabilities and forced termination probabilities for the three classes of traffic for varying mobility. Load to the system was moderate but unfair based on resource requirements for each class with $\rho_1 = 3.0$, $\rho_2 = 3.0$, and $\rho_3 = 3.0$. The number of channels in the cell was strictly partitioned proportionally based on the number of channels each class of traffic requires. Threshold values are set at $Th_1 = 8, Th_2 = 32$, and $Th_3 = 80$, but the thresholds were set up for originating calls only. This figure illustrates that cut-off priority work best at high loads. Fig. 8 shows the same scenario, but with handoffs also partitioned. Since the sum of the channels in the partitions is 240 channels, sharing occurs at moderate load, and control policies are enacted at high loads.

4.5. Strategy 4: Cut-off with Partial Blocking

Fig. 9 shows originating call loss probabilities and forced termination probabilities for the three classes of traffic for varying parameter w with all traffic at maximum mobility. Load to the system was moderate but unfair based on resource requirements for each class with $\rho_1 = 3.0$, $\rho_2 = 3.0$, and $\rho_3 = 3.0$. The number of channels in the cell was strictly partitioned proportionally based on the number of channels each class of traffic requires. Threshold values are set at $Th_1 = 8$, $Th_2 = 32$, and $Th_3 = 80$, with originating and handoff traffic each getting half that number. This figure illustrates the effects of the parameter w_i on loss

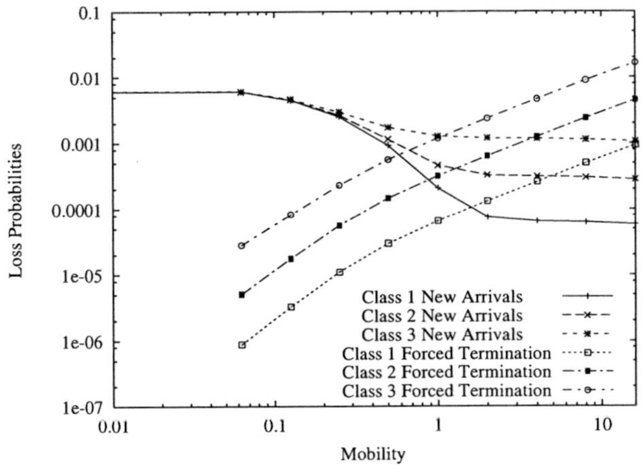

Figure 7. Call blocking for partitioned bandwidth for originating calls only

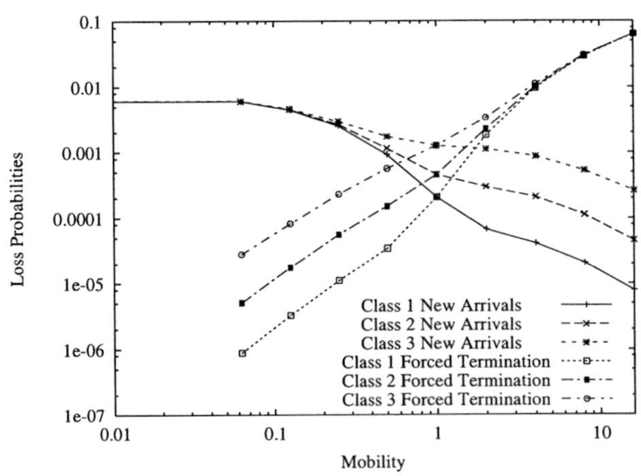

Figure 8. Call blocking for partial partitioning for all calls

probabilities. When $w_i = 0.0$, the bandwidth is completely partitioned. When $w_i = 1.0$ there is complete sharing of the channels.

Weight wParameter w

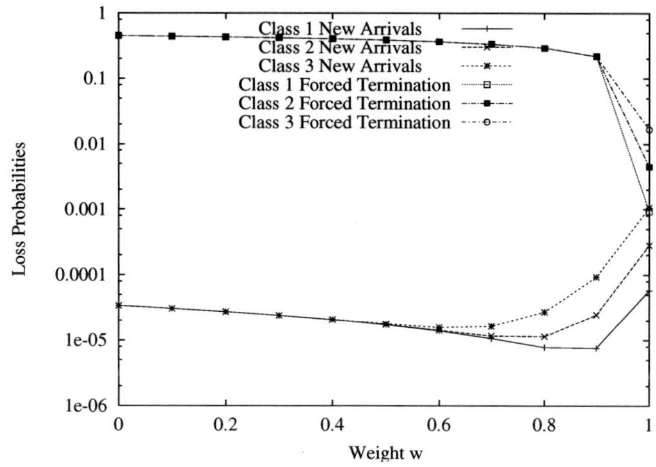

Figure 9. Call blocking for partial blocking and partitioning

4.6. Asymmetric Load

In this section we look at the effects of mobility on performance with a non-uniform load in the network. We assume utilization for each class at all cells to be $\rho_1 = 1.0$, $\rho_2 = 1.0$, and $\rho_3 = 1.0$, with the exception of a cluster of cells (cell 6, 7, and 10) where the offered load of type 1 traffic is very heavy ($\rho_1 = 20.0$). Mobility for class 2 and 3 traffic is set at 1. Fig. 10 shows the blocking probabilities at all cells for when mobility of class 1 is 0 and when mobility of class 1 is 16. The mobility rate of the class 1 traffic impacts the loss probabilities of the other classes of traffic.

4.7. Adaptive Policy for Non-stationary Traffic

The traffic load to the network may frequently exhibit non-stationary behavior such as time of day fluctuations. In [18], Srivastava and Mitchell propose an adaptive scheme for bandwidth sharing between control and data channels in a mobile cellular network that can be easily generalized to multiple traffic classes. Whenever a call of class i is blocked, the partition is adjusted such that the number of channels in partition i increases by 1 with a corresponding decrease by 1 in the partition that has not experienced blocking for the longest amount of time. For non-stationary traffic where traffic load varies proportional to amplitude α, Figs. 11 and 12 show the adaptive scheme is far superior to fixed partition schemes. For a detailed explanation see [18].

Mu1 Class 1 $\alpha_1 = 0$ Mu2 Class 2 $\alpha_1 = 0$ Mu3 Class 3 $\alpha_1 = 0$ Mu4 Class 1 $\alpha_1 = 16$ Mu5 Class 2 $\alpha_1 = 16$ Mu6 Class 3 $\alpha_1 = 16$

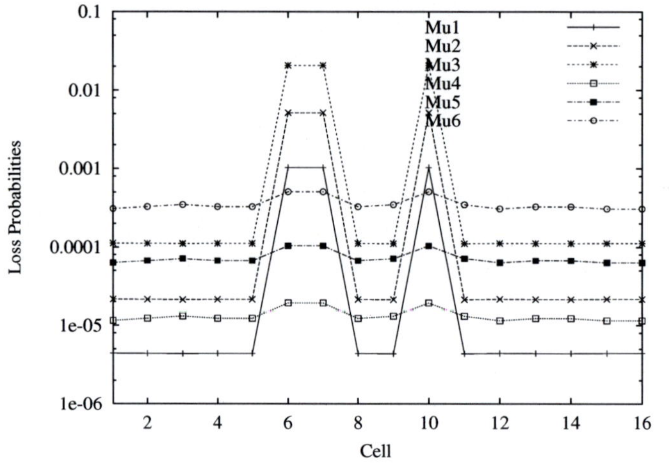

Figure 10. Losses in sixteen cell model for mobility α_1 of class 1 traffic

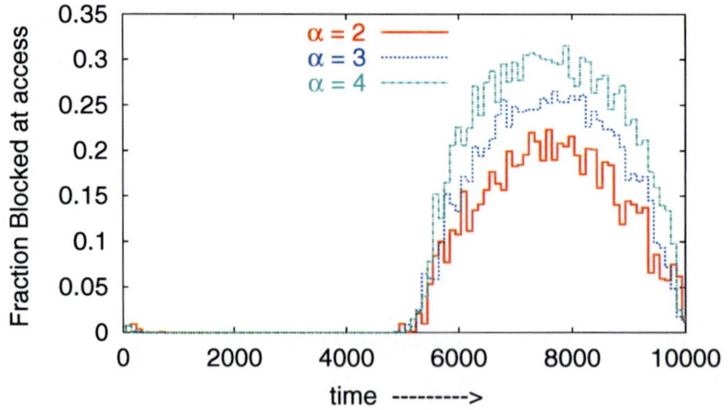

Figure 11. Fraction of blocked attempts for fixed approach (from [18])

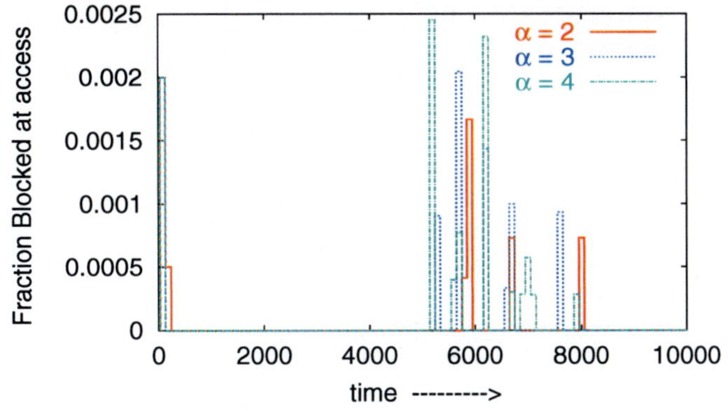

Figure 12. Fraction of blocked attempts for adaptive approach (from [18])

5. Generalized Mobility and Call Service Times

In many instances, subscribers do not travel in a random fashion. Subscribers may move along fixed paths which may be roadways, hallways, commuter rail lines, walking paths, etc., from a starting point to a destination point. At any time along this path, a user may begin or terminate a call, and may pass through a single cell or many cells depending on the mobility rate.

For small urban networks, each cell may have a small number of entry and exit points. For example, in a Manhattan grid, each cell has only four paths connecting with neighboring cells. Traffic entering from each direction will with a certain probability either terminate in the cell or exit at one of the four access points. In a similar manner to what was presented in [15], we can break down spatial dependencies in mobility into 16 mobility classes with a class assigned for each source-destination flow in each cell.

Cell dwell times and call holding times are not necessarily exponentially distributed. The difficulty in modeling non-exponential call holding times is that one must take into account the state of the call holding process at the time of a handoff between adjacent cells. In many cases this quantity is difficult or impossible to calculate, as it represents the internal state of a particular call process after an arbitrary number of handoffs. Here we propose a simple approximation to the residual time vector that is applicable to networks with heterogeneous traffic patterns.

5.1. System Description

For the development of this system we make the following assumptions about the processes involved.

- The call arrival process for each class r is Poisson with mean rate λ_r.
- The handoff process for each class r into a cell from adjacent cells is also assumed to be Poisson.
- The dwell process may be any renewal process that has a rational Laplace Stieltges transform and has a matrix exponential distribution representation $(\boldsymbol{p}, \boldsymbol{B}, \boldsymbol{e})$

- The call holding process may also be any renewal process that has a rational Laplace Stieltges transform and also has a matrix exponential distribution representation $(\boldsymbol{p}, \boldsymbol{B}, \boldsymbol{e})$

5.1.1. Matrix Exponential Distribution

A matrix exponential (ME) distribution [12] is defined as a probability distribution with representation $(\boldsymbol{p}, \boldsymbol{B}, \boldsymbol{e})$, i.e.,

$$F(t) = 1 - \boldsymbol{p}\exp(-\boldsymbol{B}t)\,\boldsymbol{e}',\ t \geq 0, \tag{31}$$

where $\boldsymbol{p}$ is the starting operator for the process, $\boldsymbol{B}$ is the process rate operator, and $\boldsymbol{e}'$ is a summing operator. The order of the representation is indicated by the dimension of the the matrix $\boldsymbol{B}$, and the degree of the distribution $F(t)$ is the minimal order of all its representations. The n-th moment of the matrix exponential distribution is given by

$$E[X^n] = n!\boldsymbol{p}\boldsymbol{V}^n\boldsymbol{e}', \tag{32}$$

where $\boldsymbol{V}$ is the inverse of $\boldsymbol{B}$.

The class of matrix exponential distributions is identical to the class of distributions that possess a rational Laplace-Stieltjes transform. Phase-type distributions form a proper subset of matrix exponential distributions. Also, the class of matrix exponential distributions is dense in the set of all distributions, so *any* distribution can be approximated arbitrarily close with a matrix exponential distribution. A matrix exponential distribution may have an underlying probabilistic interpretation (i.e., a phase-type distribution), but is not necessarily limited to such an interpretation.

A representation is not unique. If $F(t)$ has representation $(\boldsymbol{p}, \boldsymbol{B}, \boldsymbol{e})$, then $(\boldsymbol{p}\boldsymbol{X}, \boldsymbol{X}^{-1}\boldsymbol{B}\boldsymbol{X}, \boldsymbol{X}^{-1}\boldsymbol{e})$ is also a representation for any non-singular matrix $\boldsymbol{X}$. The only limitations on $(\boldsymbol{p}, \boldsymbol{B}, \boldsymbol{e})$ stem from the requirement that $F(t)$ must form a distribution function. Consequently, $(\boldsymbol{p}\boldsymbol{e}')$ is a real number between zero and one and the eigenvalues of $\boldsymbol{B}$ must either be positive reals or must come in complex conjugate pairs with a positive real part. As there are no prescribed structural or domain restrictions on the components $(\boldsymbol{p}, \boldsymbol{B}, \boldsymbol{e})$, one can choose a physically based representation (i.e., phase-type), or an algorithmic representation.

5.1.2. Single Cell Model

Again, consider a network of M cells, where each cell numbered from $m = 0, 1, \ldots, M$ has C_m channels. Traffic arriving at each cell is partitioned into L separate streams numbered from $l = 0, 1, \ldots, L$ based on the entry and exit points of approaching and departing streams in a cell. As in the previous model, each stream of subscribers, depending on bandwidth requirements, may be divided into i substreams, numbered from $i = 1, 2, \ldots, i$, each requiring c_i units of bandwidth. Additionally, handoff traffic may have priority over new call arrivals. Therefore, new calls and handoff calls may be treated as separate classes, creating an additional $k \in \{1, 2\}$ subclasses for each stream of subscribers. We drop the subscript k, so each class has representation (i, l). An applications class i is based on the number of channels required by a call, and a mobility class l is based on the entry and exit point in a cell. Each class has two priorities (handoffs or new calls). There are a total of $2r$ classes where r is now expressed as $r = IL$.

The rate of handoff from cell j to cell k for class (i,l) traffic is denoted as $\alpha_{(i,l)_{jk}}$. Hence, the time a subscriber of class (i,l) spends at cell j is matrix exponentially distributed with mean $1/\sigma_{i_j}$ where each active subscriber will hold cell resources (channels) according to a distribution which is defined as $\min(W,S)$ where W is a random variable with distribution $F_w(t)$ describing the time a subscriber is physically located in a cell (dwell time) and S is a random variable with distribution $F_s(t)$ describing the call holding (service) time.

Recall equation (1) which describes the mean rate a subscriber departs a cell $\sigma_{i_j} = \mu_i + \sum_{k \neq j} \alpha_{i_{jk}}$. For the single cell model with mobility classes and non-exponential service times, the mean time a subscriber remains in a cell is

$$\left(\sigma_{(i,l)}\right)^{-1} = \boldsymbol{p}_s \widehat{\boldsymbol{p}}_{(i,l)w} \left(\widehat{\boldsymbol{B}}_s + \widehat{\boldsymbol{B}}_{(i,l)w}\right)^{-1} \widehat{\boldsymbol{e}}'_{(i,l)w} \boldsymbol{e}'_s, \tag{33}$$

where

$$\widehat{\boldsymbol{p}}_{(i,l)w} = \boldsymbol{I}_s \otimes \boldsymbol{p}_{(i,l)w}, \tag{34}$$

$$\widehat{\boldsymbol{B}}_s = \boldsymbol{B}_s \otimes \boldsymbol{I}_{(i,l)w}, \tag{35}$$

$$\widehat{\boldsymbol{B}}_{(i,l)w} = \boldsymbol{I}_s \otimes \boldsymbol{B}_{(i,l)w}, \tag{36}$$

$$\widehat{\boldsymbol{e}}'_{(i,l)w} = \boldsymbol{I}_s \otimes \boldsymbol{e}'_{(i,l)w}, \tag{37}$$

and $\boldsymbol{I}_s$ and $\boldsymbol{I}_{(i,l)w}$ are identity matrices of the same dimension as $\boldsymbol{B}_s$ and $\boldsymbol{B}_{(i,l)w}$ respectively. Replacing equation (1), with equation (33) for each class (i,l) where $r = il$, equations (2) through (13) can be used to derive the performance metrics for a single cell.

Assuming that all new calls and handoff calls are Poisson, with probability that a dwell time completes before a call holding time, the call will be handed off to the next cell m as determined by the mobility class of the call

$$P[W < S] = \int_0^\infty (1 - F_s(t))\, f_w(t)\, dt. \tag{38}$$

5.1.3. Multi-cell Analysis

The fixed point approximation for non-exponential service times follows from our previous multi-class model. Since the approximation is applied to each class separately, we will drop the subscript (i,l) for purposes of clarity.

When using the fixed point approximation with non-exponential service times, one must take into account the residual time of a call in progress upon handoff. The observation is that each iteration of the fixed-point approximation represents the progression of the call holding process (i.e., all additional load offered to the cell has undergone an additional handoff). Therefore, in addition to the mean handoff rate which is calculated to neighboring cells, the vector for the residual call service time denoted as $\boldsymbol{p}_{rs}$ at each handoff must also be calculated.

In the first iteration, all traffic is considered to be new arrivals to the cell, and therefore the vector for any arriving call is the starting vector for the call process $\boldsymbol{p}_s$. In the second iteration, the additional load presented to the cell is traffic that has received exactly one handoff. Assuming all arrivals are Poisson, the starting vector for this additional traffic at the $k+1$ iteration is

$$\boldsymbol{p}_{rs}(k+1) = \frac{\boldsymbol{p}_{rs}(k)\widehat{\boldsymbol{p}}_w \boldsymbol{H}_w \widehat{\boldsymbol{e}}'_w}{\boldsymbol{p}_{rs}(k)\widehat{\boldsymbol{p}}_w \boldsymbol{H}_w \widehat{\boldsymbol{e}}'_w \boldsymbol{e}'_{rs}}, \tag{39}$$

where $\boldsymbol{H}_w$ is defined as $\left(\widehat{\boldsymbol{B}}_s + \widehat{\boldsymbol{B}}_w\right)^{-1} \widehat{\boldsymbol{B}}_w$.

Since new call arrivals and handoff arrivals are assumed to be Poisson, they will see time averages. The expression $\boldsymbol{p}_{rs}(k)\widehat{\boldsymbol{p}}_w \boldsymbol{H}_w \widehat{\boldsymbol{e}}'_w$ represents the conditional internal state of the call holding process conditioned on the completion of a call handoff at steady state.

At each iteration of the fixed point approximation, the average of all the vectors weighted by the mean arrival rate from each source is used for the starting vector $\boldsymbol{p}_{rs}(k)$ of the call service process

$$\boldsymbol{p}_{rs}(k) := \frac{\sum_{i=1}^{\alpha} \boldsymbol{p}_{rs,i}(k-1)\boldsymbol{\lambda}_{d,i} + \boldsymbol{p}_s \boldsymbol{\lambda}_a}{\sum_{i=1}^{\alpha} \boldsymbol{\lambda}_{d,i} + \boldsymbol{\lambda}_a}. \tag{40}$$

The fixed point approximation for arrivals is calculated in the same manner as described in section 3.

5.1.4. Forced Termination Probabilities

In this subsection we derive general expressions for forced termination for general mobility patterns with multiple classes and non-exponential dwell and call service times. Forced termination probabilities may be calculated separately for each class (i, l), so the subscript (i, l) is dropped for purposes of clarity. The handoff blocking probabilities from cell j to k, denoted as $P_{H_{jk}}$ are used to calculate forced termination probabilities. Once a new call has been successfully initiated in cell j, there are three possibilities for a transition out of cell j. The first possibility is that the call completes successfully within cell j with probability

$$\Pr[\text{success}] = \boldsymbol{p}_{\bar{r}s}\widehat{\boldsymbol{p}}_w \boldsymbol{H}_{\bar{r}s} \widehat{\boldsymbol{e}}'_w \boldsymbol{e}'_{\bar{r}s}. \tag{41}$$

The call may also successfully complete a handoff to cell k with probability

$$\Pr[j \text{ to } k] = \boldsymbol{p}_{\bar{r}s}\widehat{\boldsymbol{p}}_w \boldsymbol{H}_w \widehat{\boldsymbol{e}}'_w \boldsymbol{e}'_{\bar{r}s} \alpha_{jk}\left(1 - P_{H_{jk}}\right), \tag{42}$$

or the call can be dropped when a handoff occurs between cell j and k with probability

$$\Pr[\text{dropped}] = \boldsymbol{p}_{\bar{r}s}\widehat{\boldsymbol{p}}_w \boldsymbol{H}_w \widehat{\boldsymbol{e}}'_w \boldsymbol{e}'_{\bar{r}s} \alpha_{jk} P_{H_{jk}}. \tag{43}$$

We define the single step probability transition matrix $\boldsymbol{P}$ with dimension $m+2$ where there are two absorbing states. The absorbing state representing the probability of a call being dropped is denoted as d, and the state representing the probability of a successful completion of a call is denoted as s. Transition probabilities are defined by

$$[\boldsymbol{P}]_{jk} = \boldsymbol{p}_{\bar{r}s}\widehat{\boldsymbol{p}}_w \boldsymbol{H}_w \widehat{\boldsymbol{e}}'_w \boldsymbol{e}'_{\bar{r}s} \alpha_{jk}\left(1 - P_{H_{jk}}\right), \tag{44}$$

$$[\boldsymbol{P}]_{jd} = \sum_{k \neq j} \boldsymbol{p}_{\bar{r}s}\widehat{\boldsymbol{p}}_w \boldsymbol{H}_w \widehat{\boldsymbol{e}}'_w \boldsymbol{e}'_{\bar{r}s} \alpha_{jk} P_{H_{jk}}, \tag{45}$$

$$[\boldsymbol{P}]_{js} = \boldsymbol{p}_{\bar{r}s}\widehat{\boldsymbol{p}}_w \boldsymbol{H}_{\bar{r}s} \widehat{\boldsymbol{e}}'_w \boldsymbol{e}'_{\bar{r}s}, \tag{46}$$

$$[\boldsymbol{P}]_{dd} = 1, \tag{47}$$

$$[\boldsymbol{P}]_{ss} = 1. \tag{48}$$

The vector $\boldsymbol{\pi}$ is the steady state distribution of successful call arrivals (new calls which have not been blocked). The jth element of $\boldsymbol{\pi}$ is

$$[\boldsymbol{\pi}]_j = \frac{\lambda_j \left(1 - P_{B_j}\right)}{\sum_i \lambda_i \left(1 - P_{B_i}\right)}, \tag{49}$$

$$[\boldsymbol{\pi}]_d = [\boldsymbol{\pi}]_s = 0. \tag{50}$$

The expression for the call forced termination probability in the network is the steady state solution to the discrete Markov chain described by the single step probability transition matrix $\boldsymbol{P}$

$$\Pr[\text{dropout}] = \lim_{i \to \infty} \left[\boldsymbol{\pi} \boldsymbol{P}^i\right]_d. \tag{51}$$

5.2. Numerical Results

In this section we study the effects of mobility on performance with a non-uniform load in a 16 cell Manhattan grid network as depicted in Fig. 13. We assume the distribution of time that time that a subscriber remains in a cell to be distributed according to an Erlang 2 distribution with a squared coefficient of variation $c^2 = 1/2$ with varying mean according to the speed of the mobile in each cell. Call service times are modeled using either an exponential distribution or a hyper-exponential distribution with $c^2 = 10$.

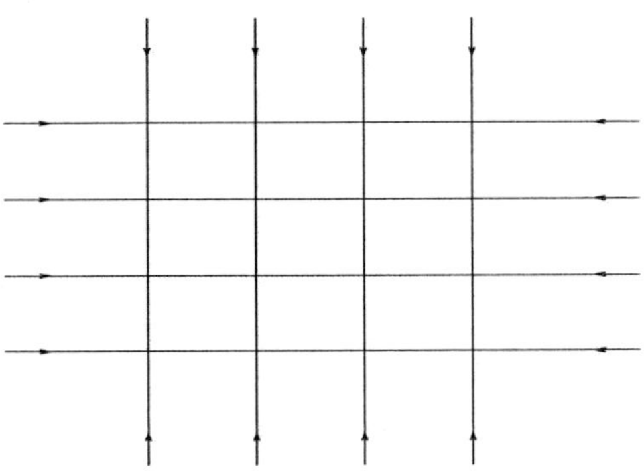

Figure 13. Sixteen cell Manhattan grid model

Fig. 14 shows blocking at the different cells where the rate of new call arrivals steadily increases according to cell number. The different dwell times in each cell is reflected in the differences in blocking. Note that blocking is considerably less for the the bursty session lengths than for exponentially distributed service lengths. This illustrates the importance of incorporating non-exponential service times in the analysis.

Service c2 = 10Service $c^2 = 10$

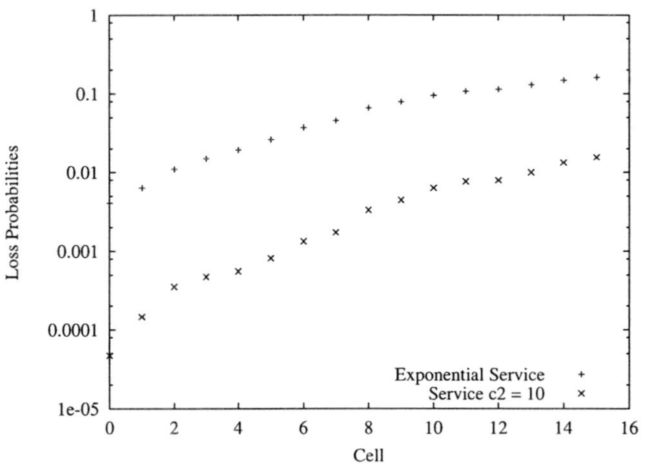

Figure 14. Call blocking in Manhattan grid model

6. Conclusions

In this work, we have presented a multi-cell, multi-class analytic model for cellular wireless networks. The model can be a very powerful analytic tool in cellular network design as it can be used to derive many performance parameters of interest, yet uses the well known multi-rate, multi-class Erlang loss formulas. In general, complete sharing of channels results in lower overall loss probabilities and higher channel utilization than using forms of bandwidth partitioning. However, this strategy is unfair to subscribers requesting high bandwidth. Fairness can easily be achieved by using cut-off priority strategies at the expense of higher overall blocking in a cell. Additionally, for non-stationary traffic, an adaptive policy has been proposed. Our studies have also shown that the mobility rate of a class of subscribers can adversely affect the QoS of other classes of subscribers in a non uniform environment. A combination of multiple cut-off thresholds and partial blocking can generate complex control policies which can be easily analyzed. Additionally, dependencies in the mobility patterns of subscribers can be modeled by the use of mobility classes that describe a stream of subscribers from an ingress to an egress point in each cell. Utilizing the Erlang insensitivity property, non-exponential dwell and call service times are also incorporated into the model.

References

[1] C.-J. Chang, T.-T. Su, and Y.-Y. Chiang, Analysis of a cutoff priority cellular radio system with finite queueing and reneging/dropping, *IEEE/ACM Transactions on Networking,* **2** (1994), pp. 166–175.

[2] S.-H. Choi and K. Sohraby, Analysis of a mobile cellular system with hand-off priority and hysteresis control, in *Proceedings of the Conference on Computer Commu-*

nications (IEEE Infocom), Tel Aviv, Israel, 2000.

[3] G. L. Choudhury, K. K. Leung, and W. Whitt, An algorithm to compute blocking probabilities in multi-rate multi-class multi-resource loss models, *Advances in Applied Probability,* **27** (1995), pp. 1104–1143.

[4] J. N. Daigle and N. Jain, A queueing system with two arrival streams and reserved servers with application to cellular telephone, in *Proceedings of the Conference on Computer Communications (IEEE Infocom)*, Florence, Italy, Apr. 1994.

[5] R. Guerin, Channel occupancy time distribution in a cellular radio system, *IEEE Transactions on Vehicular Technology,* **35** (1987), pp. 88–99.

[6] ——, Queueing-blocking system with two arrival streams and guard channels, *IEEE Transactions on Communications*, **COM-36** (1988), pp. 153–163.

[7] D. Hong and S. Rappaport, Traffic model and performance analysis for cellular mobile radio telephone systems with prioritized and nonprioritized handoff procedures, *IEEE Transactions on Vehicular Technology,* **vt-35** (1986), pp. 77–92.

[8] F. P. Kelly, Reversibility and Stochastic Networks, Wiley, New York, 1979.

[9] F. P. Kelly, Fixed point models of loss networks, *Austral. Math. Soc. Ser. B,* **31** (1989), pp. 204–18.

[10] C.-T. Lea and A. Alyatama, Bandwidth quantization and states reduction in the broadband ISDN, *IEEE/ACM Transactions on Networking,* **3** (1995), pp. 352–360.

[11] Y. Lin, S. Mohan, and A. Noerpel, Queueing priority channel assignment strategies for PCS hand-off and initial access, *IEEE Transactions on Vehicular Technology,* **43** (1994), pp. 704–712.

[12] L. Lipsky, Queueing Theory: A Linear Algebraic Approach, MacMillan, New York, 1992.

[13] D. McMillan, Delay analysis of a cellular mobile priority queueing system, *IEEE/ACM Transactions on Networking,* **3** (1995), pp. 310–319.

[14] K. Mitchell and K. Sohraby, An analysis of the effects of mobility on bandwidth allocation strategies in multi-class cellular wireless networks, in *IEEE Infocom 2001,* Anchorage, AK, Mar. 2001, pp. 1005 –1011.

[15] K. Mitchell, K. Sohraby, A. van de Liefvoort, and J. Place, Approximation models of wireless cellular networks using moment matching, in *IEEE Infocom 2000*, Tel Aviv, Israel, Mar. 2000, pp. 187 –197.

[16] A. A. Nilsson, M. Perry, A. Gersht, and V. B. Iversen, On multi-rate Erlang-B computations, in 16th *International Teletraffic Congress, Edinburgh, Scotland, June 1999* , pp. 1051–60.

[17] M. Sidi and D. Starobinski, New call blocking versus handoff blocking in cellular networks, in *Proceedings of the Conference on Computer Communications (IEEE Infocom)*, San Fransisco, California, Mar. 1996.

[18] A. Srivastava, S. Srivastava, and K. Mitchell, Determining traffic and control channels for a packet based cellular system having time-varying traffic, in *Proceedings of the IEEE Vehicular Technology Conference,* June 2005.

[19] W. Whitt, Blocking when service is required from several facilities simultaneously, *AT&T Technical Journal,* **64** (1985), pp. 1807–56.

In: Adaptation Techniques in Wireless Multimedia Networks ISBN 1-59454-883-8
Editors: Wei Li and Yang Xiao, pp. 307-326

Chapter 14

DYNAMIC CHANNEL ALLOCATION SCHEMES IN GPRS NETWORKS

Yan Zhang[1,*] ***B.H. Soong***[2†] ***and Yuming Jiang***[3‡]
[1]Wireless Communications Laboratory, National Institute of Information and Communications Technology, Singapore
[2]School of Electrical and Electronic Engineering, Nanyang Technological University, Singapore
[3]Department of Telematics, Norwegian University of Science and Technology, Norway

Abstract

In this chapter, we study the performance of GSM/GPRS networks with the emphasis on the adaptive characteristics exhibited by multi-slot capability. The analytical model generalizes the GPRS data channel requirement assumption (M channels). In addition, we propose a channel re-allocation scheme (RAS) with or without buffer mechanism to efficiently utilize the limited bandwidth. Numerical investigation is performed to demonstrate the advantage of RAS as well as the integration of buffer mechanism. The results indicate that RAS not only significantly decreases the voice call blocking probability and the average GPRS packet transmission time, but also slightly increases the channel utilization, with a negligible degradation on the GPRS packet blocking probability. The conventional assumption for small M (e.g.$M = 2$) underestimates the improvements obtained by RAS. The buffer mechanism provides a flexible manner in voice/data integrated wireless networks in helping to tune the system performance tradeoff, and an optimum buffer length is able to provide a good balance among the different performance metrics.

Keywords: GSM, GPRS, UMTS, multi-slot, adaptive capability, dynamic channel allocation, dynamic channel de-allocation, dynamic channel re-allocation, buffer mechanism

[*]E-mail address: yanzhang@ieee.org
[†]E-mail address: ebhsoong@ntu.edu.sg
[‡]E-mail address: ymjiang@ieee.org

1. Introduction

General Packet Radio Service (GPRS) network [1] [2] [3] [4] [5] [6], together with the popular Global System for Mobile Communications (GSM) [7], represent a major evolvement toward the next generation wireless multimedia network such as Universal Mobile Telecommunication System (UMTS) [8] [9] [10] [11]. Fig. 1 illustrates the simplified GPRS network architecture. The principle motivation to introduce GPRS network is to significantly enhance and concurrently simplify the wireless Internet access (e.g. web-browsing, email, ftp, mobile database and instant messenger). Traditionally, GSM is a circuit-switched network, mainly designed for voice communication and to support low data rate service (e.g. short text message). With the increasing demand of multimedia services, GPRS, as a packet-based data bearer service in enabling any existing IP applications over GSM network, is deployed to support high-data rate services. To achieve this, two components are introduced to the core network: serving GPRS support node (SGSN) and gateway GPRS support node (GGSN). The SGSN connects the GPRS core network and the radio access network, and is responsible for delivering packets between the Mobile Station (MS) and the core data network. The GGSN acts similarly as a gateway between GPRS and the external IP networks such as the Internet and enterprise intranets.

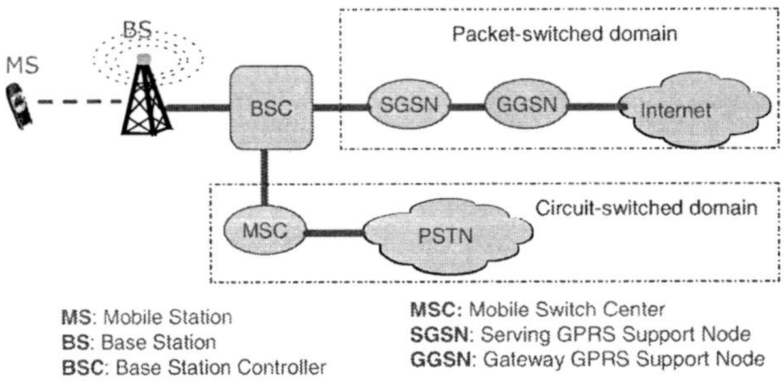

Figure 1. Integrated GSM/GPRS network architecture

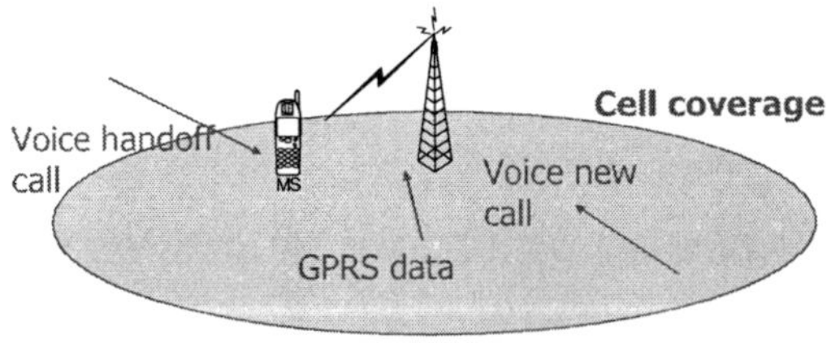

Figure 2. Data and voice traffic processes to BS within its cell coverage area

Fig. 2 shows that the various traffic processes to the Base Station (BS) in the GPRS network include voice and data call requests. The characteristics of voice and data are in strong contrast. The voice traffic is delay-sensitive and loss-insensitive while the data traffic is delay-insensitive and loss-sensitive. As a consequence, due to the diverse Quality of Service (QoS) requirements of voice and data and the motivation for efficient bandwidth utilization, the channel allocation scheme in the voice/data integrated GSM/GPRS network is one of the most crucial issues to achieve the best system performance and has been studied extensively.

The previous literatures have proposed and studied different channel allocation schemes, e.g. channel reservation, non-preemptive or preemptive queueing priority, dynamic channel allocation and channel de-allocation schemes. Lin and Lin [13] compared the performance of several proposed channel allocation schemes and the authors suggested that a dynamic channel allocation scheme with buffering mechanisms can provide flexible and better performance. Based on the work [13], Lin subsequently proposed several different buffering policies by assigning different priorities to voice or data calls [14]. Chen *et al.* [15] also compared the dynamic channel scheme with queueing and observed that employing buffer for GPRS data can greatly reduce its blocking probability. Quershi and Ween [16] studied the issue for GSM-GPRS over a Low Earth Orbit (LEO) satellite system. In [17], Rajaratnam and Takawira employed an analytical approach and simulation to study the performance of channel allocation schemes. Stuckmann and Muller [18] developed a simulation tool called GPRSsim to investigate the radio capacity induced by the integration of GPRS with fixed or on-demand channel allocation. Tripath and Sivarajan [19] proposed two algorithms for channel allocation, i.e., First Fit (FF) and Best Fit (BF), and used the simulation approach to study the performance under simplified assumptions. Via simulation, Ferng and Tsai [20] investigated dynamic channel schemes employing various techniques such as channel reservation, service priority and queueing strategies. Based on stochastic Petri net formalism, Cao [21] analyzed a queueing mechanism in GPRS/EGPRS systems. [22] examined a hybrid radio resource management strategy on the basis of a developed simulator and found that the number of dedicated channels for GPRS traffic may seriously degrade the voice QoS. [23] presented an analytical Markov model and simulation results in studying a hybrid dynamic channel allocation scheme. Specifically, the scheme permanently allocates part of the channels to GPRS data and makes the remaining channels for equal competition between GSM voice call and GPRS data call. The goal of their study is to investigate the number of guard channels for GPRS data with guaranteed QoS for the voice and data service. There are also relevant works in the framework of voice/data integrated wireless networks with buffering mechanism, reserved channel and preemptive or non-preemptive priority schemes [24] [25] [26].

However, the adaptive characteristics implied by the mutli-slot capability has not attracted much attention in the aforementioned channel allocation scheme design. Multi-slot capability refers to the feature of a flexible channel allocation scheme in which not single but up to eight time slots per frame can be simultaneously assigned for a single MS. It is one of the key features of GPRS network enabling the high data rate service up to 115 kbps and also providing an efficient mechanism for voice/data to share the precious radio resources [6]. Ni and Haggman [12] showed that employing fixed multi-slot service in GPRS will result in higher blocking probability and longer delay than using the single-slot service, and these effects can be alleviated by implementing the resource allocation scheme with dynamic multi-slot service. Lin *et al.* [13] [14] employed the multi-slot capability to assign channels to a GPRS data. The number of assigned channels may be less than the required quantity if the resource is not sufficient. De-allocation scheme (DAS), recently proposed and studied in [27], is an approach specifically employing the multi-slot property upon voice call arrival to avoid the voice call from rejection by de-allocating a channel from a multi-slot GPRS data packet and then allocate this channel to the voice call request while the rate of the GPRS data is reduced. Chen *et al.*[27] observed that, with DAS, higher channel utilization can be achieved without tremendous packet delay. For the sake of analytical simplicity and tractability, the aforementioned studies employing the multi-slot capability assume that only 1 or 2 or 3 maximum channels are required by a GPRS data request. This simple assumption is not realistic since one MS can use up to 8 slot in a frame [3] [4] [5] [30] [31], which motivates the study on the general channel requirement by GPRS data. The studies [27], [28] investigated the dynamic channel allocation with realistic channel requirement of GPRS data, and hybrid dynamic channel allocation scheme composing of DAS, guard channel for data service and buffer for voice service based on the fact that voice service can endure the reneging time. In this chapter, we will also relax the assumption by generalizing the channel requirement by GPRS data as M channels.

Next, we will describe the motivation for the proposed channel re-allocation scheme (RAS). Note that with the references [13] [14] [27], upon channel release, the freed channels are left idle. We propose that these idle channels can be re-allocated to the *degraded GPRS call*, terming the GPRS data using less than M channels, with the potential to reduce the GPRS packet transmission duration and to fully utilize the available resources. We call the proposed scheme as channel re-allocation scheme (RAS). Fig. 3 shows an example for the RAS motivation with $M = 8$. Upon the moment of GPRS data G2 completion, there are four degraded GPRS calls in service, i.e. G1, G3, G4 and G5. G1 uses 3 channels, G3 uses 6 channels, G4 uses 6 channels and G5 uses 4 channels. It is unreasonable that there are free channels while these degraded GPRS calls are not able to use them. Hence, the released 5 channels from G2 can be re-allocated to the degraded GPRS calls, i.e. 3 channels to G1 and 2 channels to G3 in this example. Then, the number of channels used by G1 is doubled and hence its transmission time is approximately halved, resulting in more GPRS packets being able to complete the service and more channels available for voice call to access.

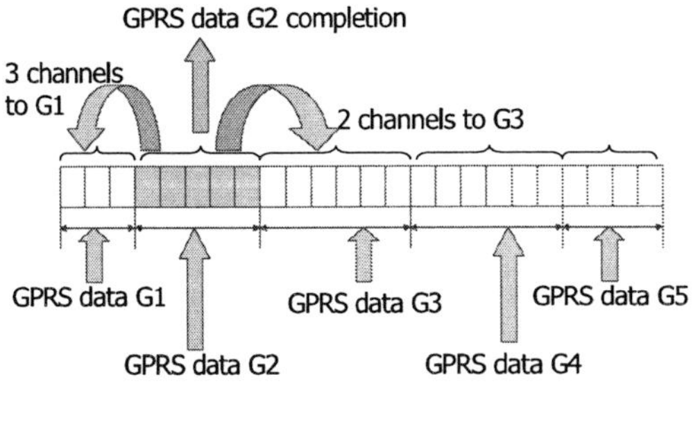

Figure 3. RAS motivation

From this intuitive example, it is evident that at least the voice call blocking probability and GPRS data transmission time can benefit from the re-allocation scheme upon such channel release. However, the general channel requirement of GPRS data considerably complicates the channel re-allocation scheme design. When GPRS data channel requirement is 2 channels, the released channels can solely be re-allocated to the GPRS calls using 1 channel. For general case, i.e. the GPRS data channel requirement is not restricted to 2 channels but M channels, one natural problem arises: how to re-allocate the released channels to degraded GPRS calls or which GPRS calls should be selected to upgrade. Fig. 3

only depicts a simple re-allocation/upgrading strategy. We will describe the specific channel allocation policies corresponding to these two questions in the next section.

In this chapter, we focus on studying the adaptive characteristics in voice/data integrated GSM/GPRS network with the elimination of the conventional assumption of GPRS data channel requirement. A channel re-allocation schemes is proposed to fully utilize the available resource. Multi-dimensional Markov chain is presented to study the performance. An iterative algorithm to compute the stationary state probability is given. The performance metrics in terms of voice blocking probability, GPRS packet blocking probability, channel utilization and GPRS packet transmission time are given.

2. Traffic Model and Dynamic Channel Allocation Scheme

In this section, we consider a Base Station (BS) with C channels for the voice call and GPRS packet sharing in a homogeneous GSM/GPRS network. We term the GPRS packet using $m(m \in \mathcal{M} \equiv \{1, 2, \cdots, M\})$ channels as *type-m GPRS call*. We assume that the voice new call, voice handoff call and GPRS packet follow the Poisson Process with the arrival rate λ_{vn}, λ_{vh} and λ_g, respectively. Let $\lambda_v = \lambda_{vn} + \lambda_{vh}$ be the total voice call arrival rate. Although a GPRS session can last for a long period, the transmission duration of a single packet is very short. Hence, it is reasonable to ignore the GPRS packet handoff traffic. The call holding time defines the time duration of a requested call connection time. The cell residence time terms the period an MS resides in the radio coverage of a cell. We assume that the call holding time and the cell residence time of the voice call are independently exponentially distributed with rate μ_c and μ_{crt}, respectively. After a channel has been assigned to a voice call, the channel can be released due either to call completion or to the handoff procedure to the neighboring cell. As a consequence, the channel holding time, defined as the time duration a call actually occupies a channel, is the minimum value of call holding time and cell residence time. Thereafter, the channel holding time also follows the exponential distribution with the parameter $\mu_v = \mu_c + \mu_{crt}$. The service time of each GPRS packet with one channel is exponentially distributed with mean $1/\mu_g$. Evidently, the service time of each GPRS packet with m channel is exponentially distributed with mean $1/(m\mu_g)$.

2.1. Channel Allocation Scheme without Buffer (CASwoB)

There are 4 events triggering the system state transition: voice call arrival, voice call normal completion/handoff, GPRS data arrival and GPRS data completion. Note that the voice call normal completion event and voice call handoff event have been categorized into a single event due to the identical consequences of the system state after the execution of these two events.

For a GPRS data arrival, it requests M channels. The network will allocate at most M channels to the data request. Assume there are C_{idle} available channels in the system upon a GPRS data request arrival. If $C_{idle} \geq M$, M channels are allocated to the data request. If $0 < C_{idle} < M$, C_{idle} channels are allocated to the data request. If $C_{idle} = 0$, the data request is blocked.

For a voice call arrival, it requests one channel. If an idle channel is available, it will be assigned to the voice request. In the case when there are no free channels, if there exists a type-q ($q > 1$) GPRS data packet, the type-q GPRS data packet will de-allocate one channel to serve the voice call request, and the type-q GPRS data call will degrade to type-$(q-1)$ (the details about the policy on which GPRS should be selected to degrade will be discussed in the next section). This process is called channel de-allocation scheme (DAS). In the case when there are no free channels and all GPRS packets are type-1, the voice call request will be blocked.

For a voice call completion or handoff, one channel will be released. If there exists a degraded GPRS data packet, the released channel will be allocated to the degraded GPRS data packet (the details upon the policy about which degraded GPRS is selected for upgrading will be addressed in the next section). This process is called channel re-allocation scheme (RAS). In the case when all the GPRS data packets are type-M, the freed one channel will be kept idle.

For a type-m GPRS data packet completion, m channels will be released. If there are degraded GPRS data packets, the freed m channels will be re-allocated to the degraded GPRS data packets based on RAS scheme. In the case when all GPRS data packets are type-M, the released m channels will be kept idle.

2.2. Channel Allocation Scheme with Buffer (CASwB)

Based on the fact that there is an overlapped area between adjacent cells and hence a time duration residing in this area before call termination [32], we introduce a buffer mechanism for voice calls (new calls and handoff calls) with finite length K. The extension of any other buffer mechanism (e.g., buffering only new calls or handoff calls, or buffering GPRS data packets) is straightforward. For a voice call waiting in the buffer, it leaves the buffer with rate η_v due to the expiration of the exponentially distributed maximum waiting time.

In this proposed scheme, RAS and DAS, together with a buffer mechanism for voice calls, are adopted to dynamically share the channels for voice/data integration and efficiently utilize the available resources as follows.

For a GPRS data packet arrival, it requests M channels. The network allocates at most M channels to the data request. If the number of available channels C_{idle} upon the data request arrival is greater than M, M channels are allocated to the data request. If $0 < C_{idle} < M$, C_{idle} channels are allocated to the data request. If $C_{idle} = 0$, i.e. all channels are occupied, the data request is blocked. The policy for GPRS data in CASwB is identical to that in CASwoB.

For a voice call arrival, it requests one channel. If an idle channel is available, it will be assigned to the voice request. In the case when there are no free channels, if there exists a type-q ($q > 1$) GPRS data packet, the type-q GPRS data packet will de-allocate one channel to serve the voice call request, and the type-q will degrade to the type-$(q-1)$ GPRS data. In the case when there are no free channels and all GPRS packets are type-1, if there is a waiting space available in the buffer, the voice call request will be placed in the buffer until it gains a channel for service or until the maximum waiting time expires. In the case when there are no free channels and all GPRS packets are type-1, if the buffer is full, the voice call request will be blocked.

For a voice call completion or handoff, one channel will be released. If there is a voice call in the buffer, the freed one channel will be allocated to the voice call. In the case when the buffer is empty, if there exists a degraded GPRS data packet, the released channel will be allocated to the degraded GPRS data packet. In the case when the buffer is empty, if all the GPRS data packets are type-M, the freed one channel will be kept idle.

For a type-m GPRS data packet completion, m channels will be released. If there is a voice call in the buffer, which implicitly suggests that all GPRS data packets are type-1 (otherwise DAS process would have been applied and there would be no voice call waiting in the buffer), then only one channel is released and it will be assigned to a voice call in the buffer. In the case when the buffer is empty, if there are degraded GPRS data packets, the freed m channels will be allocated to the degraded GPRS data packets. In the case when the buffer is empty, if all GPRS data packets are type-M, the released m channels will be kept idle.

3. Performance Analysis of CASwoB

3.1. State Definition and State Space

Let the BS state be denoted as $\mathbf{x} = (i, j_1, j_2 \cdots j_M)$ with i representing the number of voice calls in service and $j_m (m \in \mathcal{M})$ the number of type-m ongoing GPRS calls. It is clear that the total number of ongoing voice and GPRS calls is not greater than C and the maximum number of type-m GPRS in service is $\lfloor \frac{C}{m} \rfloor$, where $\lfloor x \rfloor$ stands for the maximum integer less than or equal to x. The state space of BS, $\mathcal{X}$, is hence given by

$$\mathcal{X} \equiv \left\{ \mathbf{x} = (i, j_1, j_2 \cdots j_M) | 0 \leq i + \sum_{m=1}^{M} m \cdot j_m \leq C; \right. \\ \left. 0 \leq j_m \leq \left\lfloor \frac{C}{m} \right\rfloor, m = 1, 2 \cdots M \right\} \tag{1}$$

Denote S_{max} as the number of states in the state space. Let $r_{\mathbf{x} \to \mathbf{x}'} (\mathbf{x} \in \mathcal{X}, \mathbf{x}' \in \mathcal{X})$ be the transition rate from the current legal state $\mathbf{x} = (i, j_1, j_2 \cdots j_M)$ to the destination legal state $\mathbf{x}' = (i', j_1', j_2' \cdots j_M')$. If $\mathbf{x} \notin \mathcal{X}$ or $\mathbf{x}' \notin \mathcal{X}$, then $r_{\mathbf{x} \to \mathbf{x}'} = 0$. In state $\mathbf{x}$, denote

$$\xi = C - i - \sum_{m=1}^{M} m \times j_m \tag{2}$$

as the number of free channels in the BS. Define the indicator function $\mathbf{1}_E$ equal to 1 when the event E is true, and zero otherwise.

3.2. State Transition

Case 1: GPRS data arrival. Upon a GPRS data arrival (denoted as ARR_g), if the number of idle channels ξ is greater than or equal to M, M channels are allocated to ARR_g. If

$1 \leq \xi < M$, ξ channels are allocated to ARR_g. Otherwise, i.e. $\xi = 0$, ARR_g is blocked. Hence, the state transition can be expressed as follows

$$r_{\mathbf{x}\to\mathbf{x}'} = \begin{cases} \lambda_g \cdot \mathbf{1}_{\xi \geq M} & i' = i, j'_M = j_M + 1, j'_q = j_q \; (q \in \mathcal{M} - \{M\}), \\ \lambda_g \cdot \mathbf{1}_{1 \leq \xi < M} & i' = i, j'_\xi = j_\xi + 1, j'_q = j_q \; (q \in \mathcal{M} - \{\xi\}). \end{cases}$$

Case 2: Voice call arrival. Let j_α denote the first non-zero value in the ordered sequence $(j_M, j_{M-1}, \cdots, j_2)$.

$$\alpha = \max\{q : j_q > 0, 2 \leq q \leq M\}, \tag{3}$$

If α does not exist, then all GPRS data packets (if available) are utilizing 1 channel. In contrast, the α existence shows that $j_q = 0$ (where $q = M, M-1, \cdots, \alpha+1$). For instance, $\alpha = 3$ and $j_\alpha = j_3 = 4$ in the sequence $(j_6, j_5, j_4, j_3, j_2) = (0, 0, 0, 4, 8)$ with $M = 6$. The parameter α is introduced to determine the type of a GPRS data packet to de-allocate a channel for voice call connection request.

For a voice call arrival (denoted as ARR_v), if $\xi > 0$, one channel is assigned to ARR_v. If $\xi = 0$ and α exists, then a channel from a type-α GPRS data packet is de-allocated for ARR_v and the type-α GPRS data packet becomes type-$(\alpha - 1)$. If $\xi = 0$ and α does not exist, then all channels are occupied upon ARR_v arrival and all GPRS data packets (if available) are type-1. In this case, ARR_v is blocked. Thus, we obtain the state transition as follows

$$r_{\mathbf{x}\to\mathbf{x}'} = \begin{cases} \lambda_v \cdot \mathbf{1}_{\xi > 0} & i' = i+1, j'_q = j_q \; (q \in \mathcal{M}). \\ \lambda_v \cdot \mathbf{1}_{\xi = 0}\mathbf{1}_{(\text{exist }\alpha)} & i' = i+1, j'_{\alpha-1} = j_{\alpha-1} + 1, j'_\alpha = j_\alpha - 1, \\ & j'_q = j_q \; (q \in \mathcal{M} - \{\alpha - 1, \alpha\}). \end{cases}$$

Case 3: Voice call normal completion or handoff. Let j_β denote the first non-zero value in the sequence $(j_1, j_2, \cdots, j_{M-1})$.

$$\beta = \min\{q : j_q > 0, 1 \leq q \leq M - 1\}, \tag{4}$$

The non-existence of β implies that all GPRS data packets (if available) are type-M while its existence suggests that $j_q = 0$ (where $q = 1, 2, \cdots, \beta - 1$). For instance, $\beta = 3$ and $j_\beta = j_3 = 8$ in the sequence $(j_1, j_2, j_3, j_4, j_5) = (0, 0, 8, 1, 2)$ with $M = 6$. The parameter β is defined to determine the GPRS data packet type to upgrade at the moment of RAS execution.

Upon the channel release due to voice call completion or handoff, if β exists, which indicates that there are degraded GPRS data, then the released single channel will be re-allocated to upgrade a type-β GPRS data packet into a type-$(\beta + 1)$ GPRS data packet. If β does not exist, which suggests that all GPRS data packets (if any) are type- M, then the released channel keeps idle for future voice or packet request. Therefore, we have

$$r_{\mathbf{x}\to\mathbf{x}'} = \begin{cases} i\mu_v \cdot \mathbf{1}_{(\text{exist }\beta)} & i' = i-1, j'_\beta = j_\beta - 1, j'_{\beta+1} = j_{\beta+1} + 1, \\ & j'_q = j_q \; (q \in \mathcal{M} - \{\beta, \beta+1\}). \\ i\mu_v \cdot \mathbf{1}_{(\text{not exist }\beta)} & i' = i-1, j'_M = j_M, j'_q = 0 \; (q \in \mathcal{M} - \{M\}). \end{cases}$$

Case 4: GPRS data completion. We now consider the channel reallocation upon the channel release due to a type-m GPRS data packet completion. In this case, the number of

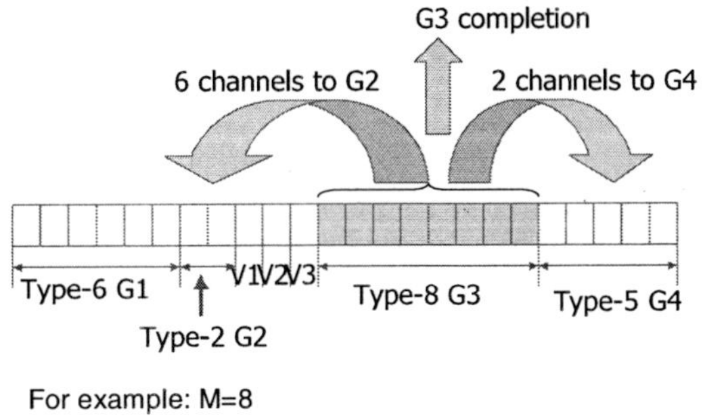

Figure 4. Worst degraded first upgrading policy upon GPRS data completion

type-q ($q \in \mathcal{M}$) GPRS data becomes $(j_1, j_2, \cdots, j_m - 1, \cdots, j_M)$ and the m freed channels are available to upgrade the degraded GPRS data packets. We adopt the worst degraded first upgrading (WDFU) policy, i.e. type-m GPRS data packet can be upgraded if and only if type-$q(q = 1, 2, \cdots, m-1)$ have been already upgraded. In upgrading a specific type-m GPRS data packet, the type-m GPRS data packet will attempt to upgrade to type-M if sufficient channels are available; and will upgrade to the possible type by using up the idle channels otherwise. If there are remaining channels after all type-m GPRS data packets have been successfully upgraded to type-M, the type-$(m+1)$ will attempt to upgrade to type-M. This process will be repeated till the released channels are used up for re-allocation or all degraded GPRS data packets have been fully upgraded to type-M. To be concrete, we define $\mathbf{F}(k)$ as the number of channels required to fully upgrade type-$q(q = 1, 2 \cdots k)$ into type-M provided that there are degraded GPRS data packets. We have,

$$\mathbf{F}(k) = \sum_{q=1}^{k} j_q(M - q) \tag{5}$$

with $\mathbf{F}(0) = 0$. Denote the index q^* $(1 \leq q^* \leq M-1)$ satisfying

$$\mathbf{F}(q^* - 1) \leq m < \mathbf{F}(q^*), \tag{6}$$

which shows that the type-1 to type-(q^*-1) can be fully upgraded to type-M, and the type-q^* can be partially upgraded to type-M under the condition of m channel availability. In the special case $\mathbf{F}(M-1) \leq m$, showing that all the degraded GPRS data packets in service can be fully upgraded to type-M GPRS data packet, we designate $q^* = M$. Non-existence of q^* in the set $\mathcal{M}$ suggests that there is no GPRS data packet or all GPRS data packets are type-M after the type-m GPRS data packet completion.

For the sake of easy understanding of the WDFU policy, we illustrate a simple example in Fig. 4. Upon the service completion instant of type-8 GPRS data G3, 8 channels are released and available for re-allocation. There are three voice calls V_i $(i = 1, 2, 3)$ and one type-6 G1, one type-2 G2 and one type-5 G4. The GPRS data calls sequence the order with an ascending number of channels, i.e. (G2, G4, G1), which is also the upgrading sequence during RAS. The GPRS data with the smallest number of channels, i.e. type-2 G2, is firstly selected for upgrading. In the upgrading procedure, G2 attempts to upgrade to type-8, which equivalently uses 6 channels out of the released 8 channels. After the full upgrading of G2, 2 channels are left for the upgrading of the remaining degraded GPRS data. Then, G4 is chosen since G4 is the second degraded GPRS call. Unfortunately, G4 cannot be fully upgraded to type-8 since there are only 2 channels for its upgrading. Hence, after the upgrading, G4 becomes type-7 and at this moment all released 8 channels are used up. As a consequence, after the RAS execution, there are three voice call V_i $(i = 1, 2, 3)$ and one type-6 G1, one type-8 G2 and one type-7 G4.

Therefore, upon the channel release due to type-m GPRS data packet completion, if q^* exists, then the type-1 to type-$(q^* - 1)$ can be fully upgraded to type-M while type-q^* can be partially upgraded to type-M. If q^* does not exist, which implies that all GPRS data (if available) are type-M, then the released channels keep idle for future voice or packet use. Hence, the transition state can be obtained as follows

$$r_{\mathbf{x}\to\mathbf{x}'} = \begin{cases} j_m(m\mu_g)\cdot\mathbf{1}_{(\text{exist } q^*)} & \mathbf{x}'_{GPRS}; (m \in \mathcal{M}) \\ j_m(m\mu_g)\cdot\mathbf{1}_{(\text{not exist } q^*)} & i' = i, j'_M = j_M, j'_q = 0(q \in \mathcal{M} - \{M\}), \\ & (m \in \mathcal{M}) \end{cases}$$

where the component in the destination state $\mathbf{x}'_{GPRS} = (i', j'_1, j'_2 \cdots j'_M)$ due to a type-m GPRS data packet completion is expressed as

1. $1 \le q^* \le M - 1$

$$\begin{aligned} & i' = i, j'_q = 0(q = 1, 2 \cdots q^* - 1), j'_{q^*} = j_{q^*} - \Delta - 1, \\ & j'_{q^*+\Delta_r} = j_{q^*+\Delta_r} + 1, j'_M = \sum_{q=1}^{q^*-1} j_q + \Delta + j_M, \\ & j'_q = j_q \text{ (other index q in set } \mathcal{M}) \end{aligned} \tag{7}$$

where $\Delta = \left\lfloor \frac{m - \mathbf{F}(q^*-1)}{M - q^*} \right\rfloor$ is the maximum number of type-q^* GPRS data can be upgraded. $\Delta_r = m - \mathbf{F}(q^* - 1) - \Delta \cdot (M - q^*)$ represents the number of channels used to upgrade one type-q^* GPRS data packet to type-$(q^* + \Delta_r)$.

2. $q^* = M$

$$i' = i, j'_q = 0 \; (q \in \mathcal{M} - \{M\}), j'_M = \sum_{q=1}^{M} j_q \tag{8}$$

Fig. 5 shows an example of the state transition diagram assuming $C = 4$ and $M = 3$. For instance, in the current state (1,1,1,0) having one voice call, one type-1 GPRS data call and one type-2 GPRS data call, the total service rate of GPRS data is $3\mu_g$. In case the type-1 GPRS data is completed, one channel is released. The released channel is then re-allocated to the type-2 GPRS data call and this call becomes type-3. The reasoning is similar when the type-2 GPRS data is completed. Consequently, triggered by the event GPRS data completion, the next consecutive state is (1,0,0,1). From the transit rate, the generator matrix, $\mathbf{R}$, governing the evolution of the multi-dimensional Markov chain, can be easily obtained. The steady-state probability of the state $\mathbf{x}$ is denoted as $\pi(\mathbf{x})$. Let $\mathbf{\Pi}$ be

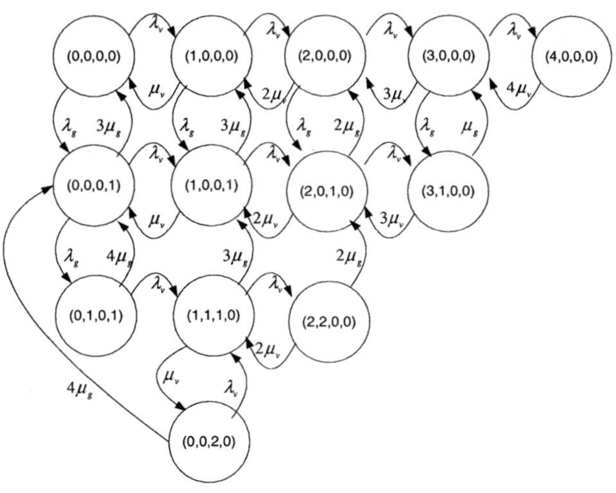

Figure 5. Multi-dimensional Markov Chain when $C = 4$, $M = 3$

the stationary state probability vector of the cell and be ordered in the lexicographic order based on the state $\mathbf{x}$. Then, $\mathbf{\Pi}$ is the solution of linear equation $\mathbf{\Pi R} = 0$ and $\mathbf{\Pi e} = 1$ with $\mathbf{e}$ representing the column vector with all ones [33]. The set of linear equations can be solved by using an iterative method called *successive over-relaxation* (SOR) [34] [35].

Voice call request is characterized as delay-intolerant and its major performance metric is the voice call blocking probability. GPRS data packet is delay-tolerant and its principal performance indicator is the average GPRS packet transmission time. The other important performance metrics include the GPRS packet blocking probability and the channel utilization.

A voice call is blocked when all channels are occupied by voice call and type-1 GPRS data packets. Hence, the voice call blocking probability P_{bv} is given by

$$P_{bv} = \sum_{i+j_1=C, j_2=0, j_3=0, \cdots, j_M=0} \pi(\mathbf{x}) \tag{9}$$

A GPRS data packet is blocked when there is no channel available upon its arrival. Thereafter, the GPRS packet blocking probability P_{bg} is given by

$$P_{bg} = \sum_{i+\sum_{q=1}^{M} q \times j_q = C} \pi(\mathbf{x}) \tag{10}$$

The GPRS packet transmission time T_g is defined as the duration from the time the GPRS data packet is receiving the service to the moment that the GPRS data packet ends the service. Its expected value is given as

$$T_g = \frac{\sum_{\mathbf{x} \in \mathcal{X}} \left[\sum_{m=1}^{M} j_m \cdot \pi(\mathbf{x}) \right]}{\lambda_g (1 - P_{bg})} \tag{11}$$

The channel utilization γ is defined as the ratio of the average busy channels to the total channels. Thus, the channel utilization γ is expressed as

$$\gamma = \frac{\sum_{\mathbf{x}\in\mathcal{X}} \left(i + \sum_{m=1}^{M} m \cdot j_m\right) \cdot \pi(\mathbf{x})}{C} \tag{12}$$

It is noteworthy that the voice handoff call arrival rate is not an input parameter, but dependent on the network architecture, MS mobility pattern and system configuration. In system steady state, the voice call arrival rate is equal to the voice call rate leaving out of the cell. Hence, the voice handoff call arrival rate is described as

$$\lambda_{vh} = \left(\sum_{\mathbf{x}\in\mathcal{X}} i \cdot \pi(\mathbf{x})\right) \cdot \mu_{crt} \tag{13}$$

Since the voice handoff call arrival rate and steady state probability are mutually dependent, a fixed point algorithm `FPA` (see Appendix) should be employed.

4. Performance Analysis of CASwB

Comparing with the scheme CASwoB, an additional dimension should be employed to represent the finite buffer dynamics in CASwB. Let the Base Station (BS) state denote as

$$\mathbf{z} = (i, j_1, j_2 \cdots j_M, k) \tag{14}$$

with i representing the number of voice calls in service, j_m the number of type-m GPRS data packets in progress and k the number of voice call in the buffer. The state space of BS, $\mathcal{Z}$, is defined as

$$\mathcal{Z} \equiv \left\{ \mathbf{z} | 0 \leq i + \sum_{m=1}^{M} m \cdot j_m \leq C; 0 \leq j_m \leq \left\lfloor \frac{C}{m} \right\rfloor, m = 1, 2 \cdots M; 0 \leq k \leq K \right\} \tag{15}$$

The total number of states is denoted as S_{max}. In state $\mathbf{z}$, denote $\xi = C - i - \sum_{m=1}^{M} m \cdot j_m$ as the number of free channels in the BS. Let $r_{\mathbf{z}\to\mathbf{z}'}(\mathbf{z} \in \mathcal{Z}, \mathbf{z}' \in \mathcal{Z})$ be the transition rate from the current legal state $\mathbf{z} = (i, j_1, j_2 \cdots j_M, k)$ to the destination legal state $\mathbf{z}' = (i', j_1', j_2' \cdots j_M', k')$. When the source state $\mathbf{z} \notin \mathcal{Z}$ or destination state $\mathbf{z}' \notin \mathcal{Z}$, then $r_{\mathbf{z}\to\mathbf{z}'} = 0$.

4.1. State Transitions

The state transition can be triggered by the following four events: GPRS data arrival, voice call arrival, voice call normal completion or handoff, and GPRS data completion. We describe these events in details as follows.

Case 1: GPRS data arrival. Upon a GPRS data arrival (denoted as ARR_g), if the number of idle channels ξ is greater than or equal to M, M channels are allocated to ARR_g. If $1 \leq \xi < M$, ξ channels are allocated to ARR_g. Otherwise, i.e. $\xi = 0$, ARR_g is blocked. Hence, the state transition can be expressed as follows

$$r_{\mathbf{z}\to\mathbf{z}'} = \begin{cases} \lambda_g \cdot \mathbf{1}_{\xi \geq M} & i' = i, j_M' = j_M + 1, j_q' = j_q \ (q \in \mathcal{M} - \{M\}), k' = k = 0; \\ \lambda_g \cdot \mathbf{1}_{1 \leq \xi < M} & i' = i, j_\xi' = j_\xi + 1, j_q' = j_q \ (q \in \mathcal{M} - \{\xi\}), k' = k = 0; \end{cases}$$

Case 2: Voice call arrival. For a voice call arrival (denoted as ARR_v), if $\xi > 0$, one channel is assigned to ARR_v. If $\xi = 0$ and α exists, then a channel from a type-α GPRS data packet is de-allocated for ARR_v and the type-α GPRS data packet becomes type-$(\alpha - 1)$. If $\xi = 0$ and α does not exist, then all channels are occupied upon ARR_v arrival and all GPRS data packets (if available) are type-1. In this case, ARR_v is stored in the buffer provided that the buffer is not full and is blocked otherwise. Thus, we obtain the state transition as follows

$$r_{\mathbf{z}\to\mathbf{z}'} = \begin{cases} \lambda_v \cdot \mathbf{1}_{\xi>0} & i^{'} = i+1, j_q^{'} = j_q\ (q \in \mathcal{M}), k^{'} = k = 0; \\ \lambda_v \cdot \mathbf{1}_{\xi=0}\mathbf{1}_{(\text{exist}\ \alpha)} & i^{'} = i+1, j_{\alpha-1}^{'} = j_{\alpha-1}+1, j_\alpha^{'} = j_\alpha - 1, \\ & j_q^{'} = j_q\ (q \in \mathcal{M} - \{\alpha-1, \alpha\}), k^{'} = k = 0; \\ \lambda_v \cdot \mathbf{1}_{\xi=0}\mathbf{1}_{(\text{not exist}\ \alpha)}\mathbf{1}_{k<K} & i^{'} = i, j_q^{'} = j_q\ (q \in \mathcal{M}), k^{'} = k+1; \end{cases}$$

Case 3: Voice call normal completion or handoff. Upon the channel release due to voice call completion or handoff, if the buffer is not empty, which implies that all GPRS are type-1(if available), then the freed channel will be allocated to the voice call at the head of the buffer. Note that the departure rate of a waiting voice call in the buffer is η_v. If the buffer is empty and β exists, which indicates that there are degraded GPRS data, then the released single channel will be re-allocated to upgrade a type-β GPRS data packet into a type-$(\beta+1)$ GPRS data packet. If the buffer is empty and β does not exist, which suggests that all GPRS data packets (if any) are type-M, then the released channel remains idle for future voice or packet request. Therefore, we have

$$r_{\mathbf{z}\to\mathbf{z}'} = \begin{cases} (i\mu_v + k\eta_v) \cdot \mathbf{1}_{k\neq 0} & i^{'} = i, j_q^{'} = j_q\ (q \in \mathcal{M}\}), k^{'} = k-1; \\ i\mu_v \cdot \mathbf{1}_{k=0}\mathbf{1}_{(\text{exist}\ \beta)} & i^{'} = i-1, j_\beta^{'} = j_\beta - 1, j_{\beta+1}^{'} = j_{\beta+1}+1, \\ & j_q^{'} = j_q\ (q \in \mathcal{M} - \{\beta, \beta+1\}), k^{'} = k = 0; \\ i\mu_v \cdot \mathbf{1}_{k=0}\mathbf{1}_{(\text{not exist}\ \beta)} & i^{'} = i-1, j_M^{'} = j_M, j_q^{'} = 0\ (q \in \mathcal{M} - \{M\}), \\ & k^{'} = k = 0; \end{cases}$$

Case 4: GPRS data completion. We consider the channel reallocation upon the channel release due to a type-m GPRS data packet completion. Upon the channel release due to type-m GPRS data packet completion, if the buffer is not empty, which implicitly suggests that all GPRS are type-1 (if available), then one channel is freed and assigned to the voice call at the head of the buffer. If the buffer is empty and q^* exists, then the type-1 to type-(q^*-1) can be fully upgraded to type-M while type-q^* can be partially upgraded to type-M. If the buffer is empty and q^* does not exist, which implies that all GPRS data (if available) are type-M, then the released channels keep idle for future voice or packet use. Hence, the transition state can be obtained as follows

$$r_{\mathbf{z}\to\mathbf{z}'} = \begin{cases} j_1\mu_g \cdot \mathbf{1}_{k\neq 0}\mathbf{1}_{j_1>0} & i^{'} = i+1, j_1^{'} = j_1 - 1, \\ & j_q^{'} = j_q = 0\ (q \in \mathcal{M} - \{1\}), k^{'} = k-1; \\ j_m(m\mu_g) \cdot \mathbf{1}_{k=0}\mathbf{1}_{(\text{exist}\ q^*)} & \mathbf{z}_{GPRS}; k^{'} = k = 0; (m \in \mathcal{M}) \\ j_m(m\mu_g) \cdot \mathbf{1}_{k=0}\mathbf{1}_{(\text{not exist}\ q^*)} & i^{'} = i, j_M^{'} = j_M, j_q^{'} = 0\ (q \in \mathcal{M} - \{M\}), \\ & k^{'} = k = 0; (m \in \mathcal{M}) \end{cases}$$

where the component in the destination state $\mathbf{z}_{GPRS}^{'} = (i^{'}, j_1^{'}, j_2^{'} \cdots j_M^{'}, k^{'})$ occurs under the condition of an empty buffer ($k = 0$) and a type-m GPRS data packet completion. Hence, we have $k^{'} = k = 0$, and

1. $1 \leq q^* \leq M-1$

$$i^{'} = i, j^{'}_q = 0 (q = 1, 2 \cdots q^* - 1), j^{'}_{q^*} = j_{q^*} - \Delta - 1,$$
$$j^{'}_{q^*+\Delta_r} = j_{q^*+\Delta_r} + 1, j^{'}_M = \sum_{q=1}^{q^*-1} j_q + \Delta + j_M,$$
$$j^{'}_q = j_q \text{ (other index q in set } \mathcal{M}) \tag{16}$$

where $\Delta = \left\lfloor \frac{m - \mathbf{F}(q^*-1)}{M-q^*} \right\rfloor$ is the maximum number of type-q^* GPRS data packets can be upgraded. $\Delta_r = m - \mathbf{F}(q^* - 1) - \Delta \cdot (M - q^*)$ represents the number of channels used to upgrade one type-q^* GPRS data packet to type-$(q^* + \Delta_r)$.

2. $q^* = M$

$$i^{'} = i, j^{'}_q = 0 \; (q \in \mathcal{M} - \{M\}), j^{'}_M = \sum_{q=1}^{M} j_q \tag{17}$$

A voice call is blocked when either all channels are occupied or the buffer is full, or the actual waiting time for a voice call in the buffer is larger than the maximum allowed queueing time. Hence, the voice call blocking probability P_{bv} is given by

$$P_{bv} = \sum_{i+j_1=C, j_2=0, j_3=0, \ldots, j_M=0, k=K} \pi(\mathbf{z}) + \sum_{i+j_1=C, j_2=0, j_3=0, \ldots, j_M=0, 0 \leq k \leq K-1} \frac{(k+1)\eta_v}{i\mu_c + j_1\mu_g + (k+1)\eta_v} \pi(\mathbf{z}). \tag{18}$$

In addition, the GPRS packet blocking probability P_{bg}, the expected GPRS packet transmission time T_g and the channel utilization γ are expressed as

$$P_{bg} = \sum_{i+\sum_{q=1}^{M} q \times j_q = C, 0 \leq k \leq K} \pi(\mathbf{z}). \tag{19}$$

$$T_g = \frac{\sum_{\mathbf{z} \in \mathcal{Z}} (\sum_{m=1}^{M} j_m) \cdot \pi(\mathbf{z})}{\lambda_g (1 - P_{bg})}. \tag{20}$$

$$\gamma = \frac{\sum_{\mathbf{z} \in \mathcal{Z}} \left(i + \sum_{m=1}^{M} m \cdot j_m \right) \cdot \pi(\mathbf{z})}{C}. \tag{21}$$

Similarly, the voice handoff call arrival rate is not an input parameter and is expressed as

$$\lambda_{vh} = \left(\sum_{\mathbf{z} \in \mathcal{Z}} i \cdot \pi(\mathbf{z}) \right) \cdot \mu_{crt} \tag{22}$$

Similarly, a fixed point algorithm, e.g. FPA, should be employed to calculate the steady-state probability and performance metrics in (18)-(21).

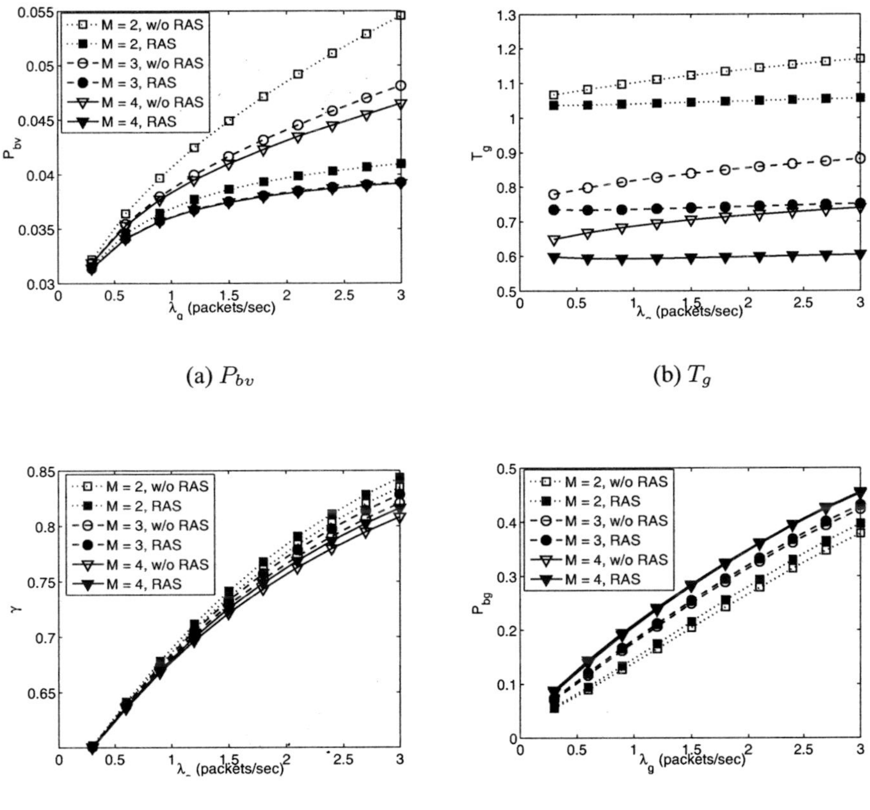

(a) P_{bv} (b) T_g

(c) γ (d) P_{bg}

Figure 6. Performance metrics v.s. GPRS packet arrival rate λ_g (packets/sec) with CASwoB

5. Numerical Results

In this section, we concentrate on showing the illustrative numerical results upon the proposed RAS benefits as well as the buffering mechanism advantages. The following set of parameters is chosen: $C = 12$, $1/\mu_c = 180.0$ sec, $1/\mu_{crt} = 120.0$ sec, $1/\mu_g = 2.0$ sec, $\lambda_{vn} = 0.04$ calls/sec and $\eta_v = 5\mu_c$.

Fig.6 shows the performance metrics in terms of GPRS packet arrival rate with the scheme CASwoB. It is clear that RAS can substantially decrease P_{bv} and T_g. This is because, when using RAS, more channels are accessible for GPRS packet, leading to the great reduction in the GPRS packet transmission time. As a consequence, in a fixed interval, more GPRS packet can complete the service and more channels are available for potential voice call connections, which results in smaller P_{bv} and T_g. Additionally, there is a slight increase in the channel utilization when using RAS due to the tendency of RAS to fully utilize all idle channels. We can observe that the penalty on P_{bg} is negligible. From the perspective of GPRS data request, RAS reduces its available channels. Hence, P_{bg} becomes larger with the involvement of RAS. However, the substantially reduced GPRS packet transmission time, leading to more available channels for GPRS packet access, compensates such dete-

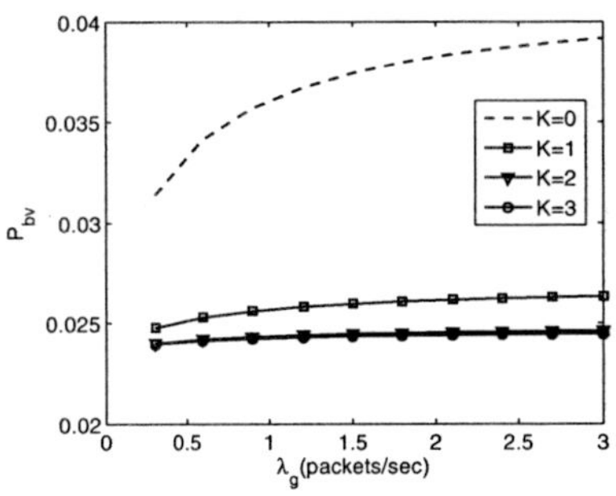

(a) P_{bv}

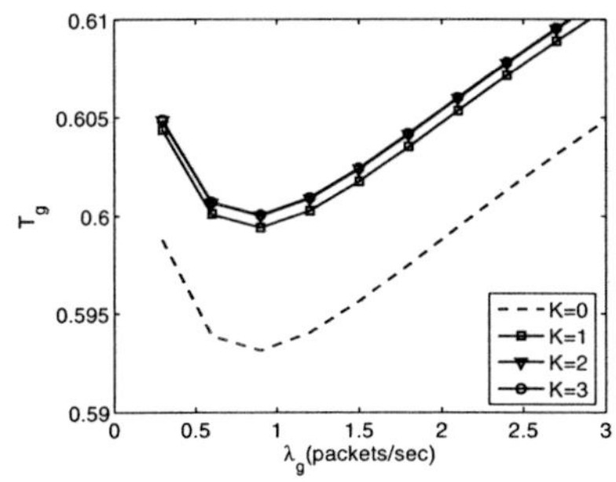

(b) T_g

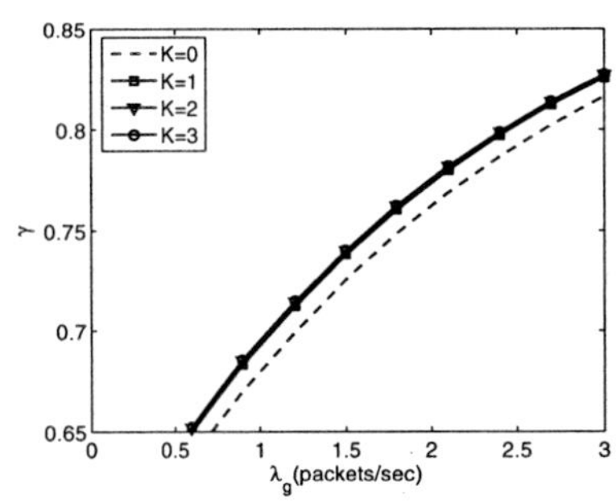

(c) γ

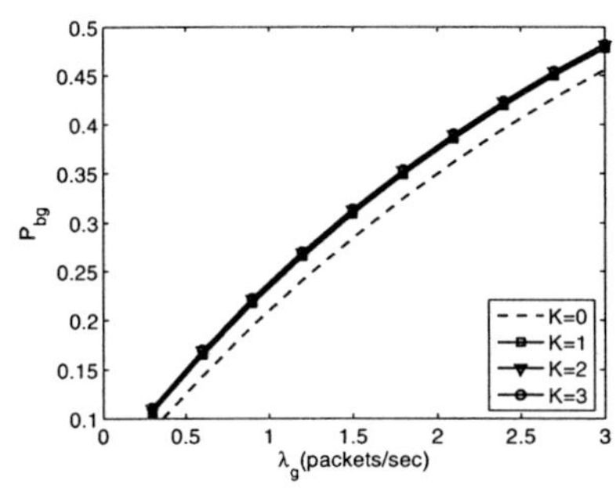

(d) P_{bg}

Figure 7. Performance metrics v.s. GPRS packet arrival rate λ_g (packets/sec) with CASwB ($M = 4$)

rioration and therefore makes this issue insignificant. Furthermore, better performance can be achieved with larger M and such benefit becomes more apparent with a burstier GPRS packet. Hence, the commonly used assumption, i.e. two channels requirement by a GPRS data, underestimates the actual achievement of dynamic channel allocation. It is also observed that continuously increased M is not always advantageous. Fig. 6(a) indicates that the voice call blocking probability under the case of $M = 3$ is almost the same as that under the case of $M = 4$. Hence, the choice of the multiple-channel requirement by a GPRS data request is dependent on the operating context.

Next we demonstrate the advantage of buffer mechanism in CASwB. Our study is implemented by comparing four different cases when choosing the buffer length K as 0, 1, 2 and 3 respectively, under the scenario M set as 4. Fig. 7 shows the corresponding results. When the buffer strategy is employed, a voice call will be placed in the buffer rather than being blocked, when there is no free channel and no available GPRS data packet to deallocate a channel. Hence, the buffer mechanism provides the voice call more opportunities to access a channel, and subsequently, the buffer mechanism significantly reduces the voice call blocking probability. Notice that there is a slight improvement on the channel utilization, as well as a slight degradation on both the GPRS packet blocking probability and the average GPRS packet transmission time. However, due to the constraint of the maximum waiting time, a voice call will be dropped when leaving out the overlapped area of cells. The consequence of this constraint is the existence of an optimal value for buffer length, e.g $K = 2$ in this example.

6. Conclusion

In this chapter, we generalize the channel requirement of GPRS data and propose the channel re-allocation scheme (RAS) with or without buffer strategy in voice/data integrated GSM/GPRS networks. The RAS scheme employs the adaptive characteristics inferred by multi-slot capability. The result shows that RAS can considerably increase the system performance, and the buffer policy provides a flexible manner in tuning the performance matrices tradeoff. It is noteworthy that the similar principle as well as the analytical model is applicable to the next generation of wireless multimedia networks specifically targeting the integrated voice/data services.

Appendix

Fixed Point Algorithm in Section 3.2

Note that we may call a $(M+1)$-dimensional state in $\mathcal{X}$ simply as state $k \in [1, S_{max}]$ without confusion in the algorithm.

Fixed point algorithm FPA:

1. Input $C, M, \lambda_{vn}, \mu_c, \mu_{crt}, \lambda_g, \mu_g$;
2. Construct the matrix $\mathbf{R}$;
3. Set initial $\lambda_{vh}^{(0)}(= 0.1)$. Set the convergence criteria ϵ_1 and ϵ_2, set the relaxation factor $\omega(1 \le \omega < 2)$;
4. Repeat

   (a) • Set initial state probability $\mathbf{\Pi}^{(0)}$ with each element equal to $\frac{1}{S_{max}}$;
   • Repeat and calculate

$$\mathbf{\Pi}^{(n)} = \omega\mathbf{\Pi}^{(n-1)}\mathbf{R} + (1-\omega)\mathbf{\Pi}^{(n-1)}$$

   until $\sum_{k=1}^{S_{max}} \frac{|\pi_k^{(n)} - \pi_k^{(n-1)}|}{|\pi_k^{(n)} + \pi_k^{(n-1)}|} < \epsilon_1$.

   (b) Calculate the handoff call arrival rate l^{th} iteration

$$\lambda_{vh}^{(l)} = \left(\sum_{k=1}^{S_{max}} \text{(voice call in state } k) \cdot \pi_k \right) \cdot \mu_{crt}$$

   until $\left| \lambda_{vh}^{(l)} - \lambda_{vh}^{(l-1)} \right| < \epsilon_2$.

5. Output the steady state probability $\mathbf{\Pi}$.

6. Calculate the performance metrics P_{bv}, P_{bg}, T_g and γ.

References

[1] G.Brasche and B.Walke, Concepts, services and protocols of the new GSM phase 2+ general packet radio service, *IEEE Communications Magazine*, vol.35, pp. 94-104, 1997.

[2] J. Cai and D.J.Goodman, General packet radio service in GSM, *IEEE Communications Magazine*, vol.35, pp. 122-131, 1997.

[3] ETSI Digital cellular telecommunications system (Phase 2+); General Packet Radio Service (GPRS); Service description; Stage 1 (GSM 02.60 version 7.2.0 release 1999), ETSI/TC, *Tech. Rep. Rec. GSM* 02.60, 1999.

[4] ETSI Digital cellular telecommunications system (Phase 2+); General Packet Radio Service (GPRS); Service description; Stage 2 (GSM 03.60 version 7.0.0 release 1999) ETSI/TC, *Tech. Rep. Rec. GSM* 03.60, 1999.

[5] 3GPP, 3rd Generation Partnership Project; Technical Specification Group Services and Systems Aspects; General Packet Radio Service (GPRS); Service Descripton; Stage 2, 2000, Technical Specification 3G TS 23.060 version 3.6.0 (2001-01).

[6] E. Seurre, P. Savelli and P. J. Pietri, *GPRS for Mobile Internet*, Artech House, UK, Dec.2002.

[7] Y.-B. Lin and I. Chlamtac, *Wireless and Mobile Network Architectures*. New York: Wiley, 2001.

[8] UMTS Forum. (2000) The UMTS third generation marketStructuring the service revenues opportunities. *Tech. Rep.* **9**, UMTS. [Online]. Available: www.umts-forum.org.

[9] UMTS Forum. (2000) Shaping the mobile multimedia futureAn extended vision from the UMTS forum. *Tech. Rep.* **10**, UMTS. [Online]. Available: www.umts-forum.org.

[10] UMTS Forum. (2000) Enabling UMTS/third generation services and applications. *Tech. Rep.* **11**, UMTS. [Online]. Available: www.umtsforum. org.

[11] H. Kaaranen, A. Ahtiainen, L. Laitinen, S. Naghian and V. Niemi, *UMTS Networks: Architecture, Mobility and Services*, John Wiley & Sons LTD, UK, 2001.

[12] S. Ni and S. G. Haggman, GPRS performance estimation in GSM circuit switched services and GPRS shared resource systems, in *Proceedings of the IEEE Wireless Communications and Networking Conference (WCNC'99)*, vol. 3, pp. 1417-1421, Sep. 1999.

[13] P.Lin and Y.B. Lin, Channel allocation for GPRS, *IEEE Transaction on Vehicular Technology*, vol.50, no.2, pp. 375-387, 2001.

[14] P.Lin, Channel allocation for GPRS with buffering mechanisms, *ACM/Baltzer Journal of Wireless Networks (WINET)*, vol.9, pp. 431-441, 2003.

[15] W.-Y. Chen, J.-L.C. Wu and H.-H. Liu, "Performance analysis of radio resource allocation in GSM/GPRS networks", in *Proceedings of IEEE Vehicular Technology Conference (VTC'02-Fall)*, vol.3, pp. 1461-1465, Sep. 2002.

[16] A. G. Qureshi and A. S. Ween, Dynamic resource allocation for GSM/GPRS services over a LEO satellite system, in *Proceedings of the 1st International Conference on Information, Communications and Signal Processing (ICICS'97)* , pp.20-24, Nov. 1997.

[17] M. Rajaratnam and F. Takawira, A single cell model for the performance analysis of the radio layer in GSM Phase 2+ (GPRS) networks under voice and data traffic, in *Proceedings of the 12th IEEE Personal Indoors and Mobile Radio Communications (PIMRC'01)*, pp. 118-123, Sep. 2001.

[18] P. Stuckmann and F. Muller, "GPRS radio network capacity and quality of service using fixed and on-demand channel allocation techniques," in *Proceedings of the 51th IEEE Vehicular Technology Conference (VTC 2000)*, pp. 440-444, 2000.

[19] A. Tripathi and K. N. Sivarajan, Resource allocation in GPRS wireless networks, in *Proceedings of the IEEE International Conference on Personal Wireless Communications (ICPWC'00)*, pp. 388-394, Dec. 2000.

[20] H.-W. Ferng and Y.-C. Tsai, Channel allocation and performance study for the integrated GSM/GPRS system, in *Proceedings of the IEEE Wireless Communications and Networking Conference (WCNC'2003)*, vol.3, pp. 1861-1865, Mar. 2003.

[21] Y.Cao, H.R.Sun, K.S.Trivedi, "Performance analysis of reservation media-access protocol with access and serving queues under bursty traffic in GPRS/EGPRS", *IEEE Transactions on Vehicular Technology*, vol.52, no.6, pp.1627-1641, Nov. 2003.

[22] H.Araujo, J.Costa and L.Correia , Analysis of a traffic model for GSM/GPRS, in *Proceedings of the 12th IEEE Personal Indoors and Mobile Radio Communications (PIMRC'01)*, vol.1, pp. 124-128, Sep. 2001.

[23] C. Lindemann and A.Thümmler, Performance analysis of the general packet radio service, *Elsvier Journal of Computer Networks*, vol.41, no.1, pp.1-17, Jan. 2003.

[24] B. Li, Q-A. Zeng, K. Mukumoto, and A. Fukuda, A preemptive priority handoff scheme in integrated voice and data cellular mobile systems, *IEICE Transactions on Communications*, vol. E82-B, no.10, pp. 1633-1642, Oct. 1999.

[25] J. Wang, Q-A Zeng, and D. P. Agrawal, Performance analysis of integrated wireless mobile networks with priority reservation and preemptive priority handoff scheme, *IEEE Transactions on Mobile Computing*, vol. 2, no.1, Jan.-Mar. 2003.

[26] Q-A Zeng and D. P. Agrawal, Modeling and efficient handling of handoffs in integrated wireless mobile networks, *IEEE Transactions on Vehicular Technology*, vol. 51, no. 6, pp. 1469-1478, Nov. 2002.

[27] W.-Y. Chen, J.-L.C. Wu and L.-L Lu, Performance comparisons of dynamic resource allocation with/without channel de-allocation in GSM/GPRS networks, *IEEE Communications Letters*, vol. 7, no. 1, pp. 10-12, Jan. 2003.

[28] H.Y. Shin, J.C. Wu and W.-Y. Chen, Performance analysis of an integrated voice/data wireless network with voice buffer, *IEICE Transaction on Communincations*, vol. E87-B, no. 1, pp. 36-45, Jan. 2004.

[29] J. Ho, Y. Zhu, and S. Madhavapeddy, Throughput and buffer analysis for GSM general packet radio service, in *Proceedings of the IEEE Wireless Communications and Networking Conference (WCNC'99)*, pp. 1427-1431, Sep. 1999.

[30] Yan Zhang and Boon-Hee Soong, Performance evaluation of GSM/GPRS networks with channel re-allocation scheme, *IEEE Communincations Letters*, vol.8, no.5, pp. 280-282, May. 2004.

[31] Yan Zhang and Boon-Hee Soong, Multi-slot capability increases GPRS network performance, in *Proceedings of the IEEE 1st International Symposium on Wireless Communication Systems (ISWCS'2004)*, pp. 154-158, Sep. 2004.

[32] Yan Zhang and Boon-Hee Soong, Handoff dwell time distribution effect on mobile network performance, *IEEE Transactions on Vehicular Technology*, to appear

[33] L. Kleinrock, *Queueing Systems.* New York, NY: John Wiley and Sons, 1975.

[34] R. Cooper, *Introduction to Queueing Theory.* New York: North-Holland, 1981.

[35] W.J. Stewart, *Introduction to the Numerical Solution of Markov Chains* . Princeton, NJ: Princeton University Press, 1995.

In: Adaptation Techniques in Wireless Multimedia Networks ISBN: 1-59454-883-8
Editors: Wei Li and Yang Xiao, pp. 327-346

Chapter 15

MULTIMEDIA DELIVERY OVER WIRELESS MULTI-HOP NETWORKS

Qian Zhang[1*], Fan Yang[2**], Haitao Wu[2] and Zhi-Li Zhang[3****]***
[1] Department of Computer Science, Hong Kong University of Science and Technology, Hong Kong, China,
[2] Microsoft Research Asia, No. 49, Zhichun Road, Haidian District, Beijing, 100080, China
[3] University of Minnesota, USA

Abstract

A wireless multi-hop network is a collection of wireless nodes that dynamically form a temporary network without a fixed infrastructure. With the increase in the bandwidth of wireless channels, and in the computing power of devices, video applications are expected to become available in wireless multi-hop networks in a near future. However, limited network resource, severe interference/ contention among neighbor traffic, dynamic changing route, etc. pose many challenges for supporting video communication over wireless multi-hop networks. In this paper, we will give an overview for the state-of-art work on multimedia over multi-hop networks. More specifically, we review the recent progress in efficient and resilient video coding and protection solutions, the network routing and transport protocols, link-layer QoS support enhancement, and cross-layer design for media over multi-hop environment.

1. Introduction

As various wireless networks evolve into the next generation to provide better services, a key technology, wireless multi-hop networks has emerged recently. In multi-hop networks, each node operates not only as a host but also as a router, forwarding packets on behalf of other nodes that may not be within direct wireless transmission range of their destinations. A multi-

[*] E-mail address: qianzh@cs.ust.hk
[**] E-mail address: fanyan@microsoft.com
[***] E-mail address: hwu@microsoft.com
[****] E-mail address: zhzhang@cs.umn.edu

hop network is dynamically self-organized and self-configured, with the nodes in the network automatically establishing and maintaining mesh connectivity among themselves (creating, in effect, an ad hoc network). This feature brings many advantages to multi-hop networks such as low up-front cost, easy network maintenance, robustness, and reliable service coverage.

Wireless multi-hop network is a promising wireless technology for numerous applications, e.g., broadband home networking, community and neighborhood networks, enterprise networking, building automation, etc. It is gaining significant attention recently. More and more people are now working on the voice, image, and video communication over multi-hop networks. To support high quality multimedia delivery over multi-hop wireless networks, there are several fundamental challenges.

1. It is well known that bit error rate (BER) of wireless network is much higher than that in wired-line links. The shared wireless media and contention from neighbor traffic further exacerbates the limitations of bandwidth and the presence of channel errors in multi-hop wireless network. While today's video coders, such as MPEG and H.26x, can compress video efficiently, the compressed bitstream is fragile in the face of channel loss. This fragility is a direct consequence of the prediction-based coding that underlies these codecs, since to decode the current frame correctly, all previous frames need to have been received correctly. Packet losses thus lead to loss of synchronization or "drift" between encoder and decoder, i.e. the predictors at the encoder and decoder are not the same.
2. In wireless network, congestion is not the only reason (nor even the main reason) for packet losses. Many losses cause by the random channel error can be measured over multi-hop wireless networks [1]. Moreover, route change/break that frequently occurred in multi-hop networks can also cause the packet loss. For multimedia streaming applications, being aware of the reason of packet losses is critical to perform correct error control and resource allocation. For example, when the packet loss rate is increasing, if the loss rate increment is due to wireless channel errors, it's proper for a multimedia streaming application to increase the error control level; if the rate increment is due to congestion, increasing the error control level might be of no use and reducing the sending rate should be the right decision; if the loss rate increment is due to route change or break, to stop transmitting data until a new path is found (or the old route is reconnected) is the correct measure of the multimedia streaming application.
3. Lack of QoS support mechanism in multi-hop wireless networks. The IEEE 802.11 carrier sense multiple access with collision avoidance (CSMA/CA) protocol has become the de-facto medium access control (MAC) standard [2] for the current generation multi-hop networks. Clearly 802.11 MAC, while generally available and simple to use, has serious deficiencies in a multi-hop environment in terms of throughput, delay variance and fairness (see for example, [3]). It is well known that the recent 802.11e enhanced DCF [4] is attempts to provide limited QoS control in WLAN's operated in the infrastructure mode. However, this solution does not apply directly to the multi-hop 802.11 scenario, because of the hidden terminal effects and content from neighbor traffics.
4. Because of the direct coupling between the physical layer and the upper layers, the traditional "layered" protocol stack is not sufficient for wireless networks. In a

wireless network, physical layer, MAC layer, and routing layer together contend for the network resource. The physical layer affects MAC and routing decisions by its transmission power and rate. The MAC layer is responsible for scheduling and allocating the wireless channel, which finally will determine the available bandwidth of the transmitter and the packet delay. This bandwidth and packet delay also can affect the decision at the routing layer to select the link. The routing layer chooses the wireless links to relay the packets to the destination. The routing decision will change the contention level at the MAC layer, and accordingly the physical layer parameters.

Target at the above challenges, the existing research address how to improve the quality for multimedia delivery over wireless multi-hop networks from different aspects. In this paper, we will give an overview for the state-of-art effects on multimedia over multi-hop networks. Basically, we can classify the existing work into several categories. More specifically,

- To handle the impact of channel error to compressed media, the efficient and resilient video coding and protection solutions are first presented.
- Target at dealing with the dynamic changing network condition (e.g., packet loss, delay, route) and also utilizing the uniqueness of multi-hop networks (e.g., multiple paths available due to the mesh topology), the network and transport protocols are reviewed.
- To provide QoS support for multimedia delivery, research studies related to link layer, including power control, scheduling, enhanced MAC, are summarized.
- Finally, to improve the overall wireless network performance, cross-layer approaches that jointly consider different layers, including multimedia application, routing and transport protocol, link layer scheduling, and physical layer power control are surveyed.

2. Efficient and Resilient Video Coding and Protection

While coding efficiency is the most important aspect in the design of any video coding scheme, the transmission of compressed video through noisy channels has always been a key consideration. Lots of research works have been there for developing various error-resilience tools. In [5], a good summary has been given, which basically classify these proposed schemes into the following categories, which are not mutually exclusive.

- Localization – removes the spatial/temporal dependency between segments of the video to reduce error propagation. The basic localization tools include resynchronization marker, adaptive Intra refresh, reference picture selection, and multiple reference pictures.
- Data Partitioning – groups coded data according to relative importance to allow for unequal error protection or transport prioritization. In MPEG-2, data partitioning divides the coded bitstream into a high-priority partition and a low-priority partition.

In MPEG-4, the data partitioning is achieved by separating the motion and macroblock header information away from the texture information.

- Redundant Coding – codes segments of the video signal or syntactic elements of the bitstream with added redundancy to enable robust decoding. Reversible Variable Length Coding (RVLC), Multiple Description Coding (MDC), and redundant slice [6] belong to this category.
- Concealment-Driven – enables improved error concealment after decoding using additional information embedded into the coded stream or by uniquely ordering segments of the video signal. The basic tools include concealment motion vectors and flexible macroblock order.

Obviously, all of these solutions for error-resilience indirectly lead to an increase in the bit-rate and loss of coding efficiency. Many research works are trying to address the best tradeoff between error resilience and coding efficiency under different conditions.

In the following, we will introduce the recent progress for error resilience related work from video standard and individual research point of view, respectively.

2.1. Error-Resilience Mechanisms on H.264

Rather flexible error-resilience mechanisms are provided in H.264 [7, 8], where it makes error resilience mechanisms available both on the encoder and on the decoder side.

In the encoder side, there are several tunable parameters for the trade-off between compression rate and error resilience under heterogeneous environments. *Random intra macroblock refreshes* and the insertion of intra-coded pictures (I frames) are the most commonly used methods to stop the temporal propagation of errors when no feedback channel is available. Considering intra frame generally generates high bandwidth cost and causes severe bit rate variations, the use of *random intra macroblock refreshes* is more effective. The *macroblock line Intra update* is another robustness option where a group of blocks will be intra coded every *N* frames. It is another form of macroblock updating. *Multiframe compensation prediction* is another tool targeting to increase both compression performance and error resilience, since the loss of an entire reference frame will have less critical effects on later predicted frames.

The use of slices is yet another method commonly used to improve robustness by stopping spatial error-propagation. The macroblocks belonging to a slice can be decoded independently from other slices since no inter-slice dependencies are allowed. Flexible Macroblock Ordering (FMO) aims essentially at dealing with packet loss bursts, whereby the sender can transmit macroblocks in non-scan order. This method, although similar to slice interleaving, provides much greater flexibility and can be tuned to be more effective in terms of error resiliency. SP slices make use of motion-compensated predictive coding to exploit temporal redundancy in the sequences, like P slices do. Unlike P slices, however, SP slice coding allows identical reconstruction of a slice even when different reference pictures are being used. They aim essentially at bit stream switching, splicing, random access, VCR functionalities and error resilience issues.

Another tool that integrates the H.264 codec is rate distortion optimization. Distortion can arise due to either quantization errors or prediction from concealed blocks. If prediction does not provide good compression, intra compression for single macroblocks is allowed. Considering the transmission errors in the network, a third mode is available where the encoder takes into account the expected packet loss rate of the network, as well as the decoder's methods to cope with errors in order to decide weather to intra or inter code a block.

Concerning the decoder, it also plays a fundamental role in error resilience since it is responsible for error concealment tasks. With that purpose it keeps a status map for macroblocks which indicates for each frame being decoded weather a certain macroblock has been correctly received, lost or already concealed. The methods used vary between intra and inter frames. For intra frames the task mainly consists on of performing a weighted pixel averaging on each lost block in order to turn it into a concealed one. For inter frames the task performed consists mainly of guessing the adequate motion vector for lost macroblocks, although intra-style methods can also be used.

The decoder also has other tasks like handling multiple reference frames or entire frame losses. The reference decoder for H.264 does not incorporate bit error resilience features, thus bit error detection and handling has to be processed externally.

2.2. Other Research Efforts Related to Error Resilience for Wireless Networks

Error Resilience Entropy Coding (EREC) [9] is one of the earliest error-resilience schemes, in which the incoming bit-stream is reordered without adding redundancy such that longer VLC blocks fill up the spaces left by shorter blocks in a number of VLC blocks that form a fixed-length EREC frame. Such fixed-length EREC frames of VLC codes are then used as synchronization units, where only one EREC frame, rather than all the codes between two synchronization markers, will be dropped when transmission errors occur.

In [10], a rate-distortion framework with analytical models that characterize the error propagation of a corrupted video bit-stream subjected to bit errors was proposed. The models were used to guide the use of spatial and temporal localization tools: synchronization markers and intra-refresh, respectively, to compute the optimal bit-allocation among spatial error-resilience, temporal error-resilience, and the source rate. Align with this direction, in [11] optimal error resilience insertion is divided into two sub-problems: optimal mode selection for MB's and optimal resynchronization marker insertion. In [12], a method to recursively compute the expected decoder distortion with pixel-level precision to account for spatial and temporal error propagation in a packet loss environment is proposed for optimal MB coding mode selection.

The above work is target for general network. Interesting to note that, one characteristic of a wireless multi-hop network is that it is usually possible to establish more than one path between a source and a destination given their mesh topology. Recently people start thinking about leveraging multi-path to further improve the efficiency. Three representative schemes are described more as examples in the following.

Feedback Based Reference Picture Selection (RPS) technique is proposed in [13]. Specifically, a time domain partitioning method (i.e., sending coded even and odd frames separately in different paths) is used in this work. Moreover, a network-aware coding method is used, which selects the reference picture based on feedback and estimated path status.

Assume that the decoder sends a positive feedback (ACK) and negative feedback (NACK), respectively, for a received or lost frame. The encoder can then estimate the status of the paths and infer which of the previous frames are damaged. Based on the estimation, for a picture to be coded, the closest picture for which itself as well as its reference pictures have been transmitted on the better path is selected as reference picture. The RPS scheme offers a good trade-off between coding efficiency and error resilience. That is, when both paths are good, RPS uses the neighboring frame as reference, thereby achieving the highest possible prediction gain and coding efficiency; when one path is bad, the encoder avoids using any frames that are affected by that path, thereby minimizing the error propagation period. Note that the RPS scheme is only applicable for on-line coding, because it adapts the encoding operation based on channel feedback.

Layered Coding with Selective ARQ is the second option to leverage multipath characteristic. When the layered video is transmitted over multiple paths (e.g., two paths), the traffic allocator sends the base layer packets on one path and the enhancement layer packets on the other path. The path with a lower packet loss rate is used for the base layer if the two paths have different quality. The receiver returns selective ARQ requests to the sender to report base layer packet losses. When the sender receives such a request, it retransmits the requested base layer packet on the enhancement layer path. The transmission bit rate for the enhancement layer will be reduced correspondingly according to the bandwidth reallocated for base layer retransmissions.

The third option is to use multiple description coding (MDC). MDC is a technique that generates multiple *equally important* descriptions. The decoder reconstructs the video from *any* subset of received descriptions, yielding a quality commensurate with the number of received descriptions. Compared with layered coding, MDC does not require the network or channel coder to provide different levels of protection. Neither does it require any receiver feedback. Acceptable quality can be achieved even when both descriptions are subject to relatively frequent packet loss, as long as the losses on the two paths do not occur simultaneously and sufficient amount of redundancy is added by appropriately choosing the predictor coefficient and the mismatch signal quantizer.

3. Effective Network and Transport Protocol for Video Delivery

Without any infrastructure support, in multi-hop wireless networks, the changing network topology results in highly dynamic path conditions and frequently changed routing. Meanwhile, as we mention in the introduction section, congestion is not the only reason causing packet losses any more. Instead, transmission errors in wireless channels and mobility induced route change/break can also contribute to packet losses. Multimedia streaming over multi-hop networks faces even bigger challenges because it has more stringent requirements on bandwidth, delay, and loss than best effort services. However, the mesh topology of multi-hop networks provides the existence of multiple routes between two endpoints which can be utilized to perform multipath streaming to better support real time multimedia applications. Figure 1 illustrates the basic idea of multimedia delivery over multi-hop networks.

In general, multipath multimedia streaming in multi-hop networks has the following advantages over single path streaming. First, it can potentially provide higher aggregate bandwidth to real-time multimedia applications (given that the bottleneck is not shared by the paths). Second, data partitioning over multiple paths can reduce the short term correlation in real-time traffic, thus improve the performance of multimedia streaming application [14]. Third, the existence of multiple paths can help to reduce the chance of interrupting the streaming service due to node mobility.

In the rest of this section, we will review the recent progress of multi-path media delivery for unicast and multicast service over multi-hop networks, respectively.

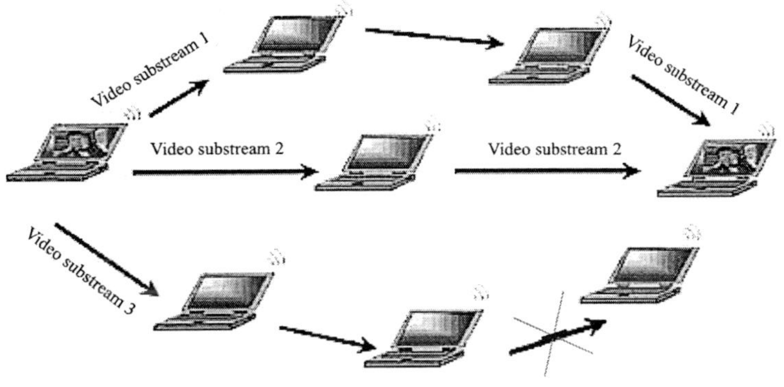

Figure 1. Basic illustration of media over multi-hop networks.

3.1. Multi-path Video Delivery for Unicast Service

To better support multipath streaming over wireless multi-hop networks, the two important issues need to be addressed are listed as follows. Firstly, packet losses due to different causes should be differentiated since incorrect information of different packet loss ratio may decrease the end-to-end multimedia streaming quality (more discuss can be found in the next section). This relies on the accurate discrimination among congestion, channel error, and route change/break states. Secondly, streaming protocols need to choose multiple maximally disjointed paths to achieve good streaming quality. Although multipath routing protocols can provide multipath routing information to the streaming protocol, we still need an efficient way to select multiple paths with best quality and assign appropriate sending rate to the new path and smoothly switch to a new path in case that the current path is about to break.

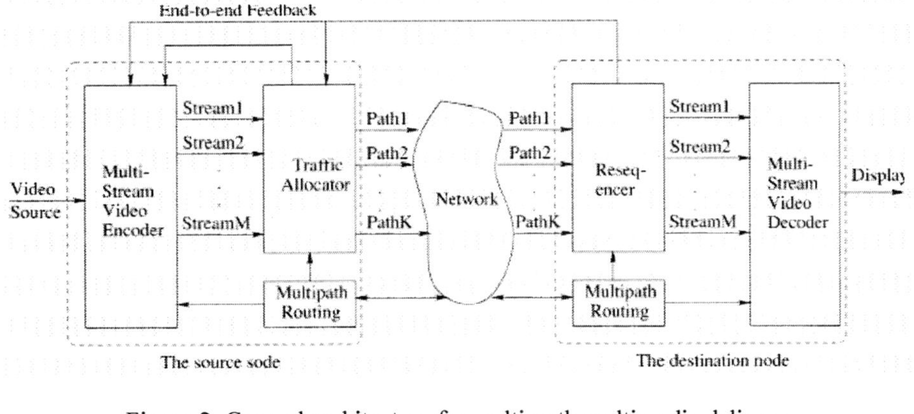

Figure 2. General architecture for multi-path multimedia delivery.

A number of approaches have been proposed to enable multimedia streaming over multiple paths in multi-hop networks. They share the same general architecture [13] as shown in Figure 2. More specifically, the multipath routing layer sets up N possible paths between the source and destination, and monitors the network condition of each path. Based on path quality information, the encoder generates M substreams for the target multimedia. Packets from the M substreams are dispersed by the traffic allocator among the N paths. At the receiver, packets arriving from all the paths are put into a resequencing buffer. The decoder will reconstruct a video sequence from the received substreams. Next we will overview the individual proposals in detail.

Recently Ad Hoc TCP Friendly Rate Control (ADTFRC) protocol has been proposed to detect and be adaptive to the varying network conditions in mobile ad hoc networks and peacefully coexist with TCP traffic [15]. In contrast to traditional streaming protocols which use packet loss as a signal of congestion, ADTFRC proposes to use multiple metrics, Short-Term Throughput (STT), Inter-packet Delay Difference (IDD), Packet Out-of-order Ratio (POR) and Packet Loss Rate (PLR), to detect network congestion, channel error, and route change/break. By jointly using them, ADTFRC can accurately identify congestion state. The accuracy of detecting channel error and route change/break states needs to be improved when both mobility and channel error are present.

The Multi-flow Realtime Transport Protocol (MRTP) was designed for supporting multipath transport architectures for realtime multimedia applications [14]. Its companion control protocol is Multi-flow Realtime Transport Control Protocol (MRTCP). MRTP/MRTCP is a natural extension of RTP/RTCP (Real-time Transport Protocol/Real-time Transport Control Protocol) and facilitates multipath transport to combat frequent link failures and congestion in multi-hop networks. An MRTP session should be established first by MRTCP, where two end nodes exchange information such as available paths, session/flow identifiers, and initial sequence numbers. During data transmission, a new path may be added to the session when a better path is found, and a degraded path may be removed from the session based on QoS reports. The flows could then be switched from degraded paths to new paths. With MRTP, a traffic allocator partitions and disperses the real time multimedia traffic to multiple flows.

Robust multipath source routing protocol (RMPSR) ([16]) is a multipath extension to DSR, which applies several new rules to address requirements of video communication applications. The basic idea of PMPSR is disjoint routes, where two routes are nearly disjoint only when the ratio of the number of shared nodes to the number of the nodes of the shorter route is smaller than a threshold value. To provide robustness for video applications, RMPSR builds and maintains multiple nearly disjoint route sets for the video communication almost all the time. Here, a route set consists of one primary route and several alternative routes. The primary route connects the source node and the destination node; alternative routes connect intermediate nodes and the destination node.

RMPSR uses a per-packet allocation scheme to distribute video packets over two primary routes of two route sets. When one transmitting primary route is broken, the intermediate node that corresponds to the broken link will send a Route ERRor (RERR) packet to the source node. Upon receiving the RERR packet, the source node removes the broken primary route from its route cache, and switches the transmission to another primary route. Two new features of RMPSR that related to video streaming can be summarized as follows.

- When the transmission route is broken, alternative routes in the same route set are used to salvage midway packets. The reason is that routes in the same route set are correlated, so if the primary route is broken, it is likely that alternative routes have been broken or will be broken shortly.
- RMPSR triggers new route request process before the connectivity is entirely lost in order to reduce the number of temporary network outages during the transmission.

In [17], the authors address the routing issues for MPEG4 Fine-Granular-Scalability (FGS) multimedia stream over wireless multi-hop networks. The FGS encodes video traffic into two different layers, base layer (BL) and enhancement layer (EL). The BL provides basic information for video sequence and the EL provides better display quality of video sequence by adding more detailed information to the BL. Handling the low-level link quality and the dynamically reconfigurable wireless network property in multi-hop networks, layered video traffic is transmitted through multiple paths. Since BL traffic carries basic information for the video sequence, retransmission is needed for the loss of BL traffic. To minimize the retransmission time, BL traffic is routed on a path with the minimum transmission delay. If multiple such paths exist, the path with largest available bandwidth should be selected. The routing problem for BL traffic was been modeled as a shortest maximum-bandwidth problem. On the other hand, since EL traffic should arrive before BL data is decoded in the destination, the path routed for EL traffic should be delay-bounded. If multiple such paths exist, the path with maximum available bandwidth is selected. The routing problem for EL traffic was modeled as a delay-bounded maximum-bandwidth problem.

For multipath video streaming over wireless multi-hop networks, received video quality is influenced by both the encoder performance and the delayed packet arrivals due to limited bandwidth. To achieve satisfactory performance, it is essential to choose the best video transmission rate among multiple routes. In [18], the objective function, expected distortion at the decoder, which incorporates both the influence of encoder performance and network congestion, is introduced. The authors propose a rate allocation scheme to optimize the expected received video quality based on simple models of encoder rate distortion performance and network rate-congestion trade-offs. Given the embedded bitstreams of a group of video frames and the cross traffic condition on a given set of paths, a gradient descent method is used to determine the amount of bits to be transmitted over the multiple paths for each frame. In that paper, two rate allocation strategies, the layered representation mode and the partial repetition mode are also investigated. The basic conclusion is that while layered representation is more efficient for mild delay constraints, partial repetition is more robust against packet losses.

To help multimedia streaming applications to perform error control and resource allocation correctly, AMTP, an Ad hoc Multipath sTreaming Protocol, is proposed [19]. AMTP can accurately differentiate the cause of packet losses by detecting the network states (errors, congestion and route change) and report those states to multimedia applications. Specifically for route change/break status, when a route is about to break due to mobility or signal attenuation, AMTP can find a new path before the old route breaks and informs the path conditions (such as available bandwidth and path error rate) to upper layer multimedia application so as to decide a proper error protection level and a fitting streaming rate.

In this paper, M-PDSR, a Multipath extension of the Preemptive Dynamic Source Routing Protocol [20] is proposed to discover and maintain multiple paths for real time streaming applications. By measuring the radio signal strength of wireless channels, M-PDSR monitors the quality of each route. When a route is about to break, it informs AMTP and suggests several candidate routes. According to AMTP's path selection algorithm, M-PDSR preemptively switches to a new qualified route before the old route breaks. AMTP sender selects multiple paths with best quality to transmit multimedia data through the underlying multi-hop network to the AMTP receiver. To accurately detect network state, it collects the cross-layer information such as packet loss rate due to errors, congestion, and route change/break and current available bandwidth along each path from intermediate nodes and the feedbacks from the AMTP "network monitor" on the receiver side. Then AMTP sender adjusts its transmission rate according to the congestion status along each of the selected paths in a TCP-friendly manner. When preemptively switching to a new path, AMTP assigns a proper rate to the path according to the path condition. The cross-layer information is also periodically (or preemptively in case of route change/break) reported to upper layer multimedia applications to perform error control and resource allocation [21].

3.2. Multi-path Video Delivery for Multicast Service

Multicast is an essential technology for many applications, such as group video conferencing and video distribution, and results in bandwidth savings as compared to multiple unicast sessions. Although there are some works address the unicast video delivery topic, so far, multicast video communication over wireless multi-hop networks has not been studied extensively.

In [22], the authors propose multiple tree video multicast communication to provide robustness for video multicast applications. The proposed multiple tree multicast video communication system consists of two parts: a multicast routing protocol to construct multiple disjoint trees, and a scheme to distribute video packets into different trees.

The core idea of this paper is the proposed distributed serial multiple disjoint trees multicast routing protocol called Serial MDTMR, which builds two disjoint multicast trees in a distributed way, for multiple tree multicast streaming. The Serial MDTMR achieves reasonable tree connectivity while maintaining disjoint-ness of two trees. For interactive video applications, Multiple Description Coding (MDC) is used to form multiple video streams, and transmit different video streams through different trees simultaneously. The basic protocol of MDTMR is described as follows. First, a shortest path multicast tree is constructed. Then after requiring all the middle nodes in the first tree not to be middle nodes of the second tree, another shortest path tree is constructed. Since these two trees do not share middle nodes at all, they are node disjoint.

In [23], the authors present a practical video multicast scheme using multiple trees, where video is coded into multiple descriptions (MD), each sent on a different tree. Due to highly diverse wireless links, the end-to-end bandwidths from the sender to the receivers (called *path bandwidth*) are also highly diverse. It would be beneficial to code each description into a number of layers, such that a receiver with a high path bandwidth can receive more layers and achieve an improved video quality, while a receiver with a low path bandwidth can at least receive the base layer for an acceptable video quality. Many sophisticated schemes for multiple description coding have been investigated over the years. A particularly efficient and practical scheme used in the paper is based on the time-domain partitioning coding, where multiple descriptions are generated by separating the video frames and coding them separately.

In this paper, the authors take an application-centric, cross-layer approach to formulate MD video multicast routing as a combinatorial optimization problem. The objective is to optimize the application layer performance metric, i.e., video distortion. The problem formulated for the cross-layer multicast routing is highly complex and is expected to be NP-complete. Therefore, efficient heuristic algorithms would be most useful in practice. A *Genetic Algorithm* (GA) [4] that is eminently suitable for addressing problems of this type is proposed in this paper, which has complex objective functions and large, unstructured combinatorial solution spaces. A GA-based solution procedure is constructed, which demonstrates its efficacy through extensive performance studies.

4. Link Layer Centric Solutions for Video Delivery

4.1. Power Adaptation for Fairly Video Delivery

In a multi-hop wireless network, devices autonomously share a common broadcast channel and collaborate to transport information. Transmitting multimedia presents many challenges related to throughput and fairness. This is mainly due to capture effect stemming from the MAC protocol in IEEE 802.11, the distributed coordination function (DCF) that based on CSMA/CA, as used in multi-hop networks. To transmit data, a node employs a four-way handshake messaging mechanism given by request-to-send (RTS), clear-to-send (CTS), DATA, and ACK. During the handshake, the node reserves the channel by advertising to others in the network the duration of its transmission as indicated by its network allocation vector (NAV). Despite the aforementioned collision avoidance mechanism, IEEE 802.11 cannot eliminate collisions completely. Moreover, due to the existence of hidden terminal and a lack of central control, channel capture phenomenon will occur in wireless multi-hop networks, where the common channel is monopolized by a single or a few nodes [24]. Capture leads to starvation in some nodes, thereof degrading the network fairness and throughput.

In [25], the authors classified the channel capture effects into three categories, i.e., "receiver-capture", *source capture, and* "power capture". Several examples are given to illustrate that all the three channel-captures can not be broken by fixed power assignment, static (the minimum) power assignment, and the maximum power level assignment schemes. To break the capture, therefore, a dynamic power algorithm is needed. In this paper, a distributed power adaptation algorithm termed power adaptation for starvation avoidance (PASA) is proposed to dynamically adjust the transmission power in each node to break capture and achieve higher spatial reuse. The main idea of the scheme is to adjust the transmission power in each node according to its current condition so that all the mobile nodes in the network can share the medium channel more efficiently. In the algorithm, a node starts its transmission with the maximum power RTS. After it succeeds in sending a packet, it "politely" uses a lower power level (while maintaining connectivity) to continue its communication to give other nodes opportunity to use the channel. If a node finds itself in the starvation state (as indicated by, for example, repeated failures in transmission), it increases its power level so that its transmission attempt can be made known to the others in the network.

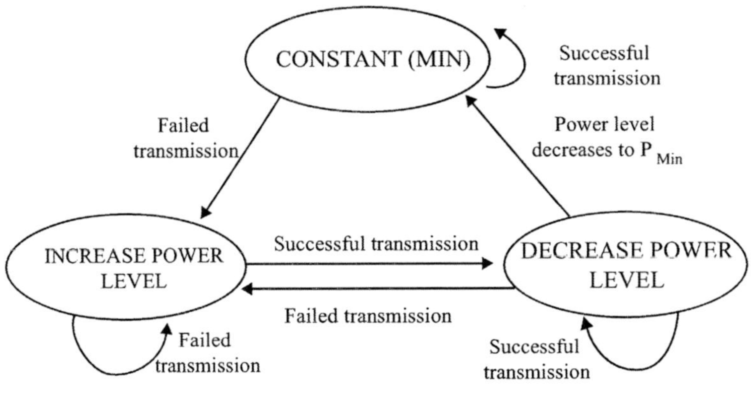

Figure 3. State transition diagram of the power adaptation algorithm.

There are three states in the algorithm: constant, increase and decrease. The state transition diagram is shown in Fig. 3. Briefly speaking, when there is no transmission failure, a node stays in the "constant" state and uses its minimum power level to transmit data. If there is certain number of consecutive failures, a node enters the "increase" state and increases its power level to retry. After a number of successful transmissions, it goes to the "decrease" state and decreases its power. The PASA algorithm is simple, efficient, autonomous, and without control message overhead.

4.2. Link-Layer Scheduling for Video Delivery

Scheduling is the core function in the link layer. In [26], the authors discuss how to allocate rate optimally to the source and take fully advantage of that rate via smart scheduling. The congestion-distortion optimized scheduling (CoDiO) is proposed for the packets delivery. This schedule indicates when the packets of the video stream will be sent from the server to the client given a discrete set of transmission times. To limit the exponential number of possible schedules, the time horizon covered by the schedule is limited. The aim of CoDiO is to determine a schedule minimizing the expected Lagrangian cost $D+\lambda\Delta$, where D is the distortion of the received video stream and Δ is the end-to-end delay which serves as the congestion metric. To minimize this objective function, CoDiO selects the most important packets in terms of video quality, and transmits them in an order which minimizes the average backlog of the bottleneck queue. For example, the I frames of a video stream will be transmitted in priority whereas B frames might be dropped. In addition, CoDiO will avoid transmitting packets in large bursts as this has the worse effect on the queuing delay. The core of CoDiO is to evaluate the expected end-to-end delay and distortion corresponding to a given transmission schedule of a set of packets.

In wireless LANs or multihop networks, it is usual that node concurrently communicates with several neighbors. Since channel quality is normally time-varying and independent across different neighbors, this provides the node an opportunity to choose one of its neighbors with good channel quality to transmit data before those with bad channel quality. In [27], the authors investigate how to schedule the transmissions of packets to its neighbors so that it can improve the performance and find out the significance of such multi-user diversity in contention-based multi-hop networks. In this paper, they present the Opportunistic packet Scheduling and Auto Rate (OSAR) protocol to exploit the channel variations. The basic idea of OSAR is rather than just matching the channel condition for a node pair in communications, the proposed protocol takes advantage of the multi-user diversity as much as possible and adapt the rate accordingly, i.e., based on the channel conditions to its neighboring nodes, the sender chooses the neighboring node with channel quality better than certain level to schedule the transmissions of packets in its queue, then the overall system throughput may be increased. The key mechanisms of OSAR are channel aware media access, rate adaptation and packet bursting.

Supporting real-time traffic in the multi-hop network environment is difficult because of the lack of a centralized infrastructure. [28] targets to investigate the distributed 802.11 MAC protocol extensions that can support flow QoS in multi-hop wireless networks. The distributed link scheduling multiple access (D-LSMA) protocol is proposed to extend 802.11 with a flexible distributed scheduling mechanism that permits nodes to reserve periodic time slices for real-time flows, i.e. each node can request a specified flow bandwidth on links to neighboring nodes. The proposed scheme's distributed scheduling algorithm also eliminates the exposed node problem in 802.11 by permitting parallel scheduling of traffic on links which do not interfere with each other. Scheduling in D-LSMA is loosely synchronized through channel events rather than with more rigid TDMA framing methods.

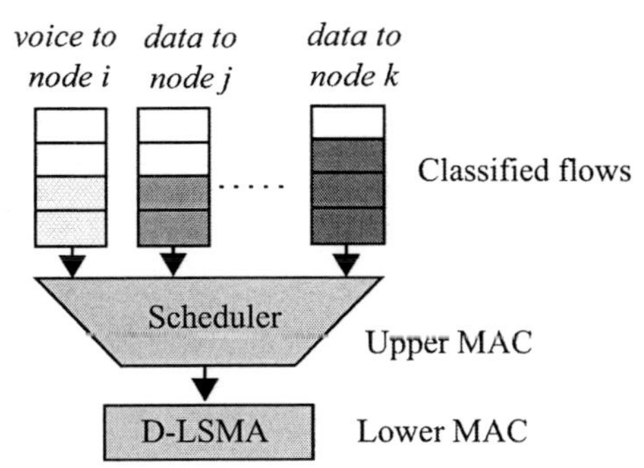

Figure 4. New MAC architecture: Scheduler + basic MAC.

In the proposed D-LSMA protocol, the scheduling in 802.11 channels is generalized by introducing the concept of segregation between an "upper MAC" and "lower MAC" (see in Figure 4). The lower MAC in D-LSMA is similar to that in the 802.11 standard with extensions to the control syntax necessary to support more general scheduling policies. The upper MAC can be thought of as an intelligent scheduler which monitors flows to neighboring nodes and makes decisions on "when to reserve?" and "how much to reserve?" depending on traffic volume and flow QoS requirements at that node. The upper MAC can incorporate algorithms that are aware of network topology and traffic load in order to maintain flow QoS. As shown in Fig 4, the upper MAC can be designed to support a mix of traffic types with separate packet queues and different scheduling policies. Note that this type of segregation by traffic type is an important feature in multi-hop networks where radio nodes have to handle both local and cross traffic with different service/bandwidth requirements.

4.3. Layer 2.5 MAC for Video Delivery

There are several challenges in effectively realizing real-time services over multi-hop wireless networks. First of all, in an 802.11-based wireless network, the transmissions from neighboring nodes may interfere with each other, causing collision. IEEE 802.11 (in DCF mode) employs CSMA/CA based MAC to reduce collisions, which is not good enough for multi-hop wireless network [29], where collisions happen due to the transmitters is out the carrier sense range of each others. Without carefully controlled traffic volume to reduce collision, a multi-hop wireless network deploying DCF can be easily overwhelmed. Furthermore, since real-time services are likely to co-exist with data services over a multi-hop wireless network, using 802.11 DCF MAC real-time (RT) traffic must compete with best-effort (BE) data traffic for access to shared media. Although QoS-enhanced 802.11 MAC mechanism, eDCF, has been developed for WLAN, it does not provide adequate service differentiation for support of real-time traffic in multi-hop wireless networks, because of the hidden terminal and other interference problems. Thus, the prioritized traffic volume regulation is necessary to provide service differentiation to RT traffic, e.g., reduce contention of RT by cutting down interfering BE traffic.

In [30], a novel software framework, called Layer 2.5 SoftMAC, is proposed to support RT services over multi-hop wireless networks utilizing off-the-shelf commercial 802.11 a/b/g NICs and standard MAC DCF. Layer 2.5 SoftMAC lies between 802.11 MAC layer and IP layer: it leverages the (limited) coordination by 802.11 MAC DCF, and extends/enhances it via software mechanisms for support of RT services. The key idea behind SoftMAC is to employ "coarse-grained" control mechanisms (e.g., distributed admission control and rate regulation) to coordinate and regulate network load and packet transmission of both RT and BE traffic among neighboring nodes in a distributed manner. The objective is to keep the channel busy time and collision rate below appropriate levels, and thus ensure acceptable RT quality.

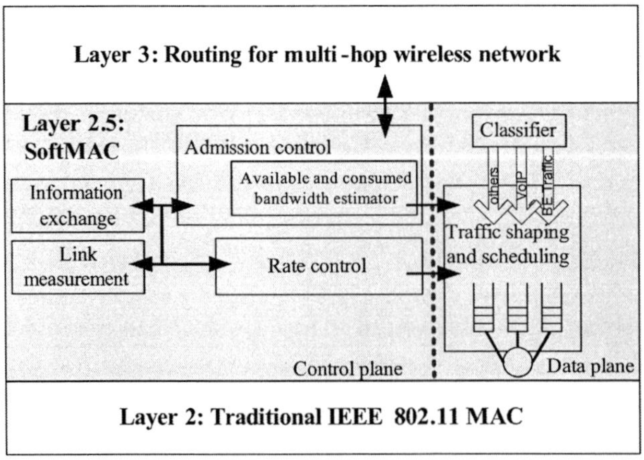

Figure 5. SoftMAC architecture and components for RT delivery.

Fig.5 provides a schematic depiction of the Layer 2.5 SoftMAC architecture, which is divided into two planes: the control plane and data plane. There are four modules in the control plane and works as following. Each node will exchange its RT and BE traffic information explicitly to its neighbors by broadcasting, and measure the current link status such as capacity and packet loss ratio. With all the information, the distributed AC module can estimate the consumed/available bandwidth and decide whether the available resource is sufficient to support new connections. Meanwhile, the rate of BE traffic will be controlled by a rate control mechanism at each node. Based on the decisions of control plane modules, priority queueing and traffic shaping are used in the data plane to regulate the rate of packet transmission.

5. Cross-Layer Design for Multi-hop Multimedia Delivery

Media streaming imposes great challenges on the underlying multi-hop wireless network, where interference among nodes limits the transmission rate over a single link. Although recently there are many researches focus on multimedia compression and transmission for multi-hop wireless networks, these solutions often do not provide adequate support for multimedia applications. This is because the resource management, adaptation and protection strategies available in the lower layers - Physical (PHY) layer, Medium Access Control (MAC) layer, and Network/Transport layers - are optimized without *explicitly* considering the specific characteristics of the multimedia applications, and conversely, multimedia compression and streaming algorithms do not consider the mechanisms provided by the lower layers for error protection, scheduling, resource management etc. This "layered" optimization leads to a simple, independent implementation, but results in suboptimal multimedia (objective and/or perceptual quality) performance. Alternatively, under adverse conditions, wireless stations need to optimally adapt their multimedia compression and transmission strategies jointly across the protocol stack in order to guarantee a predetermined quality at the receiver. This calls for the cross-layer design for video delivery over multi-hop networks.

The art of video streaming over multi-hop wireless networks is still in its infancy, especially when addressed via a cross-layer network design. Most of the recent research considers only a subset of layers of the protocol stack. In [13], path diversity in an 802.11 network combined with multi-stream coding of video is proposed and analyzed. In [31] the cross-layer design addresses the joint problem of power control and scheduling for multi-hop wireless networks with QoS. It takes SINR and minimum rate as constraints to minimize the total transmit over the links. [32] is a simple cross-layer design example. The routing information created by AODV can be shared with application programs. When a sender wants to send packets, it will check the information first. If the route changed in terms of hop counts, it will adjust encoding bit rate to adapt to the links condition.

In [33] the congestion information of mobile node is used by different layers such as network, transport and higher layer. *By locally monitoring the network interface transmission queue length and MAC layer behavior at each node, a node can establish an approximation of the degree to which the wireless medium around it is busy.* Within network layer, the congestion parameter can be used as a routing metric to select route. Within the transport layer, a node may set Explicit Congestion Notification (ECN) bits in a packet's IP header.

Within other higher layer protocols, for example, the sender may compress the data before transmission if it knows the some link in the route is very busy.

In [34], the network planning problem in wireless multi-hop networks is formulated as the problem of allocating physical and medium access layer resources or *supplies* to minimize a cost function, while fulfilling certain end-to-end communication *demands*, which are given as a collection of multicast sessions with desired transmission rates. An iterative cross-layer optimization is proposed, which alternates between: 1) jointly optimizing the timesharing in the medium access layer and the sum of max of flows assignment in the network layer and 2) updating the operational states in the physical layer.

A fundamental problem for providing high quality video delivery in wireless networks is how to map QoS criterion at different layers and optimize QoS across the layers. In [35], a cross-layer QoS mapping architecture is proposed for video delivery over wireless environment. There are several components under this architecture, including a proposal of an adaptive QoS service model that allows QoS parameters to be adaptively adjusted according to the time-varying wireless channel condition, an interaction mechanism between the priority network and video applications to provide proper QoS selection, and a resource management scheme to assign resources based on the QoS guarantee for each priority class under the time-varying wireless channel. This architecture enables to perform QoS mapping between statistical QoS guarantees at the network level to a corresponding priority class with different video quality requirements.

The robust and efficient transmission of video over WLANs is investigated in [36] that enables the joint optimization of the various protection strategies existing in the protocol stack. In existing WLAN environments, different protection strategies exist at the various layers of the protocol stack. At the physical layer, it is possible to dynamically reconfigure the modulation and channel coding techniques for each packet, based on the channel characteristics to allow for tradeoffs between throughput and reliability. Interleaving is also often applied at the physical layer to combat burst errors. In the MAC layer, retransmissions are used for protection /error control. At the application layer, FEC, ARQ, and hybrid ARQ schemes can be employed along with error resilient video coding schemes and error concealment strategies. Different packetization strategies can be adopted at the various layers (physical, MAC, transport, and application layer), leading to various tradeoffs between throughput and reliability. For an optimal performance, a joint consideration is desirable in order to provide an optimal overall performance for the transmission of video. In this paper, an analysis of the throughput efficiency for the downlink traffic is presented based on a realistic channel model, 802.11a MAC operation, and various header overheads associated with different layers of protocols. The end-to-end distortion of MPEG-4 video is modeled for various channel conditions using different unequal error protection strategies. Based on that model, a strategy for the adaptive selection of application-layer FEC, maximum MAC retransmission limit, and packet sizes depending on the channel condition to maximize the video quality under different multipath channel conditions is developed.

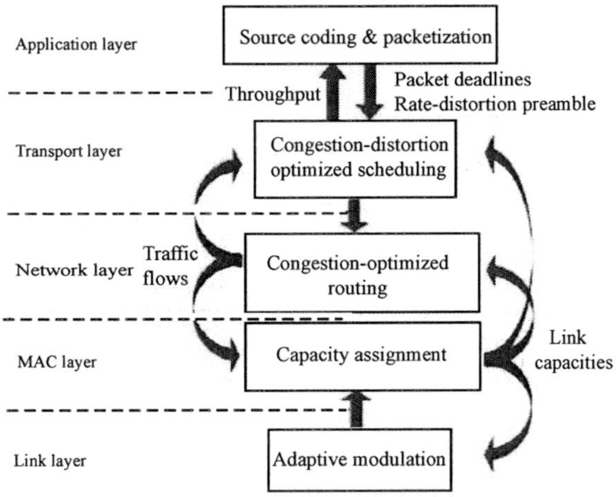

Figure 6. Cross-layer design framework for media streaming over wireless multi-hop network.

In [37], a more general framework is proposed, which encompasses the entire protocol stack. Fig. 6 shows a block diagram of a new cross-layer design framework with information exchange between the different layers. At the link layer, adaptive techniques are used to maximize the link rates under varying channel conditions. This extends the achievable capacity region of the network. Each point of this region indicates a possible assignment of the different link capacities. Based on link-state information, the MAC selects one point of the capacity region by assigning time slots, codes or frequency bands to each of the links. The MAC layer operates jointly with the network layer to determine the set of network flows that minimize congestion. To find a jointly optimal solution for the capacity assignment and the network flows, successive sub-optimal solutions are exchanged iteratively between these two middle layers which constitute the core of the cross-layer framework. At the transport layer, congestion-distortion optimized scheduling is performed to control the transmission and retransmission of video packets. Finally, the application layer determines the most efficient encoding rate.

Three pieces are discussed in that paper. First, several adaptive link layer techniques are presented to determine the maximum rates that can be transmitted between any two nodes in the network. Secondly, the authors show that how the operating network capacity can be jointly optimized with network flow to improve the end-to-end performance of video streaming. Finally, application layer information is considered to determine which packets should be transmitted and when to maximize the decoded video quality while limiting network congestion as well as how to determine the optimal streaming rate.

Although there is some work addresses cross-layer from different aspects, much work still needs to be done along these directions to identify and exploit cross-layer interactions in real-time media streaming over wireless multi-hop networks. One thing that needs to be paid much attention is that a cross-layer approach to network design can significantly increase the design complexity. The layering principle simplifies design and implementation and provides the possibility of alternative layer implementations. Thus, cross-layer design should not eliminate the design advantages of layering. Keeping some form of separation, while allowing layers to actively interact, appears a good compromise for enabling interaction between layers without eliminating the layering principle. Moreover, cross-layer design does not necessarily mean that all the layers should interact together. How to identify the layers, that can generate the most significant gains by interacting together present important areas of future research.

6. Conclusions

Enabling multimedia transport over wireless multi-hop networks is an exciting yet challenging task. Multi-hop wireless paths are highly unstable and the compressed media is susceptible to transmission errors. Meanwhile, multiple paths in a multi-hop wireless network can be exploited as an effective means to combat transmission errors. In this paper, we give an overview for the state-of-art work on multimedia over multi-hop networks. More specifically, the recent progress in efficient and resilient video coding and protection solutions, the network routing and transport protocols, link-layer QoS support enhancement, and cross-layer design for media over multi-hop environment are reviewed. Moreover, some future directions for cross-layer design is also highlighted in the paper.

References

[1] D. Aguayo, J. Bicket, S. Biswas, G. Judd, and R. Morris. Link-level measurements from an 802.11b mesh network. In *Proc. of SIGCOMM*, 2004.

[2] IEEE 802.11 specification

[3] S.Xu and T.Saadwi, "Does the IEEE 802.11 MAC protocol work well in multihop wireless ad hoc networks", *IEEE Comm.*, Jun. 2001

[4] IEEE draft for Wireless Medium Access Control (MAC) and Physical Layer (PHY) specifications: Medium Access Control (MAC) Enhancements for Quality of Service (QoS), IEEE Std 802.11e/D4.0, November 2002

[5] Vetro, J. Xin, and H. Sun, "Error-Resilience Video Transcoding for Wireless Communications", invited paper in *IEEE Wireless Communication Magazine*, 2005.

[6] Y. Wang and Q.-F. Zhu, "Error control and concealment for video communication: A review," *Proc. IEEE*, vol. 86, pp. 974–997, May 1998.

[7] ITU-T and ISO/IEC JTC 1, "Advanced Video Coding for Generic Audiovisual Services," ITU-Recommendation H.264 and ISO IEC 14996-10 AVC, 2003.

[8] G. Sullivan and T. Wiegand, "Video Compression – From Concepts to H.264/AVC Standard," *Proc. of the IEEE, Special Issue on Advances in Video Coding and Delivery*, Vol. 93, No. 1, pp. 18-31, January 2005.

[9] R. Swann, and N. Kingsbury, "Transcoding of MPEG-II for enhanced resilience to transmission errors," *Proc. IEEE Int'l Conf. Image Processing*, vol. 2, pp. 813-816, Oct 1996.

[10] G. de los Reyes, A. R. Reibman, S.-F. Chang, and J. C.-I. Chuang, "Error-resilient transcoding for video over wireless channels," *IEEE J. Selected Areas Commun.*, vol. 18, no. 6, pp. 1063-1074, June 2000.

[11] G. Cote, S. Shirani, and F. Kossentini, "Optimal mode selection and synchronization for robust video communications over error-prone networks," *IEEE J. Selected Areas Commun.*, vol. 18, no. 6, pp. 952–965, 2000.

[12] R. Zhang, S.L. Regunathan, and K. Rose, "Video coding with optimal inter/intra-mode switching for packet loss resilience," *IEEE J. Selected Areas Commun*, vol. 18, no. 6, pp. 966–976, 2000.

[13] S. Mao, S. Lin, S.S Panwar, Y. Wang, and E. Celebi, "Video transport over ad hoc networks: multistream coding with multipath transport", *IEEE Journal on Selected Areas in Communications*, Vol. 21, Issue 10, Dec. 2003, pp.1721 – 1737.

[14] S. Mao, D. Bushmitch, S. Narayanan, S. Panwar, "MRTP: a multi-flow realtime transport protocol for ad hoc networks," *Proc. IEEE VTC '03* Fall.

[15] Z. Fu, X. Meng and S. Lu, "A transport protocol for supporting multimedia streaming in mobile ad hoc networks," *IEEE JSAC*, Dec. 2003.

[16] W. Wei, A. Zakhor, "Robust multipath source routing protocol (RMPSR) for video communication over wireless ad hoc networks", *IEEE ICME '04*, Vol. 2, pp. 1379-1382, Vol.2, June 27-30, 2004.

[17] J. Wang, H. Wang, Q. Zhang, and B. Chen, "Routing for MPEG4 FGS Multimedia Stream Transmission over Ad Hoc Networks", Proceedings, 2nd IASTED International Conference on Communications, Internet & Information Technology, Scottsdale, Arizona, pp. 469–472, Nov. 2003.

[18] X. Zhu, S. Han, and B. Girod, "Congestion-Aware Rate Allocation for Multipath Video Streaming Over Ad hoc Wireless Networks," *Proc. IEEE International Conference on Image Processing, ICIP-2004,* Singapore, Oct. 2004.

[19] K. Rojviboonchai, F. Yang, Q. Zhang, H. Aida, and W. Zhu, "AMTP: A Multipath Multimedia Streaming Protocol for Mobile Ad Hoc Networks", *IEEE ICC 2005*.

[20] T. Goff, N. Abu-Ghazaleh, D. Pathak, R. Kahvecioglu, "Preemptive routing in ad hoc networks," Proc. ACM MobiCom 2001.

[21] F. Yang, Q. Zhang, W. Zhu and Y.Q. Zhang, "Bit allocation for scalable video streaming over mobile wireless Internet", *Proc. IEEE INFOCOM*, Mar. 2004.

[22] W. Wei, A. Zakhor, "Multipath Unicast and Multicast Video Communication over Wireless Ad Hoc Networks", *IEEE Broadnets*, 2004, pp. 496- 505.

[23] S Mao, X. Cheng, Y.T. Hou, and H.D. Sherali, "Multiple Description Video Multicast in Wireless Ad Hoc Networks", *IEEE Broadnets* 2004, pp. 671-680.

[24] Ware, J. Judge, J. Chicharo, and E. Dutkiewicz, "Unfairness and capture behaviour in 802.11 adhoc networks," in *Proceedings of International Conference on Communications*, pp. 159 –163, IEEE, 2000.

[25] J. Chen, S.-H.G. Chan, Q. Zhang, W. Zhu, J. Chen, "PASA: power adaptation for starvation avoidance to deliver wireless multimedia", *IEEE Journal on Selected Areas in Communications*, Vol. 21, Issue 10, pp.1663–1673, Dec. 2003.

[26] Girod, E. Setton, X. Zhu, "Congestion-Optimized Routing and Scheduling of Video Over Wireless Ad Hoc Networks," *Proc. IEEE International Symposium on Circuits and Systems, ISCAS-2005,* Kobe, Japan, May 2005. *Invited Paper.*

[27] J. Wang, H. Zhai, Y. Fang, M.C. Yuang, "Opportunistic media access control and rate adaptation for wireless ad hoc networks", *IEEE ICC 2004*, Vol. 1, pp. 154-158, June 20-24, 2004.

[28] Z. Wu, D. Raychaudhuri, "D-LSMA: distributed link scheduling multiple access protocol for QoS in ad-hoc networks", *IEEE GLOBECOM '04*, Vol. 3, pp.1670-1675, 29 Nov. 29- Dec.3, 2004.

[29] S.Xu and T.Saadwi. Does the IEEE 802.11 MAC protocol work well in multihop wireless ad hoc networks. *IEEE Comm.*, Jun. 2001

[30] H. Wu, X. Wang, Y. Liu, Q. Zhang, Z.-L. Zhang, "SoftMAC: Layer 2.5 MAC for VoIP Support in Multi-hop Wireless Networks", *MSR Technical Report*, 2005.

[31] U. C. Kozat, I. Koutsopoulos and L. Tassiulas, .A Framework for Cross-layer Design of Energy-efficient Communication with QoS Provisioning in Multi-hop Wireless Networks, Technical Research Report, The Institute for System Research, The University of Maryland, 2004

[32] H. Gharavi and K. Ban, .Dynamic adjustment packet control for video communications over ad-hoc networks,. The 2004 International Conference on Communications (ICC 2004)

[33] Y.-C. Hu and D. B. Johnson, .Exploiting Congestion Information in Network and Higher Layer Protocols in Multihop Wireless Ad Hoc Networks., 24th International Conference on Distributed Computing Systems (ICDCS'04), Hachioji, Tokyo, Japan, March 24 - 26, 2004

[34] Y. Wu, P. Chou, Q. Zhang, W. Zhu K. Jain, and S-Y. Kung, "Network planning in wireless ad hoc networks: a cross-layer approach," *IEEE Journal on Selected Areas on Communications*, submitted October 2003.

[35] W. Kumwilaisak, Y.T Hou, Q. Zhang, W. Zhu, C.-C.J. Kuo, Y.-Q. Zhang, "A cross-Layer quality-of-service mapping architecture for video delivery in wireless networks", *IEEE Journal on Selected Areas in Communications*, Vol. 21, Issue 10, pp. 1685 –1698, Dec. 2003.

[36] M. van der Schaar, S. Krishnamachari, S. Cho, and X. Xu, "Adaptive cross-layer protection strategies for robust scalable video transmission over 802.11 wlans," *IEEE Journal on Selected Areas of Communications (JSAC)*, vol. 21, no. 10, pp. 1752 -1763, Dec. 2003.

[37] Setton, T. Yoo, X. Zhu, A. Goldsmith, and B. Girod, "Cross-layer design of ad hoc networks for real-time video streaming", invited paper in *IEEE Wireless Communication Magazine*, 2005.

INDEX

A

B

C

D

E

F

G

S

T